INTRODUCTION TO
RENEWABLE ENERGY
SECOND EDITION

ENERGY AND THE ENVIRONMENT

SERIES EDITOR
Abbas Ghassemi
New Mexico State University

PUBLISHED TITLES

Introduction to Renewable Energy, Second Edition
Vaughn Nelson and Kenneth Starcher

Environmental Impacts of Renewable Energy
Frank R. Spellman

**Geothermal Energy: Renewable Energy and the Environment,
Second Edition**
William E. Glassley

Energy Resources: Availability, Management, and Environmental Impacts
Kenneth J. Skipka and Louis Theodore

Finance Policy for Renewable Energy and a Sustainable Environment
Michael Curley

Wind Energy: Renewable Energy and the Environment, Second Edition
Vaughn Nelson

Solar Radiation: Practical Modeling for Renewable Energy Applications
Daryl R. Myers

Solar and Infrared Radiation Measurements
Frank Vignola, Joseph Michalsky, and Thomas Stoffel

Forest-Based Biomass Energy: Concepts and Applications
Frank Spellman

Solar Energy: Renewable Energy and the Environment
Robert Foster, Majid Ghassemi, Alma Cota,
Jeanette Moore, and Vaughn Nelson

INTRODUCTION TO
RENEWABLE ENERGY
SECOND EDITION

Vaughn Nelson
Kenneth Starcher

CRC Press
Taylor & Francis Group
Boca Raton London New York

CRC Press is an imprint of the
Taylor & Francis Group, an **informa** business

MATLAB® is a trademark of The MathWorks, Inc. and is used with permission. The MathWorks does not warrant the accuracy of the text or exercises in this book. This book's use or discussion of MATLAB® software or related products does not constitute endorsement or sponsorship by The MathWorks of a particular pedagogical approach or particular use of the MATLAB® software.

CRC Press
Taylor & Francis Group
6000 Broken Sound Parkway NW, Suite 300
Boca Raton, FL 33487-2742

© 2016 by Taylor & Francis Group, LLC
CRC Press is an imprint of Taylor & Francis Group, an Informa business

No claim to original U.S. Government works

Printed on acid-free paper
Version Date: 20150908

International Standard Book Number-13: 978-1-4987-0193-8 (Hardback)

This book contains information obtained from authentic and highly regarded sources. Reasonable efforts have been made to publish reliable data and information, but the author and publisher cannot assume responsibility for the validity of all materials or the consequences of their use. The authors and publishers have attempted to trace the copyright holders of all material reproduced in this publication and apologize to copyright holders if permission to publish in this form has not been obtained. If any copyright material has not been acknowledged please write and let us know so we may rectify in any future reprint.

Except as permitted under U.S. Copyright Law, no part of this book may be reprinted, reproduced, transmitted, or utilized in any form by any electronic, mechanical, or other means, now known or hereafter invented, including photocopying, microfilming, and recording, or in any information storage or retrieval system, without written permission from the publishers.

For permission to photocopy or use material electronically from this work, please access www.copyright.com (http://www.copyright.com/) or contact the Copyright Clearance Center, Inc. (CCC), 222 Rosewood Drive, Danvers, MA 01923, 978-750-8400. CCC is a not-for-profit organization that provides licenses and registration for a variety of users. For organizations that have been granted a photocopy license by the CCC, a separate system of payment has been arranged.

Trademark Notice: Product or corporate names may be trademarks or registered trademarks, and are used only for identification and explanation without intent to infringe.

Library of Congress Cataloging-in-Publication Data

Nelson, Vaughn, author.
    Introduction to renewable energy / Vaughn Nelson and Kenneth Starcher. -- Second edition.
        pages cm. -- (Energy and the environment)
    Includes bibliographical references and index.
    ISBN 978-1-4987-0193-8 (hardcover : alk. paper) 1. Renewable energy sources. I. Starcher, Kenneth, author. II. Title.

TJ808.N46 2016
621.042--dc23   2015034559

Visit the Taylor & Francis Web site at
http://www.taylorandfrancis.com

and the CRC Press Web site at
http://www.crcpress.com

Contents

List of Figures

List of Tables

Preface

The big question: how do we use science and technology such that spaceship Earth will be a place for all life to exist? We are citizens of Earth, and within your lifetime, there will be major decisions over the following: energy (including food), water, minerals, space, and war (which we can state will happen with 99.9% probability). These previous statements were made over 30 years ago, when Nelson first taught introductory courses on wind and solar energy. Since then, the United States has been involved in a number of armed conflicts, so my prediction on war has been fulfilled. The era of armed conflict over resources has already started—Oil War I (Gulf War) and Oil War II (Iraq War)—and a sustainable-energy future primarily fueled by renewable energy is paramount to reduce the possibility of an Oil War III between China and the United States over dwindling supplies of petroleum. This is also the opinion of one of my Chinese colleagues working in renewable energy.

We are over 7 billion and heading toward 11 billion people, and we are all participants in an uncontrolled experiment on the effect of human activities on the Earth's environment. Renewable energy is part of the solution to the problem of finite resources of fossil fuels and the environmental impact of greenhouse gases. Renewable energy is now part of national policies with significant goals for percentage increase in generation of energy within the next decades. The reason is that there are large amounts of renewable energy in all parts of the world; in contrast to fossil fuels and minerals, renewable energy is sustainable, and it reduces greenhouse gas emissions. The growth of renewable energy has been very large, at 20% per year since 2005; however, this large growth rate is attributable to the original low levels of renewable energy generation, except for hydroelectric power generation, growth for which is around 2% per year. Hydroelectric power still remains the top source of renewable-energy generation, with an installed capacity of 1,000 GW; however, at the end of 2014, the installed capacity of wind farms was 360 GW and photovoltaics was 180 GW (a significant part of new electric plant capacity from all sources). Compare these values with the numbers from the first edition of this book (2011): installed capacity energy from wind farms was 158 GW and from photovoltaics was 23 GW.

Policies for supporting renewable energy have spread from 48 countries in 2004 to over 140 countries in 2014. Renewable energy targets along with feed-in tariffs have had the biggest impact on increasing the renewable energy market. In 2004, the majority of local governments did not consider the renewables in their energy supply, and now many of them have become leaders in advancing renewable energy, some even setting targets of 100% renewables. The future for renewable energy is very bright and you can be a part of that future by working in the field or by supporting the implementation of renewable energy.

Vaughn Nelson
and
Kenneth Starcher
West Texas A&M University

MATLAB® is a registered trademark of The MathWorks, Inc. For product information, please contact:

The MathWorks, Inc.
3 Apple Hill Drive
Natick, MA 01760-2098 USA
Tel: +1 508 647 7000
Fax: +1 508 647 7001
E-mail: info@mathworks.com
Web: www.mathworks.com

Acknowledgments

We are deeply indebted to colleagues, present and past, at the Alternative Energy Institute (AEI), West Texas A&M University (WTAMU), the Wind Energy Group at the Agricultural Research Service, and the U.S. Department of Agriculture, Bushland, Texas. The students in our classes and the students who worked at AEI provided valuable insight and feedback. We thank many others who worked with us at the AEI and the U.S. Department of Agriculture, especially the numerous international researchers and interns. We thank the Instructional Innovation and Technology Laboratory, WTAMU, for the computer drawings, and Robert Avant, Texas A&M Agri-Life Research, who reviewed Chapter 10, on bioenergy.

Vaughn: I express my gratitude to my wife, Beth, who has put up with me all these years. Dana and Vaughn Nelson (my grandchildren) assisted my efforts, especially with the PowerPoint presentations.

Ken: I credit my wife, Madeleine, with making me get up each morning and making it well worthwhile to come home each evening. I have never really had a "job," but the lifetime of involvement in renewables has been worth all the years of doing it.

Acknowledgments

Authors

Vaughn Nelson, PhD, has been involved with renewable energy, primarily wind energy, since the early 1970s. He is the author of 3 books and 4 CDs, has published over 50 articles and reports, was the principal investigator on numerous grants, and has given over 60 workshops and seminars from the local to the international level. His primary work has been on wind resource assessment, education and training, applied R&D, and rural applications of wind energy. Presently, he retired from West Texas A&M University (WTAMU). He was director of the Alternative Energy Institute (AEI) from its inception in 1977 through 2003 and then returned for another year in July 2009. He retired as the dean of the Graduate School, Research and Information Technology, WTAMU, in 2001. He served on Texas state committees, most notably the Texas Energy Coordination Council during its 11-year existence. He has received three awards from the American Wind Energy Association, one of which was the Lifetime Achievement Award in 2003; received an award as a Texas Wind Legend in 2010 from the Texas Renewable Industries Association; received an award in 2013 for Outstanding Wind Leadership in Education from Wind Powering America; and served on the board of directors for state and national renewable energy organizations. One of his projects was a renewable energy demonstration building at the AEI Wind Test Center. Dr. Nelson developed the material for a new online course in renewable energy at WTAMU in the spring of 2010, and the first edition of this book was the result. Dr. Nelson is the author of *Wind Energy* (2009, 2nd ed., 2013) and *Renewable Energy and the Environment* (2011). He received the Lifetime Achievement Award from the American Wind Energy Association in 2003.

Dr. Nelson's degrees include a PhD in physics from the University of Kansas, an EdM from Harvard University, and a BSE from Kansas State Teachers College, Emporia. He was at the Departamento de Física, Universidad de Oriente, Cumana, Venezuela, for two years and then at WTAMU from 1969 to 2003.

Kenneth Starcher began his college career and involvement with renewables in the fall of 1976. This led to a BS in physics/computer science at West Texas State University (1980). In 1980–81 he took courses in electrical engineering, electronics, and physics at Texas Tech University. He earned an MS in engineering technology at WTAMU (1995) and then took some courses in agricultural economics at WTAMU.

Starcher has been a field worker for most of the projects at the Alternative Energy Institute (AEI) since 1980. He has been the educational funnel for onsite training and public information for students and public workshops for AEI. He has served as a trainer at wind and solar training workshops locally, nationally, and internationally. He has served as a research technician, research associate, assistant director, director, and associate director (training, education, and outreach) for AEI over the past 35 years.

Starcher served as a board member of the American Wind Energy Association, is on the executive board of Class 4 Winds and Renewables, was chosen as the individual member of the year for Texas Renewable Energy Association in 2005, was chosen as the small wind educator at the Small Wind Conference in 2010, and was awarded an Outstanding Wind Leadership Education Award from Wind Powering America in 2013.

Starcher has installed and operated more than 85 different renewable energy systems, ranging in scale from 50 W to 500 kW. He has served as a consultant for wind companies in the United States and produced wind resources maps for counties, states, and Thailand and Honduras.

1 Introduction

1.1 ENERGY AND SOCIETY

Industrialized societies run on energy, a tautological statement in the sense that it is obvious. Population, gross domestic product (GDP), consumption and production of energy, and production of pollution for every country in the world are interrelated. The United States has less than 5% of the world population; however, in the world, the United States generates around 22% of the gross production and 16% of the carbon dioxide emissions and is at 18% for energy consumption (Figure 1.1). Notice that the countries listed in Figure 1.1 consume around 70% of the energy and produce around 70% of the world GDP and carbon dioxide emissions. The developed countries consume the most energy and produce the most pollution, primarily due to the increase in the amount of energy per person. On a per person basis, the United States is considered the worst for energy consumption and carbon dioxide emission.

The energy consumption in the United States increased from 34 exajoules (32 quads) in 1950 to a peak of 107 EJ (101 quads) in 2007, and because of the recession and more efficient use, consumption was 102 EJ (97 quads) in 2013. The oil crisis of 1973 showed that efficiency is a major component in gross national product and the use of energy. However, you must remember that correlation between GDP and energy consumption does not mean cause and effect.

It is enlightening to consider how the United States has changed in terms of energy use since World War II. Ask your grandparents about their lives in the 1950s and then compare the following with today:

Residential: Space heating and cooling, number of lights, and amount of space per person
Transportation: Number and types of vehicles in the family
Commercial: Space heating and cooling for buildings and lights
Industrial: Efficiency

A thought on energy and GDP: A solar clothes drying (a clothes line) does not add to the GDP, but every electric and gas dryer contributes; however, they both do the same function. We may need to think in terms of results and efficient ways to accomplish a function or process and the actual life-cycle cost. Why do we need heavy cars or sport utility vehicles with big motors that accelerate rapidly to transport people?

Now the underdeveloped part of the world, primarily the two largest countries in terms of population (China $1.3 * 10^9$ and India $1.1 * 10^9$), is beginning to emulate the developed countries in terms of consumption of energy, consumption of material resources, and greenhouse gas emissions. One dilemma in the developing world is that a large number of villages and others in rural areas do not have electricity.

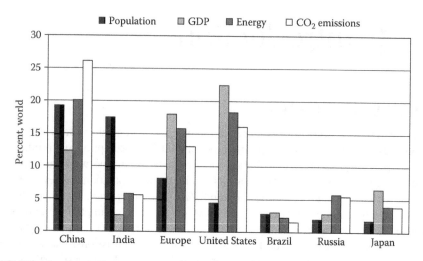

FIGURE 1.1 Comparisons, percent of world, for population (rank in world), gross domestic product, energy consumption, and carbon dioxide emission.

1.2 TYPES OF ENERGY

There are many different types of energy. *Kinetic energy* is energy available in the motion of particles, for example, wind or moving water. Potential energy is the energy available because of the position between particles, for example, water stored in a dam, the energy in a coiled spring, and energy stored in molecules (gasoline). There are many examples of energy: mechanical, electrical, thermal (heat), chemical, magnetic, nuclear, biological, tidal, geothermal, and so on.

In reality there are only four generalized interactions (forces between particles) in the universe: nuclear, electromagnetic, weak, and gravitational [1]. In other words all the different types of energy in the universe can be traced back to one of these four interactions (Table 1.1). This interaction or force is transmitted by an exchange particle. The exchange particles for electromagnetic and gravitational interactions have zero rest mass and so the transfer of energy and information is at speed of

TABLE 1.1
Information for Generalized Interactions

Interaction	Particle	Strength	Range (m)	Exchange Particle
Nuclear (strong)	Quarks	1	10^{-15}	Gluons
Electromagnetic	Charge	10^{-2}	Infinite	Photon
Weak	Leptons	10^{-6}	10^{-18}	Weakons[a]
Gravitational	Mass	10^{-39}	Infinite	Graviton

[a] My name for exchange particles (intermediate vector bosons).

light, $3 * 10^8$ m/s (186,000 miles/s). Even though the gravitational interaction is very, very, very weak, it is noticeable when there are large masses. The four interactions are a great example of how a scientific principle covers an immense amount of phenomena.

The source of solar energy is the nuclear interactions at the core of the Sun, where the energy comes from the conversion of hydrogen nuclei into helium nuclei. This energy is primarily transmitted to the Earth by electromagnetic waves, which can also be represented by particles (photons). In this course we will be dealing primarily with the electromagnetic interaction, although hydro and tides are energy due to the gravitational interaction and geothermal energy is due to gravitational and nuclear decay.

We will use exponents to indicate large and small numbers. The exponent indicates how many times the number is multiplied by itself, or how many places the decimal point needs to be moved. Powers of 10 will be very useful in order of magnitude problems, which are rough estimates.

$$10^3 = 10 * 10 * 10 = 1000$$

$$10^{-3} = \frac{1}{10^3} = 0.001$$

Note there is a discrepancy between the use of billions in the United States (10^9) and England (10^{12}). If there is a doubt, we will use exponents or the following notation for prefixes.

Factor	Name	Symbol	Factor	Name	Symbol
10^{-12}	pico	p	10^3	kilo	k
10^{-9}	nano	n	10^6	mega	M
10^{-6}	micro	μ	10^9	giga	G
10^{-3}	milli	m	10^{12}	tera	T
			10^{15}	peta	P
			10^{18}	exa	E

1 quad = 1.055 exajoules

1.3 RENEWABLE ENERGY

Solar energy is referred to as renewable and/or sustainable energy because it will be available as long as the Sun continues to shine. Estimates for the remaining life of the main stage of the Sun are another 4 to 5 billion years. The energy from the Sun, electromagnetic radiation, is referred to as *insolation or solar energy*. The other renewable energies are wind, bioenergy, geothermal, hydro, tides, and waves. *Wind energy* is derived from the uneven heating of the Earth's surface due to more heat input at the equator with the accompanying transfer of water and thermal energy by evaporation and precipitation. In this sense, rivers and dams for hydro energy are stored solar energy. The third major aspect of solar energy is the conversion of solar energy into biomass by photosynthesis. Animal products such as oil from fat and biogas from manure are derived from solar energy. *Geothermal energy* is due to heat from the Earth from decay of radioactive particles and residual heat from gravitation during formation of the Earth. Volcanoes are fiery examples of geothermal energy reaching the surface from the interior, which is hotter than the surface. *Tidal energy* is primarily due to the gravitational interaction of Earth and Moon.

Overall, 14% of the world's energy comes from *bioenergy*, primarily wood and charcoal, but also crop residue and even animal dung for cooking and some heating. This contributes to deforestation and the loss of topsoil in developing countries. Production of ethanol from biomass is now a contributor to liquid fuels for transportation, especially in Brazil and the United States.

In contrast, fossil fuels are stored solar energy from past geological ages. Even though the quantities of oil, natural gas, and coal are large, they are finite and for the long term of hundreds of years, they are not sustainable.

1.4 ADVANTAGES/DISADVANTAGES

The advantages of renewable energy are sustainable (non-depletable), ubiquitous (found everywhere across the world in contrast to fossil fuels and minerals), and essentially non-polluting. Note that wind turbines and photovoltaic panels do not need water for the generation of electricity, in contrast to steam plants fired by fossil fuels and nuclear power.

The disadvantages of renewable energy are variability and low density, which in general results in higher initial cost. For different forms of renewable energy, other disadvantages or perceived problems are visual pollution, odor from biomass, avian and bat mortality with wind turbines, and brine from geothermal. Wherever a large renewable facility is to be located, there will be perceived and real problems to the local people. For conventional power plants using fossil fuels, for nuclear energy, and even for renewable energy, there is the problem of *not in my backyard*.

1.5 ECONOMICS

Business entities always couch their concerns in terms of economics (money). We cannot have a clean environment because it is uneconomical. Renewable energy is not economical in comparison to coal, oil, and natural gas. We must be allowed to continue our operations as in the past, because if we have to install new equipment to reduce greenhouse gas emissions, we cannot compete with other energy sources, and finally, we will have to reduce employment, and jobs will go overseas.

The different types of economics to consider are pecuniary, social, and physical.

Pecuniary is what everybody thinks of as economics, *money*. On that note, we should be looking at life-cycle costs, rather than our ordinary way of doing business, low initial costs. Life-cycle costs refer to all costs over the lifetime of the system.

Social economics are those borne by everybody and many businesses want their environmental costs to be paid by the general public. A good example is the use of coal in China, as they have laws (social) for clean air, but they are not enforced. The cost will be paid in the future in terms of health problems, especially for the children today. If environmental problem(s) affect(s) someone else today or in the future, who pays? The estimates of the pollution costs for the generation of electricity by coal is $0.005 for 0.10/kWh.

Physical economics is the energy cost and the efficiency of the process. There are fundamental limitations in nature due to physical laws. Energetics, which is the energy input versus energy in the final product for any source, should be positive. For example production of ethanol from irrigated corn has close to zero energetics. So, physical economics is the final arbitrator in energy production and consumption. In the end, *Mother Nature always wins* or the corollary, pay now or probably pay more in the future.

Finally, we should look at incentives and penalties for the energy entities. What each entity wants are subsidies for themselves and penalties for their competitors. Penalties come in the form of taxes, environmental, and other regulations, while incentives come in the form of subsidies, break on taxes, do not have to pay social costs on the product, and the government pays for research and development. How much should we subsidize businesses for exporting overseas? It is estimated that we use energy sources in direct proportion to the incentives that source has received in the past. There are many examples of incentives and penalties for all types of energy production and use.

1.6 CLIMATE CHANGE

Climate change, which previously was referred to as *global warming*, is a good example that physical phenomena do not react to political or economic statements. Global warming is primarily due to human activity. "Global atmospheric concentrations of carbon dioxide, methane and nitrous oxide have increased markedly as a result of human activities since 1750 and now far exceed pre-industrial values determined from ice cores spanning many thousands of years.... The global increases in carbon dioxide concentration are due primarily to fossil fuel use and land use change, while those of methane and nitrous oxide are primarily due to agriculture" [2]. Concentrations of carbon dioxide in the atmosphere (Figure 1.2) are projected to double with future energy use based on today's trend [3,4].

The Kyoto Protocol of 1996 to reduce greenhouse gas emissions became effective in 2005 as Russia became the 55th country to ratify the agreement. The goal was for the participants collectively to reduce emissions of greenhouse gases by 5.2% below the emission levels of 1990 by 2012. While the 5.2% figure was a collective one, individual countries were assigned higher or lower targets and some countries were permitted increases. For example, the United States was expected to reduce emissions by 7%. However, this did not happen, as the United States did not ratify the treaty because the perceived economic costs would be too large and there were not enough provisions for developing countries, especially China, to reduce future emissions. Note that for the past few years, U.S. emission of carbon dioxide has decreased due to increased use of natural gas and increase in wind and solar power for the production of electricity, and because of less economic activity due to recession of 2008.

If participant countries continue with emissions above the targets, then they are required to engage in emissions trading. Notably, participating countries in Europe are

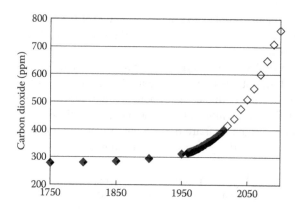

FIGURE 1.2 Carbon dioxide in the atmosphere and projected growth with no emission reductions.

using different methods for carbon dioxide trading, including wind farms and planting forests in other countries. Carbon dioxide emissions will still increase, even if nations reduce their emissions to 1990 levels, because of population growth and increase in energy use in the underdeveloped world. As the Arctic thaws, then methane, a more potent greenhouse gas than CO_2, would further increase global warming [5].

Increased temperatures and the effect on weather and sea level rise are the major consequences. Overall, the increased temperature will have negative effects compared to the climate of 1900–2000. By 2100, sea levels are projected to increase by 0.2 to 1 m, with an increase of 2 m unlikely, but physically possible. With positive feedback due to less sea ice and continued increase in carbon dioxide emissions, the melting of the Greenland ice sheets would increase the sea level by over 7 m and the West Antarctic Ice Sheet would add another 5 m. The large cities near the oceans will have to be relocated or build massive infrastructures to keep out the ocean. Who will pay for this, national or local governments?

1.7 ORDER OF MAGNITUDE ESTIMATES

In terms of energy consumption, production, supply and demand, and design for heating and cooling, estimates are needed and an order of magnitude estimate will suffice. By order of magnitude, we mean an answer to within a power of 10.

Example

How many seconds in a year. With a calculator, it is easy to determine

365 days * 24 h/day * 60 min/h * 60 s/h = 31,536,000

When you round to one significant digit, this becomes $3 * 10^7$ s.

For an order of magnitude estimate for the above multiplication, round each number with a power of 10, then multiply numbers and add the powers of 10

$$4 * 10^2 * 2 * 10^1 * 6 * 10^1 * 6 * 10^1 = 4 * 2 * 6 * 6 * 10^5$$

$$= 288 * 10^5 = 3 * 10^2 * 10^5 = 3 * 10^7 \text{s}$$

1.8 GROWTH (EXPONENTIAL)

Our energy dilemma can be analyzed in terms of fundamental principles. It is a physical impossibility to have exponential growth of any product or exponential consumption of any physical resource in a finite system. As an example, suppose Mary started employment with $1/year; however, her salary is doubled every year, a 100% increase (Table 1.2, Figure 1.3). Notice that after 30 years, her salary is one billion dollars. Also notice that for any year, the amount needed for the next period is equal to the total sum for all the previous periods plus one. The mathematics of exponential growth is given in Appendix I.

TABLE 1.2
Exponential Growth with a Doubling Time of 1 Year

Year	Salary ($)	Amount = 2^t	Cumulative ($)
0	1	2^0	1
1	2	2^1	3
2	4	2^2	7
3	8	2^3	15
4	16	2^4	31
5	32	2^5	63
t		2^t	$2^{t+1} - 1$
30	$1 * 10^9$	2^{30}	$2^{31} - 1$

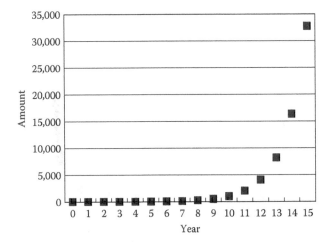

FIGURE 1.3 Salary with a doubling time of 1 year to show exponential growth.

Another useful idea is doubling time, $T2$, for exponential growth, which can be calculated by

$$T2 = \frac{69}{R} \tag{1.1}$$

where:
   R is the % growth per unit time

Doubling times for some different year rates are given in Table 1.3.

There are numerous historical examples of growth: population, 2%–3%/year; gasoline consumption, 3%/year; world production of oil, 5%–7%/year; and electrical consumption, 7%/year. If we plot the value per year for smaller rates of growth (Figure 1.4), the curve would be the same as Figure 1.3, only the timescale along

TABLE 1.3
Doubling Times for Different Rates of Growth

Growth (%/year)	Doubling Time Years
1	69
2	35
3	23
4	18
5	14
6	12
7	10
8	8
9	8
10	7
15	5

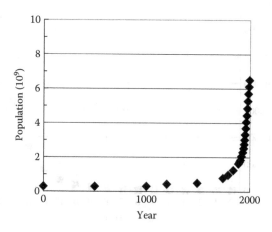

FIGURE 1.4 World population showing exponential growth.

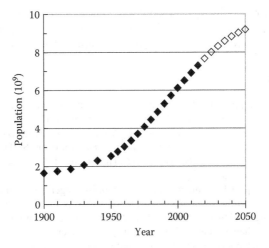

FIGURE 1.5   World population with projection to 2050 under median variant.

the bottom would be different. The United Nations projects over 9 billion people (Figure 1.5) by 2050 [6], with the assumption that the growth rate will decrease from 1.18% in 2008 to 0.34% in 2050.

However, even with different rates of growth, the final result is still the same. *When consumption grows exponentially, enormous resources do not last very long.*

This is the fundamental flaw in terms of ordinary economics ($) and announcing growth in terms of percentages. How long do they want those growth rates to continue? Nobody wants to discuss how much is enough. The theme since President Reagan is that all we need is economic development and the world's problems will be solved. However, the global economic crisis of 2008 and environmental problems have made some economists have second thoughts on continued growth. Now there are lots of books on the problems of fossil fuels, other resources such as minerals and water and environmental effects.

1.9   SOLUTIONS

We do not have an energy crisis, since you will learn energy cannot be created nor be destroyed. We have an energy dilemma because of the finite amount of readily available fossil fuels, which are our main energy source today. The problem is twofold, over population and over consumption. Population is $7.3 * 10^9$ in 2015 and is growing toward $9 * 10^9$ and maybe even larger, and developing countries want the same standard of living as developed countries. The world population is so large that we are doing an uncontrolled experiment on the Earth's environment. However, the developed countries were also major contributors to this uncontrolled experiment in terms of consumption and now increased consumption in China and India is adding to the problem.

The solution depends on world, national, and local policies and what policies do we implement and even individual actions. In our opinion, it is obvious what needs to be done for the world: reduce consumption, zero population growth, shift to renewable energy, reduce greenhouse gas emissions, reduce environmental pollution, and reduce military expenditures. What do you do as an individual? We have done things in the past to save energy and have future plans. What are yours?

REFERENCES

1. Interactions. http://hyperphysics.phy-astr.gsu.edu/hbase/forces/funfor.html.
2. IPCC. 2007. Summary for policymakers. In: *Climate Change 2007: The Physical Science Basis. Contribution of Working Group I to the Fourth Assessment Report of the Intergovernmental Panel on Climate Change.* Solomon, S., D. Qin, M. Manning, Z. Chen, M. Marquis, K.B. Averyt, M. Tignor, and H.L. Miller (eds.). Cambridge University Press, Cambridge.
3. Technical Support Document. April 17, 2009. Endangerment and cause or contribute findings for greenhouse gasses under section 202(a) of the clean air act. http://epa.gov/climatechange/endangerment/downloads/TSD_Endangerment.pdf.
4. US Climate Change Science Program. July 2007. Scenarios of greenhouse gas emissions and atmospheric concentrations; and review of integrated scenario development and application. Final Report. www.climatescience.gov/Library/sap/sap2-1/finalreport/sap2-1a-final-technical-summary.pdf.
5. S. Simpson. 2009. The peril below the ice. *Sci Am, Earth 3.0*, Vol 18, # 2, p 30.
6. United Nations. 2008. World population prospects, the 2008 revision, highlights. www.un.org/esa/population/publications/wpp2008/wpp2008_highlights.pdf.

RECOMMENDED RESOURCES

GENERAL REFERENCES

T. Flannery. 2006. *The Weather Makers: How Man Is Changing the Climate and What it Means for Life on Earth.* Atlantic Monthly Press, New York.

J. Gustave Speth. 2005. *Red Sky at Morning.* Yale University Press, New Haven, CT.

R. Heinberg. 2004. *The Party's Over.* New Society Publishers, Gabriola Island, Canada.

J. Howard Kunstler. 2005. *The Long Emergency.* Atlantic Monthly Press, New York.

M. T. Klare. 2001. *Resource Wars: The New Landscape of Global Conflict.* Metropolitan Books, New York.

M. T. Klare. 2009. *Rising Powers, Shrinking Planet.* Metropolitan Books, New York.

E. Kolbert. 2006. *Field Notes from a Catastrophe, Man, Nature, and Climate Change.* Bloomsbury, New York.

J. Lovelock. 2006. *The Revenge of Gaia: Earth's Climate Crisis and the Fate of Humanity.* Perseus Book Group, New York.

W. Youngquist. 1997. *Geodestinies: The Inevitable Control of Earth Resources over Nations and Individuals.* National Book, Portland, OR.

LINKS

Energy Information Administration, U.S. Department of Energy. www.eia.doe.gov. This site contains a lot of information on U.S. and international energy resources and production.

International energy outlook. http://www.eia.doe.gov/iea/. Data files can be downloaded, PDFs and spreadsheets.

United Nations: Information on population and projections on population. www.un.org/esa/population/unpop.htm.

U.S. Census; U.S., world population clocks. www.census.gov.

GLOBAL WARMING

Global Climate Change Impacts in the United States Report. June 2009. www.globalchange.
    gov/publications/reports/scientific-assessments/us-impacts.
Intergovernmental Panel on Climate Change. www.ipcc.ch.
Union of Concerned Scientists, Global Warming. www.ucsusa.org/global_warming.
United States Global Change Research Program. www.globalchange.gov.

PROBLEMS

1.1. What was the population of the world in 1950, 2000, this year, and projected for 2050?

1.2. What was the population of your country in 1950, 2000, this year, and projected for 2050?

1.3. List two advantages of renewable energy.

1.4. List two disadvantages of renewable energy.

1.5. Besides large hydro, what are the two most important renewable energy sources for your country?

1.6. For a sustainable society in your country, what would be the two most important policy issues?

1.7. What are the largest two sources for carbon dioxide emissions?

1.8. Besides the United States, what country consumes the most energy?

1.9. What country emits the most carbon dioxide?

1.10. The size of the European Union has increased over the years. Estimate the percentage increase in GDP and energy consumption by the addition of these new blocks of countries.

1.11. When is gravity considered a source for renewable energy?

1.12. Global warming is primarily due to what factor?

1.13. What is the predicted amount of carbon dioxide, ppm, in the atmosphere for 2050?

1.14. What two nations emit the most carbon dioxide per year (Figure 1.1)? What percent is that of the world total?

1.15. What percent of the world total of carbon dioxide emission per year is due to combustion of coal, combustion of oil, and combustion of natural gas?

1.16. What is your carbon footprint? Calculators are available on the Internet, for example, www.carbonify.com/carbon-calculator.htm or www.carbon-footprint.com/calculator1.html.

1.17. Under the Kyoto Protocol, list three participating countries and their emission levels of carbon dioxide (latest year available) compared with their levels of 1990. Remember the target levels are below 1990 levels.

1.18. The local business people want the city to grow. What rate do they want, %/year? What is the doubling time?

1.19. Suppose world population grows at 0.5% per year, what is the doubling time? After that period of time, what is the projected world population?

2 Energy

2.1 INTRODUCTION

Scientists have been successful in understanding and finding unifying principles. However, many people take the resulting technology for granted and do not understand the limitations of humans as part of the physical world. There are moral laws (or principles), civil laws, and physical laws. Moral laws have been broken (e.g., murder and adultery), civil laws have been broken (almost everybody has driven over the speed limit), *but no one breaks a physical law.* Therefore, we can only work with nature, and we cannot do anything that violates the physical world.

We have been and we will be clever in manipulating and using physical laws in terms of science and the application of science and technology. Just think what has occurred in the past century, from the first flight of airplanes (1903) to man landing on the Moon (1969), and to exploration of the solar system by robotic systems, and from the special theory of relativity (1905), which predicted the relation between mass and energy, to the atomic bomb (1945). Another major technology advance is the invention of the transistor (1947), which led to integrated circuits and myriad electronic devices; as a result, much of the population of the world (2015) now have instant mobile phones, songs, and video in their hands and via Internet access.

A major unifying concept is energy and how energy is transferred. The area that deals with heat, a form of energy, is called *thermodynamics.*

2.2 DEFINITION OF ENERGY AND POWER

To understand renewable energy, the definitions of energy and power are needed. *Work* is the result of force acting on an object, which is then moved through some distance.

$$\text{Work} = \text{force} * \text{distance}$$

or

$$W = F * D, \quad \text{joule (J)} = \text{newton (N) meter (m)} \tag{2.1}$$

A number of symbols are used, and with the easy availability of personal computers and calculators, sample calculations are used for illustration and understanding. Many people have a mental block as soon as they see mathematical symbols, but everybody uses symbols. Therefore, Equation 2.1 can be understood as a shorthand notation for the words and written concepts that preceded it.

Moving objects, doing work, and changing position between interacting particles require energy, so energy and work are measured by the same units. Some units of energy are joule, calorie, kilowatt hour (kWh), Btu, and quad.

calorie = Amount of energy required to raise
the temperature of 1 g of water by 1°C

British thermal unit (Btu) = Amount of energy required to raise the temperature
of 1 pound of water by 1°F

Some conversion factors for energy are as follows:

$$1 \text{ cal} = 4.12 \text{ J}$$

Calorie = Kilocalorie, the unit used in nutrition = 1,000 calories

$$1 \text{ Btu} = 1,055 \text{ J}$$

$$1 \text{ barrel of oil } (42 \text{ gal}) = 6.12 \times 10^9 \text{ J} = 1.7 \times 10^3 \text{ kWh}$$

$$1 \text{ ton of coal} = 2.8 \times 10^7 \text{ Btu} = 2.9 \times 10^{10} \text{ J}$$

$$1 \text{ quad} = 10^{15} \text{ Btu} = 1.055 \text{ exajoules}$$

$$1 \text{ kWh} = 3.6 \times 10^6 \text{ J}$$

$$1 \text{ ton of oil equivalent } (\text{toe}) = 4.2 \times 10^{10} \text{ J}$$

$$1 \text{ million tons of oil equivalent } (\text{Mtoe}) = 4.2 \times 10^{16} \text{ J}$$

Note: Energy content/mass for different sources of energy is fairly constant; however, energy content/volume will vary, depending on the density. Coal will vary in energy content between anthracite, bituminous, lignite, and peat (highest moisture content). This text uses metric tons (1,000 kg = 2,205 lb); however, some sources will use short ton (2,000 lb) and the previous ton or long ton (2,400 lb), so be sure what units are being used when calculating energy. The major sources on world and regional energy (Energy Information Administration [EIA], International Energy Agency, and BP world energy) may use different units, for example, quads, million tons of oil equivalent (Mtoe), millions of barrels of oil, and exajoules. Another difference is in generation of electric energy, where output is in TWh or billion kWh. In the EIA spreadsheets, they give electric generation in billion kWh or quads for some sources; however, there is not a direct conversion. If electric generation is given in terms of thermal energy, generally they account for energy input to generate the output, so there is the efficiency of the steam plants fueled by fossil fuels, nuclear, and biomass, around 38%. Therefore, one Mtoe (4.2×10^{16} J) produces around 4.4 TWh ($1.6 * 10^{16}$ J) of electricity.

Objects in motion can do work; therefore, they possess energy, kinetic energy (KE):

$$KE = 0.5mv^2 \tag{2.2}$$

where:
m is the mass of the object
v is its speed

Example 2.1

A car with a mass of 1,000 kg moving at 10 m/s has the following kinetic energy:

$$KE = 0.5 * 1,000 * 10 * 10 \text{ kg}(\text{m/s})^2 = 5,000 \text{ J}$$

Remember in calculations, units are required.

Because objects interact (e.g., by gravity), then due to their relative position they can do work or have energy, potential energy (PE). To raise a 10 kg mass a height of 2 m while standing on the Earth's surface requires 200 J of energy. You are changing the gravitational PE, so at that upper level that object has an additional 200 J of PE.

Power is the rate of energy use or production.

$$\text{Power} = \text{energy/time, joule/second} = \text{watt} \qquad (2.3)$$

If power is known, then energy can be calculated for any time period.

$$E = P * t \qquad (2.4)$$

A kilowatt (kW) is a measure of power, and a kilowatt hour (kWh) is a measure of energy. Motors (kW or horsepower) and power plants (megawatts) are rated in terms of power.

Example 2.2

A 5-kW electric motor, which runs for 2 h, consumes 5 kWh * 2 h = 10 kWh of energy.

Example 2.3

Eight 100-W light bulbs that are left on all day will consume 8 * 100 * 24 = 19,200 Wh = 19.2 kWh of energy.

2.3 HEAT

Heat is another form of energy, thermal energy. Heat is just the internal kinetic energy (random motion of the atoms) of a body. Rub your hands together, and they get warmer. As you heat your home, you are increasing the speed of the air particles. *Heat and temperature are different.* Heat is energy, and temperature is the potential for transfer of heat from a hot place to a cold place. In the transfer of this heat, work can be done. As an example of the difference between heat and temperature: Would you rather stick your finger in a cup of hot coffee, $T = 80°C$, or get hit by a high-speed proton, $T = 1,000,000°C$? One has much more energy than the other.

2.4 THERMODYNAMICS

The understanding of energy today can be embodied in the following laws or principles of thermodynamics:

1. Energy is conserved. Energy is not created or destroyed, only transformed from one form to another. In layperson's terms, this means that all you can do is break even. A number of patents have been issued for perpetual motion machines [1], a device that produces more energy than the energy needed to run the machine. A number of people have invested money in such machines, but needless to say, the money was lost since the devices contradict the first law of thermodynamics
2. Thermal energy, heat, cannot be transformed totally into work. In simpler terms, you cannot even break even. Another way of looking at it is that systems tend toward disorder, and in transformations of energy, disorder increases. As entropy is a measure of order, then in succinct terms, entropy is increasing

This means that some forms of energy are more useful than other forms. For example, the energy in a gallon of gasoline is not lost but only transformed into heat by a car. However, after the transformation, that energy is dispersed into a low-grade form (more entropy) and cannot be used to do more work in moving the car.

As an aside for the scientists, the following most famous equation says that mass is just a concentrated form of energy. Conversion of a small amount of mass gives a lot of energy (e.g., an atomic or hydrogen bomb).

$$E = mc^2$$

where:
c is the speed of light

2.5 ENERGY DILEMMA IN LIGHT OF THE LAWS OF THERMODYNAMICS

As energy cannot be created or destroyed, only transferred, we have an energy dilemma in the use of energy resources and their effect on the environment. Therefore, *the first and primary objective of any energy policy must be conservation and efficiency,* as that results in the most economic use of a barrel of oil and is less expensive than drilling for new oil.

2.5.1 CONSERVATION

Conservation means if you do not need it, do not turn it on or use it. President Carter's admonition to reduce the thermostat setting and setting a speed limit of 55 mph were conservation measures. High prices and shortages (e.g., in the California electrical crisis of 2000–2001) resulted in an increased conservation, and the high price of gasoline in 2008 made more people consider fuel efficiency before they purchase a vehicle. In general, utility companies like to sell more energy rather than have customers save energy.

2.5.2 EFFICIENCY

Efficiency is the ratio of the energy for the function or product divided by the energy input.

$$\text{Efficiency} = \text{energy out/energy in}$$

Energy can be used to do work (mechanical energy) and to heat an object or space (thermal energy) and be transformed to electrical energy or stored as PE. In each transformation, an upper limit on efficiency can be determined by the second law of thermodynamics. In thermal processes, the temperatures of the hot and cold reservoirs determine this efficiency.

$$\text{Eff} = \frac{T_H - T_C}{T_H} \tag{2.5}$$

Temperatures must be in degrees Kelvin, $T_{\text{deg K}} = T_{\text{deg C}} + 273$.

In an electrical generating plant that uses input steam at 700°C (973 K) and on the downside is cooled by water to 300°C (573 K), the maximum efficiency possible is around 0.41% or 41%. Modern thermal power plants have efficiencies of around 40%. In other words, 60% of the stored chemical (or nuclear) energy is rejected and 40% is converted into electricity.

Temperature is a measure of potential for heat transfer and is not a measure of energy. Since efficiency is always less than 1, for any system or device to continue to operate, energy must be obtained from outside the system. For every energy transformation, there is an efficiency, and the total efficiency is the product of the individual efficiencies (multiply). The efficiency of converting coal to light using incandescent light bulbs is around 2%.

Transformation	Efficiency (%)
Mining of coal	96
Transportation of coal	97
Generation of electricity	38
Transmission of electricity	93
Incandescent bulb (electricity to light)	5
Overall efficiency (coal to light)	1.6

Therefore, fluorescent lights (15%–25% efficiency) for commercial buildings and compact fluorescent lights for your home are important. Now, light-emitting diodes (LEDs) (25%–50% efficiency) are available. Countries, states, and even cities have set regulations to phase out incandescent lighting. This also says that daylighting can save money, especially during the summer as you do not need air conditioning to reduce the heat given off by lights.

As a corollary to the second law efficiency, a system for producing energy must be a net energy gainer, the *energetics* of the system, also referred to as *energy balance, energy return on energy invested, energy ratio*. In the physical world, subsidies or economics (dollars) do not change the energetics of the systems for the production of energy; all they do is tilt consumption or use in favor of different energy resources. For example, at some point in the future it will take more energy to drill for oil than the amount of energy in the oil produced. At that point, it would be foolish to subsidize the drilling for oil as an energy source. It might be that the product is so useful as a liquid fuel or as a source for other products that it could be subsidized by other energy sources. How much natural gas does it take to produce oil from tar sands in Canada? What is the energetics of producing ethanol from corn?

Prior to the oil crisis of 1973, industry and business maintained that efficiency was not cost effective and that gross domestic product (GDP) was tied directly to the use of energy. Industry changed, and the United States saved billions (10^9) of dollars since 1973 by increased efficiency in industry and higher efficiency for transportation. However, as stated, much more has to be accomplished by conservation and increasing efficiency.

An example of efficiency is cogeneration, today referred to as *combined heat and power* (CHP). In the production of electricity, the low-grade (lower-temperature) energy can be used for other processes. In most electric power plants where electricity is generated by steam (coal, oil, gas, and even nuclear), 60% of the heat is not used. Combined cycle gas turbines have higher efficiencies as heat from the first cycle is also used. In Europe, some electric power plants have heating districts associated with them.

Efficiency in transportation is an example of the difficulty in formulating a rational energy policy that would convert the world to sustainable energy within the environmental limitations. Hybrid cars entered the market in 2000, and in 2014, there are a lot of hybrid models and even a number of plug-in hybrid models and electric vehicles. A personal note, Nelson has had hybrid models for a number of years, and in 2012, he bought an electric vehicle, Nissan Leaf (average 6 km/kWh, range around 100 km). Starcher bought a Prius (Toyota hybrid) in 2014 and has averaged 45 mpg in all his travels to date.

Every U.S. president since 1973 has called for energy independence, primarily due to the high cost of imported oil. The high price for oil in 2008 and then the financial crisis with the automobile industry needing government money to operate in 2009 demonstrate the failure of past energy policies. In 2006, President G. W. Bush's energy policy was to drill for more oil and gas, and as in the past, the automobile industry fought against increasing fuel efficiency. The argument was again couched in terms of economics: Because we cannot compete with foreign manufacturers of small cars, consumers will not buy fuel-efficient cars (advertising advocates large motors, heavy sport-utility vehicles [SUVs], and large vehicles for safety). In past discussions with students, they stated that gasoline in the United States would have to be around $1/L ($4/gal) before they would buy a fuel-efficient vehicle. Of course, Europeans and people in other countries have been paying those and even higher prices for a long period. In 2008, with oil over $100/bbl, the sale of fuel-efficient vehicles increased; however, as people in the United States become accustomed to that price, sales of fuel-efficient vehicles declined. In December 2014, oil prices decreased dramatically; what will that do to the sale of fuel-efficient vehicles? Couched in terms of the safety issue, everybody should drive a semitruck (to heck with fuel efficiency), or at least we all deserve big Cadillacs. Another note is that vehicles powered by fuel cells using hydrogen are much more efficient than those with internal combustion engines. Why do we not have millions of those vehicles on the road today?

Looking back, the obvious answer was to reduce gasoline use, increase fuel efficiency, and mandate a substantial tax on gasoline after the oil crisis of 1973. There was progress on fuel efficiency as in 1975 the U.S. Congress passed laws for combined automobile fleet efficiency (CAFE) for vehicles weighing less than 3,886 kg (8,500 lb); however, pickups and large vans did not count in the CAFE requirements. This law has saved the United States billions of dollars in imported oil. The problem was that SUVs were counted as light trucks, and their fuel consumption was around 5.5 km/L (12 mpg), so the overall fuel efficiency declined as SUVs gained market share. In 1999, over half a million vehicles sold exceeded the gross vehicle weight requirements. Even with continued objections by the automobile industry and reluctant acceptance by the G. W. Bush administration, finally in 2007 CAFE was revised to eliminate the exemptions to light trucks classified as SUVs or passenger vans unless they exceed 4,500 kg (10,000 lb) gross vehicle weight, and the CAFE was increased to 15 km/L (35 mpg) by the year 2020.

Under President Obama, an agreement reached in 2011 with major automotive manufacturers (except Volkswagen) is that the CAFE should be raised to 23 km/L (54.5 mpg) for cars and light trucks by 2025. The European Union and Japan fuel economy standards for 2012 are around 19 L/km (45 mpg) and their proposed standards for 2025 will still be greater than that of the United States. U.S. federal tax credits (2009) include a new tax credit, starting at $2,500 and capped at $7,500, for plug-in hybrid and electric vehicles. The first 250,000 vehicles sold by each manufacturer get the full tax credit, and then it phases out like the hybrid vehicle tax credits were phased out with more vehicles sold.

In an interesting note, the big three automobile manufacturers in the United States received over $2 * 10^9 in research and development (R&D) funding from the government for the Partnership for New Generation of Vehicles [2]. The goal was to create a sedan for five people that would obtain 80 mpg. The manufacturers said that there is no way to reach that goal; however, they wanted money to build more efficient cars and even obtained tax breaks for a limited time period for people trading in old, inefficient cars ("cash for clunkers").

Again, the question concerns where the federal government should place these incentives. It would be cheaper to subsidize more efficient cars than to pay all the costs for imported oil. What is the additional cost for imported oil if the military costs for the Oil War I (Gulf War) and the Oil War II (Iraq War) are included? An additional $0.50 tax on a gallon of gasoline would just about pay for Oil War II and the war in Afghanistan and would help drive purchases of more efficient cars.

The Organization of Petroleum Exporting Countries (OPEC) wants to keep the price of oil in the range at which they make a lot of money, but not so high it encourages conservation and efficiency. However, *at some point the demand for oil across the world will be higher than can be supplied.* When production starts to decline, we will have higher prices, which will surpass the price of $100/bbl, even with the production from oil sands and oil shale formations.

2.6   USE OF FOSSIL FUELS

The night images of Earth (Figure 2.1) taken by satellite illustrates the tremendous amount of energy radiating into space from lights and fires [3]. In 2014, world consumption of energy was around 550 exajoules (EJ), with China surpassing the United States in 2011. With the addition of Europe, those three entities account for over 50% of the total, and with the addition of only three other nations, the amount is 70% (Figure 2.2). Most of that energy, 87%, is from fossil fuels (petroleum, natural gas, and coal) (Figure 2.3), while electricity from hydropower and nuclear power is the other major component.

The quantity of fossil fuels consumed per year is astounding; 33 Gbbls of petroleum, 3.4 Tm^3 of dry natural gas, and 7.7 Gt of coal. Then as the fuel is burned, the carbon content is emitted into the atmosphere as carbon dioxide.

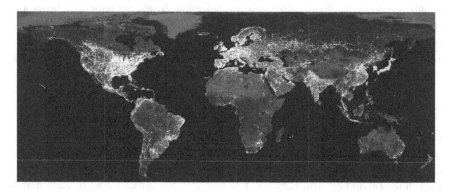

FIGURE 2.1   Nighttime image of Earth from satellites. (Courtesy of NASA, Visible Earth, http://eoimages.gsfc.nasa.gov/images/imagerecords/55000/55167/earth_lights_4800.tif.)

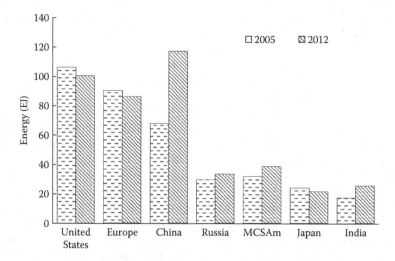

FIGURE 2.2   Consumption of energy for regions and nations with large population or large GDP. MCSAm stands for Mexico, Central, and South America.

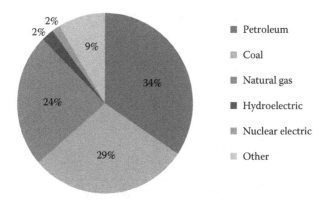

FIGURE 2.3 Consumption of world energy, 2012, by resource.

Three noticeable developments are the largest hydroelectric project in the world, Three Georges dam on the Yangtze River in China; the Fukushima nuclear disaster in Japan, which resulted in all their nuclear plants being shut down for some time; and horizontal drilling and fracking to obtain oil and gas from shale formations. Horizontal drilling and fracking of shale formations has actually increased the production of oil and gas in the United States and this technology will then be used in other parts of the world to extend the time and amount of oil and gas production.

Comparison of generation of electricity from non-fossil fuels for the major consumers of energy, population, and GDP is shown in Figure 2.4, with the United States and Europe leading in nonhydro renewables. The developed world plus China and India have a concentrated effort to increase the use of renewables for the generation of electricity. By 2014, the total for nonhydro renewables for electric generation

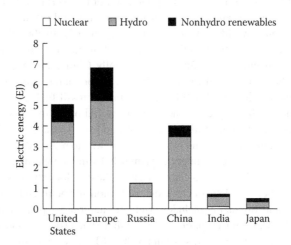

FIGURE 2.4 Electric energy from non-fossil fuels for Europe and nations with large populations or large GDP.

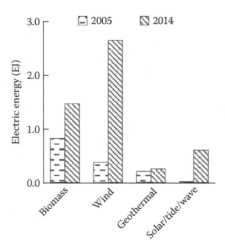

FIGURE 2.5   World electric energy from nonhydro renewables.

was around 5 EJ (Figure 2.5). The use of percent versus actual values sometimes gives a distorted view, especially if you start with small numbers. Although non-hydro renewables are still a small percentage of the total for the generation of electricity, from 2005 to 2014 they increased by around 200% (world electric generation increased by around 40%, which included new renewable generation). Generation of electricity by small wind, photovoltaic, and hydro systems does not show on this scale; however, the EIA now breaks out renewable data capacity, and estimation of energy production for all renewable energy systems is covered in detail in subsequent chapters, and more technical aspects are covered in the series *Renewable Energy and the Environment* [4].

The United States, with 4.5% of the population of the world, consumes 18% of the world's energy resources and a major portion of the mineral resources; now, demand from the developing world is increasing (see Figure 1.1). In the 1970s, the United States consumed 25% of the world's energy, which means the underdeveloped world has increased their energy consumption, especially China. The energy flow for the United States shows that fossil fuels are still the major source of energy (Figure 2.6). The energy flow for the world will be similar in that fossil fuels account for most of the energy sources; the only difference is there would be no imports. At the end-use sectors (transportation, industrial, commercial, and residential), over 50% is wasted energy. Therefore, the quickest and most economic way to reduce the use of fossil fuels and emissions is to reduce consumption, primarily by the more efficient use of energy.

The magnitude of the problem can be seen by the cost for petroleum imports in the United States. Notice that data for crude oil production and petroleum supply/consumption are different as oil supply includes crude oil, natural gas plant liquids, and other liquids. In 1973, U.S. consumption was 6.2 Gbbl (barrels)/year, and approximately 40% was imported, so the cost was around $100 * 10^9 per year for imported oil at $40/bbl (if the cost is adjusted for inflation, it would be double as oil would be $100/bbl in 2014 dollars). Even though consumption of imported oil was reduced in the 1980s,

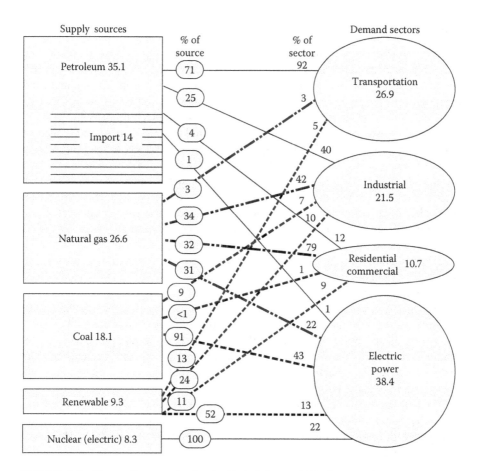

FIGURE 2.6 Energy flow, quads, for the United States, 2013, from production to end-use sectors plus percentage from supply sources and percentage to demand sectors. (Data from Energy Information Administration (EIA), U.S. Department of Energy. http://www.eia.doe.gov/.)

the cost was still expensive. In the 1990s, oil consumption and imports in the United States increased again toward the previous levels, and in 2005, consumption was over 7 Gbbl/year. Due to the world financial crisis, petroleum consumption in the United States decreased to 6.5 Gbbl in 2008, with 62% imported at $100/bbl, for a cost around $400 * 10^9$. Then, U.S. production increased due to horizontal drilling and fracking to obtain oil and gas from shale formations. By 2014, imports were below 40%; however, at $90/bbl the cost of imported oil was still quite large, $250 * 10^9$. Again note the volatility of the price of oil, as it fell from $100/bbl (July 2014) to $45/bbl (January 2015). This reduction saved consumers in the United States a large amount of money; however, even at the low price of $50/bbl, the estimated cost is $124 * 10^9$ for imported oil (estimated at 2.48 Gbbls for 2015).

The EIA reference case [5] predicts a continued increase in domestic production to the level of around 4.2 Gbbls/year by 2020. However, consumption at 6.7 Gbbls/year means imports would be still be significant, around 2.5 Gbbl/year.

2.6.1 PETROLEUM

The important concept is that crude estimates of resources give fairly good answers regarding when production for finite resources will peak. Also, predictions on the future use of the resource can be made from past production as production and consumption of a finite resource will probably be somewhat similar to the bell curve. In 1956, Hubbert [6] predicted that the U.S. oil production would peak in the mid-1970s, and he was close as the actual peak occurred in 1970. Petroleum production decreased from 1970 through 2005 (Figure 2.7), even with the production from Alaska and offshore oil fields. The increase in production starting in 2006 has reached the levels of peak production in 1970 and is expected to continue to increase until 2020, and after that date, there will again be a declining rate of domestic production. Note that the depletion rate from the shale formations will be faster than for conventional oil. In any case, because demand is large, imports will still be significant.

The prediction (logistic curve) of U.S. petroleum production in Figure 2.8 used actual production through 2014, and the prediction was calculated in a spreadsheet using the method of Deffeyes [7, Chapter 7]. The prediction curve now does not fit as well and the peak has shifted from 1970 to 1990.

Even if a larger resource base is assumed, with exponential growth the larger resource is used up at about the same time. Also, as the resource is used, it becomes more difficult to obtain the resource; that is, it takes more energy and money to obtain the resource. The amount of oil and natural gas discovered per meter of hole drilled decreases exponentially. The same type of analysis and predictions can be made for natural gas, coal, and nuclear ore.

The bell curve, also called the *normal or Gaussian curve*, will not be exact for predicting future production as advanced technology will allow us to recover more of the remaining fossil fuels and extend the time the resource is available. However, the end result is still the same.

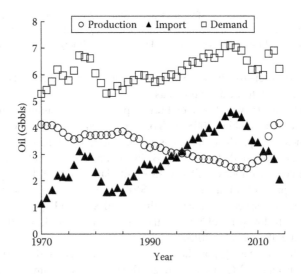

FIGURE 2.7 U.S. oil demand, production, and imports.

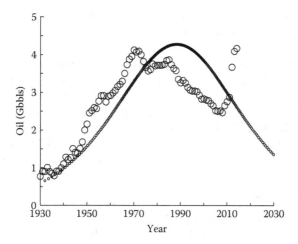

FIGURE 2.8 U.S. oil production and predicted curve.

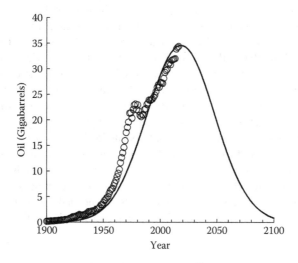

FIGURE 2.9 World oil production and predicted curve.

World oil production [8] will follow the same pattern as oil production in the United States. Notice that the bell curve shows world oil production (Figure 2.9) will peak around 2020; however, the curve should be extended to the right, so our estimated peak is from 2030 to 2050, as future production is stretched out because it includes oil from tight formations (shale), heavy oil, deep-water oil, and polar oil, all of which will be more expensive. Demand dipped in the industrial nations with the world financial crisis in 2008; however, it will not change much of the analysis. There are a number of websites on peak oil. The oil poster (http://www.oilposter.org) is well done, and it shows the world oil peak at 2010. Our estimate is world oil will peak between 2030 and 2050. Note the cost ($20 billion, $20 * 10^9$) of the BP oil leak in the Gulf of Mexico in 2010.

The reaction to the oil crises of 1973 and 1980 was increased efficiency, which shows as a dip in world production. However, as developing countries demand more energy, the demand and production will in general be approximated by the bell curve. The U.S. EIA makes low reference and high forecasts for demand, production, and prices [5]. Over the years, the range on price has been large, for example, price of oil for the reference case for 2030 was $20/bbl (forecast 2000), $45/bbl (forecast 2006), and $118/bbl (forecast 2014). The changes in the numbers show the difficulty of prediction. In 2013, the EIA is predicting a steady increase in the price for oil plus an increase in production over the next 20 years. For EIA predictions, check the international energy outlook on their website (http://www.eia.gov/forecasts/ieo/).

2.6.2 Natural Gas

Some people are touting natural gas for vehicles (compressed natural gas) because of cost for imported oil, future decline of the production of oil, and increased use of natural gas for generating electricity (now driven by cheap natural gas and natural gas has less emissions). However, over the long term, the problem is the same: A finite resource will be used fairly quickly with increasing demand.

The production of natural gas (Figure 2.10) is increasing across the world, due to advanced technology, especially for shale formations. Production of natural gas in Russia is a bit above that in the United States, with the two countries producing 50% of world production in 1995 and 38% in 2013. Total production in the United States will be less as reserves are around 9 Tm^3 compared to Russia with 31 Tm^3. Present reserves in the United States would last around 100 years at the 2014 rate of consumption; however, U.S. EIA predictions are for increased consumption to the year 2030. Anyway, peak natural gas production will probably occur in the time period of 2030–2050, although some have predicted peak production by 2020 [9–11]. Also, natural gas is an important feedstock for fertilizer and has been promoted as the feedstock for a future hydrogen economy, which would require enormous amounts of natural gas.

It is important to remember that new wells are needed for both oil and natural gas to replace decreased production from previous wells. For 2013, wells drilled in

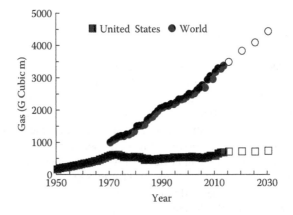

FIGURE 2.10 World and U.S. production of natural gas with predictions.

the United States accounted for around 30% of the world total. Since 1949, over 2.7 million oil and gas wells (exploratory, developmental, and dry) have been drilled in the United States, and around 850,000 are still in production. More wells have been drilled in the United States than in other regions of the world. For example, Saudi Arabia has the largest oil reserves in the world and has only drilled around 10,000 wells [12, Chapter 5].

2.6.3 COAL

Each fossil fuel industry touts the use of its product. The coal industry is promoting the sustainable development of coal and conversion of coal to liquid fuels. Clean coal, which is really stretching its total environmental impact, is the promotion of coal plants that sequester carbon dioxide. Coal provided 29% of the primary energy for the world (Figure 2.3) and 43% of global electricity.

Production of world coal (Figure 2.11) increased by 47% in the last 25 years to 7.9 Gtons in 2013, primarily due to increased production in China and now India. China is the largest producer, 46%, and consumer of world coal, 49%. Eighty-six percent of electricity in China is provided by coal, and China is constructing new coal generation plants, planned for 450 GW by 2040. Also, coal provides a major portion of energy for heating and cooking in China. The down side is the increased carbon dioxide emissions and environmental effects as every major city in China has problems with air pollution.

The world coal reserves are 891 Gt, with the United States having the largest reserves (Figure 2.12), an estimate that U.S. coal will last 200 years. Does that 200 years include increased production of coal as coal producers want to increase their share of the energy market? Of course, use of coal produces pollution and carbon dioxide emissions. For more information, go to the U.S. EIA website (http://www.eia.doe.gov/) or, for the industry viewpoint, http://americaspower.org. In the long

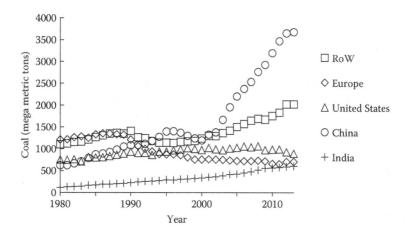

FIGURE 2.11 Production of coal in the world plus major coal-producing nations.

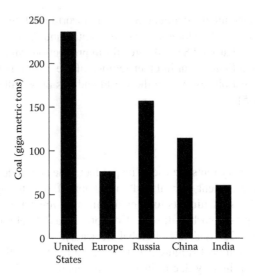

FIGURE 2.12   Major reserves of coal (anthracite, bituminous, sub-bituminous, and lignite) in the world (2013 data).

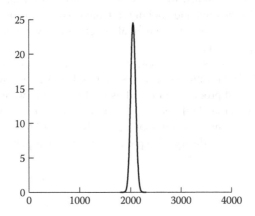

FIGURE 2.13   World use of fossil fuels on long timescale.

term, the use of fossil fuels could be called the *fickle finger of fate* (Figure 2.13). It is obvious that the world will shift toward renewable energy with the added benefit of less environmental impact. The major problem is the time frame for this shift within the restraints of reducing carbon dioxide emissions.

2.7   NUCLEAR

The first commercial nuclear power plant was built in 1957, and as of August 2014 [13–15], there were 437 nuclear power plants in operation in the world, with an installed capacity of 374.5 GW. Two plants are in long-term shutdown and 70 plants (66.8 GW) are under construction, again with China leading the way with 27 plants

(26.8 GW). However, date of operation for many plants will be some years away as the construction period is long.

Some countries in Europe are phasing out nuclear power as 6.3 GW was decommissioned in 2011. Everybody knows about the failure of the nuclear power plants in Japan due to the tsunami (May 2011), which reduced their capacity by 12 GW and resulted in all nuclear power plants being shut down for over a year. At the end of 2014, only 2 plants out of 48 that could be operational were back on line in Japan. Some of the plants are 40 years old and 6 are in the Fukushima exclusion zone, so probably up to 35 plants will be restarted by 2020. The loss of nuclear power has increased the cost of electricity because of the replacement cost for liquefied natural gas to fuel power plants.

Nuclear plants provide 2,507 TWh, 12% of global electricity (Figure 2.14), with the largest percentage in France at around 80% from 58 plants. The EIA is predicting the generation of 3,800 TWh for the world from nuclear power plants in 2030, around a 50% increase.

In the United States, there are 104 plants (installed capacity 106 GW, production 770 TWh); however, no new nuclear plants were constructed for a number of years, and the percentage of U.S. electricity from nuclear power declined from 23% to 19% as new electric plants produce electricity from natural gas and wind and solar farms. There is a revival of interest in nuclear power in the United States with 5 plants (5.6 GW) under construction, and a number of applications for licenses. Nuclear power has had a large amount of funding for R&D in the United States and continues to receive substantial federal funding and now the industry is seeking federal funding for construction. An advantage of nuclear power is that there are no carbon dioxide emissions; however thermal plants still need water, and the amount of uranium ore is finite. One solution is to construct breeder reactors.

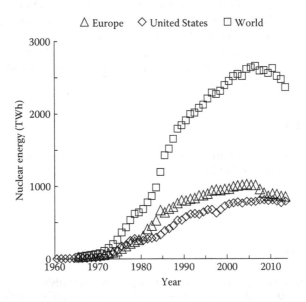

FIGURE 2.14   Production of electricity from nuclear power plants. Notice that Europe and the United States have a large portion of the total.

The proponents of nuclear energy state that a large increase in nuclear energy is a solution to global warming and the increasing need for electricity. Hansen is proposing shutting down of all coal plants [16] and replacing them with nuclear power plants. Others are for closure of existing nuclear plants and no construction of new plants, due to the problems of nuclear waste, proliferation of nuclear weapons, dirty bombs, and nuclear accidents. There have been over 30 major (official) accidents at nuclear power reactors since 1952 [17,18], with 3 significant accidents at utility reactors: Three Mile Island, Chernobyl, and Fukushima.

2.8 FINITE RESOURCE

The cumulative consumption of a resource from any initial time to any time T can be calculated by summing the consumption per year. This can be done using a spreadsheet or calculated. If the magnitude of the resource is known or estimated, then the end time T_E when the resource is used up can be estimated for growth, constant, or declining consumption using the same methods. Mathematics of resource consumption and lifetime of a finite resource are given in Appendix I.

The cumulative consumption is the area under the curve, for example, world oil production in Figure 2.9; the estimated resource is the area under the predicted curve. Note that at the peak, around half the resource has been consumed. Since those values were in a spreadsheet, the sums would give the total consumption to date and estimated total resource. Through 2014, the world had consumed 1,360 Gbbl of oil of an estimated resource of 2,800 Gbbl. In 2014, world oil reserves were estimated 1,650 Gbbl with another 335 Gbbl [19] of economically recoverable shale oil. What are the environmental impacts of shale oil? Note there is a difference between resources and reserves of oil and natural gas as reserves change as technology and economics affect the ability to extract the resource. Oil reserves increased substantially when the oil sands of Canada and heavy oil of Venezuela were included. The simplest estimate for T_E is to use estimated reserve and divide that number by the present annual consumption. Of course, as supply declines, prices will increase, and demand will decrease, so the T_E will be longer.

Example 2.4

There are around 1,650 Gbbl left of the world conventional oil, and at the present rate of consumption of 33 Gbbl/year, the lifetime is

$$T_E = \frac{1,650}{33} = 50 \text{ year}$$

Example 2.5

If you do not use the equation for T_E in the Appendix, a spreadsheet is useful for calculations as you can play with different scenarios of growth and size of the resource. A growth rate of 3% per year was used. So, at around 25 years, all the conventional oil is gone.

Year	Consumption	Cumulative
0	3.00E + 10	
1	3.09E + 10	3.09E + 10
2	3.18E + 10	6.27E + 10
3	3.28E + 10	9.55E + 10
...	...	...
23	5.92E + 10	1.00E + 12
24	6.10E + 10	1.06E + 12
25	6.28E + 10	**1.13E + 12**
26	6.47E + 10	1.19E + 12
27	6.66E + 10	1.26E + 12
28	6.86E + 10	1.33E + 12
29	7.07E + 10	1.40E + 12
30	7.28E + 10	1.47E + 12

The examples reinforce a previous statement: Exponential growth means large resources do not last long. Even with polar, deep-sea, tar sands, and oil shale, the lifetime will not be that much longer under present rates of consumption.

Similar analyses for other fossil fuels and uranium ore from estimated resources (Table 2.1) emphasize their finite lifetime under present rates of consumption. If the demand is small enough, is reduced exponentially, or is reduced at the depletion rate, a resource can essentially last a long time. However, with increased growth, T_E can be calculated for different resources, and the time before the resource is used

TABLE 2.1
Estimated Resources or Reserves, 2013*

Resource	Amount
U.S. oil	$44 * 10^9$ barrels
U.S. natural gas	$9.3 * 10^{12}$ m^3
U.S. coal	$237 * 10^9$ metric tons
U.S. uranium oxide	$2 * 10^5$ metric tons at \$110/kg
http://www.eia.gov/uranium/reserves/ures.pdf	$5 * 10^5$ metric tons at \$220/kg
World crude oil (conventional)	$1.7 * 10^{12}$ barrels
World oil—includes heavy, sands, shale, deep sea, polar oil	$2.1 * 10^{12}$ barrels
World natural gas	$186 * 10^{12}$ m^3
World coal	$891 * 10^9$ metric tons
World uranium oxide	$5.4 * 10^6$ metric tons at \$130/kg
http://www. euronuclear.org/info/encyclopedia/u/uranium-reserves.htm	$6.3 * 10^6$ metric tons at \$260/kg

* Most data are from *BP Statistical Review of World Energy 2014. Historical Data Workbook 1965–2013*. http://www.bp.com/en/global/corporate/about-bp/energy-economics/statistical-reviewof-world-energy. html.

is generally short. Remember, these are only estimates of resources, and other estimates will be higher or lower [20,21].

According to the energy companies, the continued growth in energy use in the United States is to be fueled by coal (largest fossil fuel resource) and nuclear energy. With the increase in natural gas from shale formations, the natural gas would displace some of the coal and would serve as the bridge to large renewable energy use in the future, especially in the electric sector. How long can coal last if we increase production to offset the need for importation of oil? Also, increased or even current production rates of fossil fuels will have major environmental effects as global warming has become an international political issue.

2.9 SUMMARY

Continued exponential growth is a physical impossibility in a finite (closed) system, and the Earth is a finite system. Previous calculations made about the future are just estimations, and possible solutions to our energy dilemma are as follows:

1. Reduce demand of fossil fuels to depletion rate.
2. Transition to zero population growth and begin a steady-state society.
3. Place a tax on carbon.
4. Implement more policies and incentives to increase conservation and efficiency and to decrease emissions of carbon dioxide.
5. Redefine the size of the system and colonize the planets and space; however, this will not alleviate the problem on Earth. From our present viewpoint, the resources of the solar system are infinite, and our galaxy contains over 100 billion stars.

Because the Earth is finite, there is a limit for population, availability of fresh water, fossil fuels, and minerals [22], and even a limit on the amount of food production and catch of fish from the sea. Therefore, a change to a sustainable society, which depends primarily on renewable energy, becomes imperative within this century. For the world, we will have to do the following in the transition period (next 20 years) in order of priority:

1. Implement more conservation and efficiency. Since the first energy crisis, this has been the most cost-effective mode of operation. It is much cheaper to save a barrel of oil than to discover new oil or import oil.
2. Increase substantially the use of renewable energy.
3. Reduce dependence on fossil fuels. Use of coal must include all social costs (externalities) and the sequester of carbon dioxide from coal plants producing electricity.
4. Reduce environmental impact, especially greenhouse gases.
5. Make use of nuclear energy.

Personal opinions of the authors: Nelson is ambivalent about the increased use of nuclear energy. From my viewpoint as a physicist's, taking care of nuclear waste is a technical problem, and as with any industrial operation, there will always be accidents. The question is what are predicted costs and what risks is the public willing to accept?

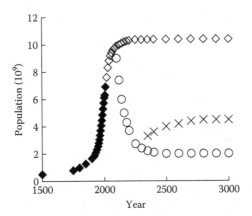

FIGURE 2.15 Population from 1500 to present with possible future populations (♦ = past, ◇ = future, steady state; ○ = catastrophe; × = revival).

However, from my view of politics, disposal of nuclear waste, proliferation of nuclear weapons, and dirty bombs will never be solved. So I am not sure whether to increase or decrease the use of nuclear energy. In any case, it is in priority five. Starcher is for any energy source that will last long term. Atomic energy has potential, and the waste/contaminated products that result from the production of electricity with atomic power are a possible source of valuable products for future generations. The trouble is the cost of construction and insurance, a purely economic consideration.

Implement policies (incentives and penalties) that emphasize items 1 and 2. State and local policies must focus on consistent goals. Efficiency can be improved in all the major sectors: residential, commercial, industrial, transportation, and even the primary electrical utility industry. National, state, and even local building codes need to be rewritten to improve energy efficiency in buildings. Finally, there are a number of things that you as an individual can do about conservation and energy efficiency. In addition, be an advocate for conservation, efficiency, renewable energy, and the environment.

For a few final comments, the possible future for human society involves conservation and efficiency, with an orderly transition to sustainable energy and a steady state with no growth, catastrophe, or catastrophe with some revival (Figure 2.15). As overpopulation and overconsumption are affecting the Earth, an uncontrolled experiment, the most probable future for the population is catastrophe or catastrophe with some revival.

REFERENCES

1. K. Adler. Summer 1986. The perpetual search for perpetual motion. *Am Herit Invent Technol,* 58.
2. Supercar. April, 1999. *Sci Am,* 46.
3. *Night Sky of Earth.* http://antwrp.gsfc.nasa.gov/apod/ap010827.html.
4. A. Ghassemi. 2013. Series Ed., *Renewable Energy and the Environment:* Vaughn Nelson, *Wind Energy, 2nd Ed,* R. Foster, M. Ghassemi, and A. Cota. 2014. *Solar Energy,* 2010; W. Glassley, *Geothermal Energy, 2nd Ed.* CRC Press, New York.

5. *Annual Energy Outlook*. 2015. http://www.eia.gov/forecasts/aeo/executive_summary.cfm.

6. M. King Hubbert. 1969. Energy resources. In: *Resources and Man*. National Academy of Sciences-National Research Council, Report of Committee on Resources and Man, Freeman, San Francisco, CA, pp. 157–242; also in Energy resources of the Earth, *Sci Am*, 60, 1971.

7. K. S. Deffeyes. 2005. *Beyond Oil: The View from Hubbert's Peak*. Hill & Wang, New York.

8. K. S. Deffeyes. 2001. *Hubbert's Peak: The Impending World Oil Shortage*. Princeton University Press, Princeton, NJ.

9. J. Darley. 2004. *High Noon for Natural Gas*. Chelsea Green, White River Junction, VT.

10. B. Powers. 2012. *Cold, Hungry and in the Dark: Exploding the Natural Gas Supply Myth*. New Society Publishers, Gabriola Island, Canada.

11. R. W. Bentley. 2002. *Global Oil and Gas Depletion: An Overview*, Energy Policy 189. http://www.oilcrisis.com/bentley/depletionOverview.pdf.

12. M. R. Simmons. 2005. *Twilight in the Desert: The Coming Saudi Oil Shock and the World Economy*. John Wiley, New Jersey.

13. International Atomic Energy Agency. Power reactor information system, comprehensive database on power reactors in operation, under construction or those being decommissioned. http://www.iaea.org/pris/. Also see nuclear power reactors in the world, Data Series No. 2, 2014 Ed. http://www-pub.iaea.org/MTCD/Publications/PDF/rds-2-34_web.pdf.

14. U.S. Energy Information Administration, Nuclear and Uranium. http://www.eia.gov/nuclear/.

15. Nuclear Energy Institute. http://www.nei.org.

16. J. Hansen. 2009. *Storms of My Grandchildren. The Truth about the Coming Climate Catastrophe and Our Last Chance to Save Humanity*. Bloomsbury, New York.

17. Y. Monget. 2007. *The World Tomorrow, Scenarios of Global Catastrophe*. Appendix 5: Major (official) nuclear accidents since 1952. Abrams, New York.

18. The Guardian. 2011. *Nuclear Power Plant Accidents: Listed and Ranked since 1952*. http://www.theguardian.com/news/datablog/2011/mar/14/nuclear-power-plant-accidents-list-rank#data.

19. *EIA/ARI World Shale Gas and Shale Oil Resource Assessment*. 2013. http://www.adv-res.com/pdf/A_EIA_ARI_2013%20World%20Shale%20Gas%20and%20Shale%20Oil%20Resource%20Assessment.pdf.

20. *BP Statistical Review of World Energy 2015*. Data Workbook-Statistical Review 2015. http://www.bp.com/en/global/corporate/about-bp/energy-economics/statistical-review-of-world-energy.html.

21. World Energy Council. Data, energy resources. http://www.worldenergy.org/data/resources/, http://www.worldenergy.org/publications/survey_of_energy_resources_2007/default.asp.

22. M. T. Klare. 2001. *Resource Wars: The New Landscape of Global Conflict*. Metropolitan Books, New York.

RECOMMENDED RESOURCES

LINKS

BP Statistical Review of World Energy 2015. http://www.bp.com/en/global/corporate/about-bp/energy-economics/statistical-review-of-world-energy.html.

Earth, a graphic look at the state of the world, a great site for overview on all aspects. http://www.theglobaleducationproject.org/earth/energy-supply.php.

Energy Information Administration, U.S. Department of Energy. http://www.eia.doe.gov. The EIA site contains a lot of information on U.S. and international energy resources and production. Reports and data files can be downloaded in both PDF and spreadsheet formats.

International Atomic Energy Agency. http://www.iaea.org.

International Energy Agency. http://www.eia.org.

Peak Oil. http://www.peakoil.com.

United Nations. *Information on Population and Projections on Population.* http://www.un.org/esa/population/unpop.htm.

U.S. Census. *Information on World Population.* http://www.census.gov.

World Energy Council. *Survey of Energy Resources 2007.* http://www.worldenergy.org/documents/ser2007_final_online_version_1.pdf; http://www.worldenergy.org/publications/survey_of_energy_resources_2007/default.asp.

Worldmapper. Shows morphed countries of the world where size depends on topical data, such as population, oil exports, oil imports, and others. http://www.worldmapper.org.

GENERAL REFERENCES

C. J. Campbell. 2005. *Oil Crisis.* Multi-Science, Essex.

R. L. Garwin and G. Charpak. 2001. *Megawatts and Megatons.* Knopf, New York.

J. Goodell. 2006. *Big Coal.* Houghton Mifflin, Boston, MA.

T. Hartmann. 1999. *The Last Hours of Ancient Sunlight.* Three Rivers Press, New York.

M. T. Klare. 2004. *Blood and Oil.* Metropolitan Books, New York.

A. McKillop, Ed., with S. Newman. 2005. *The Final Energy Crisis.* Pluto Press, London.

Y. Monget. 2007. *The World Tomorrow: Scenarios of Global Catastrophe.* Appendix 5: Major (official) nuclear accidents since 1952. Abrams, New York.

P. Tertzakian. 2006. *A Thousand Barrels a Second.* McGraw-Hill, New York.

PROBLEMS

Order of magnitude (OM) problems: provide an answer of only one or two significant digits with the proper power of 10.

2.1. A snowball, mass = 0.5 kg, is thrown at 10 m/s. How much kinetic energy does it possess? What happens to that energy after you are hit with that snowball?

2.2. For your home, estimate the power installed for lighting. Then, estimate the energy used for lighting for 1 year.

2.3. From problem 2.2, estimate the energy saved if you converted your lighting from incandescent to compact fluorescent. Fluorescent lights are more efficient, providing more light per watt.

2.4. What is the maximum power (electrical) used by your residence or home (assume all your appliances, lights, etc. are on at the same time)?

2.5. Approximately, at your residence or home, what is your average electrical energy usage per month?

2.6. What is the power rating of your vehicle (convert to kW)? What is the fuel efficiency (mpg or km/L)?

2.7. OM: Average fuel efficiency is 24 mpg in the United States. If efficiency is raised to 35 mpg, how many barrels of imported oil per year would be saved? At $100/bbl, how much money would be saved?

2.8. OM: Same as problem 2.7 but assume fuel efficiency is 50 mpg.

2.9. The Hawaii Natural Energy Institute tested a 100-kW OTEC (ocean thermal energy conversion) system. The surface temperature is 30°C, and at a depth of 1 km the temperature is 10°C. Calculate the maximum theoretical efficiency for this OTEC system. Remember that you have to use Kelvin.

2.10. Go to some websites for ethanol and obtain their numbers for energetics of producing ethanol from corn.

2.11. Efficiencies of thermal power plants can be around 40%. What is the efficiency of producing electricity from fuel cells? From combined cycle gas turbines?

2.12. Go to a website for the night sky of Earth. What is the main source of light in the Middle East?

2.13. For the latest year that data are available, what percentage of world energy was from fossil fuels?

2.14. What is the estimated time for peak production of coal in China, the United States, and the world?

2.15. OM: The Chamber of Commerce and the Board of Development are always promoting their city as the place for new industry. If a city has a population of 100,000 and a growth rate of 10% per year, what is the population after five doubling times? How many years is that?

2.16. OM: The world population in 1985 was around 4.5 * 10⁹ (the first time I taught the course) and after 40 years (2015), it is around 7.3 * 10⁹. How many people will there be on the Earth by the year 2050? Assume the present rate population growth as 1% per year.

2.17. OM: If the population growth rate could be reduced to 0.5% per year, how much longer would it take to reach the same population as in problem 2.16 for the year 2050?

2.18. OM: The most economical size of nuclear power plants is around 1,000 MW. How many nuclear power plants would have to be built in the United States over the next 50 years to meet the long-term historical growth of 7% per year in demand for electricity? In 2014, the generating capacity for the United States was around 1.1 TW.

2.19. OM: From problem 2.18, what is the total cost if the installed cost of a nuclear plant is around $5,000/kW? Suppose coal plants were installed at $1,500/kW; what is the cost?

2.20. OM: Assume new electric power plants in the United States are to be fueled by coal, and the electric growth rate is 5%. U.S. generating capacity was around 1.1 TW in 2014. How many metric tons of coal would be needed for the year 2050? Use the following conditions: Plants were operated at 95% capacity, and the efficiency of conversion is 35%.

2.21. OM: What is the efficiency of a nuclear power plant if the incoming steam is at 700°C and the outgoing steam is at 310°C. Remember that you have to use Kelvin.

2.22. OM: Use the coal reserves of the United States from Table 2.1. At today's rate of consumption, how long would that last?

2.23. OM: For problem 2.22, assume a coal consumption growth rate of 10% per year. How long will U.S. coal last?

2.24. If you could reduce your heating, cooling, and lighting bill by 50%, how much money would you save at your residence?

2.25. OM: Use natural gas reserves for the world from Table 2.1. At today's rate of consumption, how long would that last?

2.26. OM: Use uranium oxide ore reserves for the world from Table 2.1. At today's rate of consumption, how long would that last?

2.27. OM: Assume a growth rate of 7% for nuclear power plants. At today's rate of consumption, how long would that last?

2.28. OM: The population of the world is predicted to reach $11 * 10^9$. Mexico City is one of the largest cities in the world at $2 * 10^7$ people. World population in 2015 was $7.3 * 10^9$. How many new cities the size of Mexico City will have to be built to accommodate this increase in population?

2.29. OM: China has embarked on a policy regarding use of cars, from thousands in 1985 to millions on the road in 2015. Suppose China acquired the same number of cars per person as the United States. How much oil would China need per year? Compare that number with present annual oil production. Would there be a problem?

2.30. What is the world oil production? Who are the top 5 producing countries?

2.31. What is the world natural gas production?

2.32. What is the world coal production?

2.33. Go to power reactor information systems, http://www.iaea.org/pris/. How many nuclear power plants are there in the world? In your country? How many nuclear power plants in the world began operation in the last 5 years (use links, Status Change Trends)?

2.34. List at least three to five ways you are going to save energy this year.

3 Sun

3.1 SOLAR POWER

The Sun is a big ball of plasma composed primarily of hydrogen (92%), helium (8%), and small amounts of other atoms or elements. A plasma is where the electrons are separated from the nuclei because the temperature is so high (kinetic energy of nuclei and electrons is large). By the process of fusion, protons are converted into helium nuclei plus energy. The Sun (Table 3.1) is a stable main sequence star with an estimated age of $4.5 * 10^9$ years and will continue for another 4 to $5 * 10^9$ years before starting the next phase of evolution, the burning of helium. At that point, the Sun will expand and be larger than the orbit of the Earth.

Nuclei are composed of nucleons (Figure 3.1), which come in two forms: protons (which have a positive charge) and neutrons (no charge). The gravitational interaction is attractive, so in the center of the Sun the protons are close enough together for the nuclear interaction to occur, even though the protons repel one another due to their charge. At the size of nuclei (10^{-15} m), the nuclear interaction is stronger than the repulsion of the electromagnetic (EM) interaction. Protons are converted into helium nuclei, and because the mass of the helium nucleus is less than the mass of the four protons, that difference in mass (for the Sun around $5 * 10^9$ kg/s) is converted into energy. That energy is transferred to the surface of the Sun, where EM radiation and some particles (solar wind) are ejected into space.

This tremendous amount of energy is radiated into space in all directions from the surface of the Sun with a power of $3.8 * 10^{23}$ kW. The Earth only intercepts a small portion of the Sun's power; however, that is still a large amount. At the top of the atmosphere, the power intercepted by the Earth is $1.73 * 10^{14}$ kW, equivalent to 1.35 kW/m^2. Remember that this surface is perpendicular (90°) to the Sun. If a surface is at an angle to the Sun, the same amount of energy is spread over a larger area. At the surface of the Earth on a clear day, this solar insolation is around 1.0 to 1.2 kW/m^2 on a surface perpendicular to the Sun from 9 in the morning to 3 in the afternoon, depending on the amount of haze in the atmosphere and on elevation.

3.2 ELECTROMAGNETIC SPECTRUM

There are two ways of describing nature: particles and waves. Particles have mass, are localized in space, and can have charge and other properties, and no two particles can occupy the same space. Waves have no mass and are spread out over space; waves obey the principle of superposition, which means that two or more waves can occupy the same space at the same time. Many waves need a medium, for example, water and sound; however, there is no medium for EM waves. EM waves (Figure 3.2) travel at the speed of light and are described by their wavelength and frequency, which are related by

TABLE 3.1

Characteristics of Sun and Earth

	Sun	Earth
Diameter (km)	1,392,000	12,740
Mass (kg)	$1.99 * 10^{30}$	$5.98 * 10^{24}$
Surface temperature (K)	5,800	300

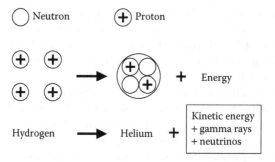

FIGURE 3.1   In the center of the Sun, protons are converted into helium nuclei plus energy.

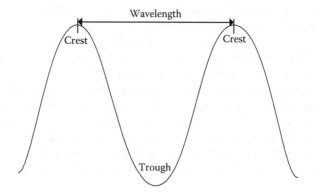

FIGURE 3.2   Wavelength is the distance from peak to peak of a wave.

$$c = \lambda * f \tag{3.1}$$

where:
  c is the speed of light in a vacuum, $3 * 10^8$ m/s
  λ is the wavelength, m, distance from peak to peak of the wave
  f is the frequency in hertz, which is the number of cycle/sec (as a wave moves by
     a point, the number of peaks or crests per second)

If you know either wavelength or frequency, you can calculate the other quantity. The speed of light will be different in a material; however, in air the speed is essentially the same as in a vacuum. EM radiation consists of paired oscillating electric and magnetic fields, perpendicular to each other and perpendicular to the direction of motion of the wave (Figure 3.3). A good applet of an EM wave is provided at http://www.walter-fendt.de/ph14e/emwave.htm.

The EM spectrum (Figure 3.4) is the range of EM radiation from very short wavelengths (high frequency) to very long wavelengths (low frequency). Sometimes, there is confusion between light, which refers to all EM waves, and the visible range (which previously was called *light* since we could see in this range). The subsections of the spectrum are labeled by how the radiation is produced and detected, and there is an overlap between the ranges. At the atomic level, the two ways of describing nature are combined, so EM waves come in units called photons; their energy is given by

$$E = h * f \tag{3.2}$$

where:

h is Planck's constant, $6.6 * 10^{-34}$ kg m^2/s

In physics texts, the symbol v is used for frequency rather than f. Large frequency corresponds to high-energy photons, such as X-rays and gamma (γ) rays, which can go through materials that absorb visible light and can cause damage to the materials. Low-frequency EM radiation can also go through materials that absorb light (e.g., radio waves go through the walls of houses).

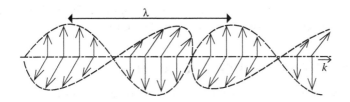

FIGURE 3.3 Diagram of electromagnetic wave showing components of electric and magnetic fields; the wave is traveling to the right.

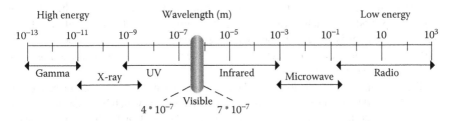

FIGURE 3.4 Electromagnetic spectrum from gamma radiation to long-wavelength radio waves.

3.2.1 VISIBLE

The range of the spectrum that we can see, visible (sometimes referred to as light), is small, with red light ($7 * 10^{-7}$ m) having a longer wavelength than blue light ($4 * 10^{-7}$ m). A rainbow is a familiar example of the colors that we can see. White light is just a superposition (combination) of all the colors. All the different colors we can see and generate are just absorption and reflections of different parts of the visible spectrum. There are some animals that see in the ultraviolet (bees) and infrared (snakes) ranges. We now have detectors for the whole range of EM radiation. With an infrared detector, you can see the thermal signatures people in the dark or the heat lost from a building. For images in the infrared spectrum, go to http://www.nationalinfrared.com/image_browser.php, and for images in the ultraviolet spectrum, go to http://www.pbase.com/kds315/uv_photos. For flowers in the visible and ultraviolet (what the bees see), go to http://www.naturfotograf.com/UV_flowers_list.html#ROSACEAX.

3.2.2 BLACKBODY RADIATION

A perfect absorber or emitter of EM radiation is a blackbody. The amount of radiation emitted per wavelength (or frequency) depends only on the temperature of the body and not on the type of material or atoms of the body. So, a blackbody curve can be generated for a specific temperature, with the peak of the curve shifting to shorter wavelengths (larger frequency) for higher temperatures. A blue flame is hotter than an orange flame. A higher-temperature object emits more radiation at all wavelengths, so the curves are a similar shape, nested within one another (Figure 3.5). Notice that the peak of the curve for the Sun is in the visible range, and it is interesting that our eyes are most sensitive to yellow–green light. The peak of the curve for the lower-temperature object is in the infrared spectrum.

3.3 ENERGY BALANCE OF THE EARTH

The energy balance of the Earth is essentially zero, except for the small amount of geothermal energy generated by radioactive decay. The Earth radiates the same amount of energy into space as the amount of EM energy absorbed from the Sun (Figure 3.6). If the energy in versus energy out were not balanced, the Earth would increase in temperature and would then radiate more energy into space to be in balance again.

The solar energy or power (remember if you know one, you know the other for any time period) interacts with the Earth's atmosphere and surface of which the major component is water. Of the incoming radiation (100 units or 100%), clouds (31%) and the surface (3%) reflect a third, and the rest is absorbed by the atmosphere (19%) and the surface (47%).

The amount of EM radiation from the Sun is primarily in the visible range, and this is absorbed and then converted primarily to thermal energy, which has a lower temperature, around 290 K, that radiates at longer wavelengths (peak at $1 * 10^{-5}$ m). The blackbody curve for Earth at 290 K would not even show on Figure 3.5a. This absorbed energy drives our weather in terms of evaporation and transportation of heat from the equator to

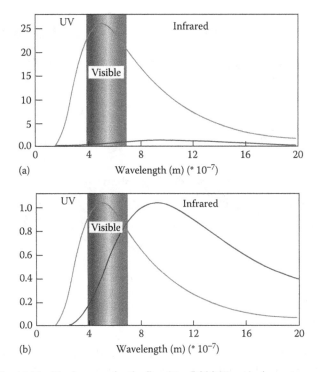

FIGURE 3.5 (a) Blackbody curve for the Sun ($T = 5,800$ K) and a lower-temperature object ($T = 3,000$ K). (b) Blackbody curve for the Sun ($T = 5,800$ K) and a lower-temperature object ($T = 3,000$ K). Curves now normalized to 1 to show similar shape.

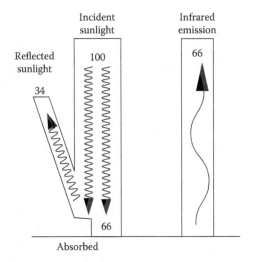

FIGURE 3.6 Energy balance of the Earth.

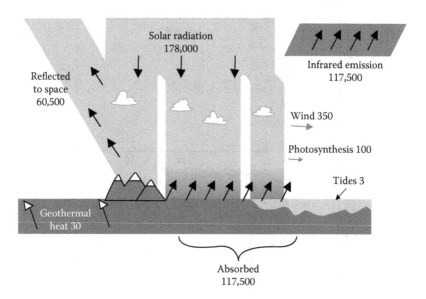

FIGURE 3.7 Transmission and absorption of electromagnetic radiation (kilowatts) and other sources of energy.

the poles and provides the energy for wind and waves and currents in the ocean; some is absorbed and stored in plants through the process of photosynthesis. Some of the infrared radiation is emitted to space (clear skies), and the rest is absorbed in the atmosphere (Figure 3.7). Of the infrared radiation absorbed in the atmosphere, some is then reradiated into space, and the rest is reradiated back to Earth. Clear nights are cooler than cloudy nights because of nighttime radiation into space, which has a temperature of 3 °K.

The atmosphere is transparent to visible and radio wavelengths but absorbs radiation in other wavelengths (Figure 3.8). Ozone in the upper atmosphere absorbs ultraviolet radiation. People in Australia, New Zealand, and Tierra del Fuego are now recipients of more high-energy ultraviolet radiation as we have been destroying the ozone layer at the poles in the upper atmosphere (Figure 3.9). Some of you may remember that industry did not want to discontinue the use of clorofluorocarbons, so their arguments were we could not replace the gas in refrigerators because of economics and the chemistry for ozone destruction was not completely certain. Some people who disagree with findings by scientists refer to this as junk science. Would people in supersonic planes high in the atmosphere have to worry about EM radiation? Do astronauts out in space have to worry about high-energy EM radiation?

3.4 EARTH–SUN MOTION

The inner planets are quite different because of their distance from the Sun and different mass (Figure 3.10), which then determines the amount of incident energy and how much atmosphere is retained. Mercury is too close to the Sun, which makes for a hot temperature, and has too small a mass to retain an atmosphere; therefore, the

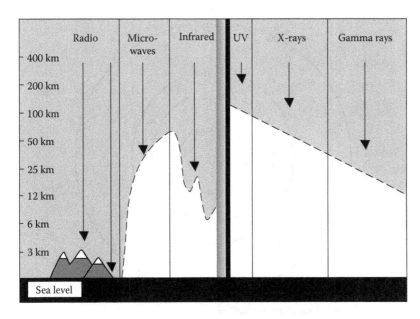

FIGURE 3.8   Visible and radio waves reach the surface, while other radiation is absorbed in the atmosphere.

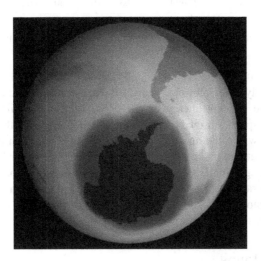

FIGURE 3.9   Large ozone hole that is three times larger than the United States over Antarctica (dark shade). Notice that the hole reaches to Tierra del Fuego in South America. (Data from NASA, Atmospheric Science Data Center, surface meteorology and solar energy. http://eosweb.larc.nasa.gov/cgi-bin/sse/sse.cgi?+s01.)

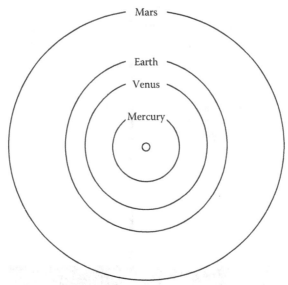

Sun	Diameter 1,390,000 km				
Inner planets	Distance from sun (000 km)	Temperature (°C) Mean	Max	Min	Atmosphere
Mercury	57,910	179	427	−173	None
Venus	108,200	482			CO_2
Earth	149,600	15	58	−89	N_2, O_2
Mars	227,490	−63	20	−140	Small CO_2

FIGURE 3.10 Relative size of orbits of the inner planets plus temperature and atmosphere.

molecules escape into space. Venus, which was once thought to be a sister planet to Earth, is completely covered by clouds and has a dense atmosphere of carbon dioxide. Venus retains the incident energy from the Sun at a higher surface temperature because the atmosphere absorbs the infrared radiation, trapping the thermal energy. Earth has liquid water and an atmosphere primarily of nitrogen and oxygen, both of which store heat and make the temperature range suitable for life. However, the amount, even though small, of greenhouse gases, primarily carbon dioxide, affects the overall temperature of Earth. Mars is too small and too far away from the Sun, so it has little atmosphere and is cold. A site for information on the solar system is http://www.solarviews.com.

3.4.1 Earth Motion

The Earth rotates (orbital motion) around the Sun once a year in an ellipse (almost a circle) and rotates (spins) on its axis once a day. The Earth is actually closest to the Sun on January 1. The orbital radius from the Sun determines how much solar energy is available, and the tilt of the spin axis (23.5°) from the plane of the orbital motion determines the seasons. The path of the Sun across sky changes during the year due to the tilt of the axis of the Earth. For example, the point where the Sun rises or sets goes from 23.5° south of east–west to 23.5° north of east–west.

Winter occurs in December in the Northern Hemisphere and in June in the Southern Hemisphere because of the tilt of the axis of the Earth (Figure 3.11). The amount of solar energy per surface area depends on the angle of the Earth's surface in relation to the Sun (Figure 3.12). Notice in Figure 3.12 that the white lines

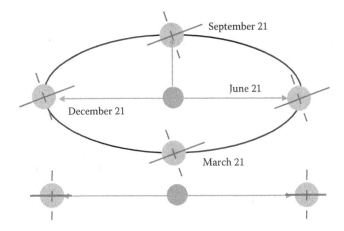

FIGURE 3.11   Seasons are due to the tilt of the axis of the Earth of 23.5° to the plane of orbit.

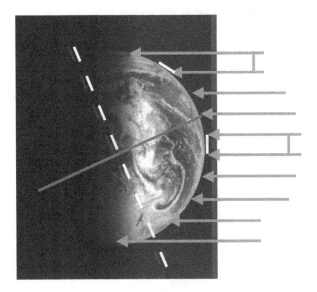

FIGURE 3.12   Effect of tilt of the axis of the Earth on incident radiation. Gray line indicates the equator, dashed white line indicates spin axis, and white lines indicate same length on the surface of the Earth.

on the surface are equal length, so that the same amount of energy in the Northern Hemisphere in winter is spread over a much larger area, plus the hours of daylight are shorter. The two obvious positions of the Sun are height above the horizon at noon and where it rises or sets. This angle changes by 47° from June 21 to December 21. At noon, the Sun is directly overhead (90°) at the equator on March 21 and September 21, and it rises and sets due east and west at the equator and other latitudes. Because sunrise and sunset are for the edge of the Sun and not the center of the Sun, at higher latitudes those angles are not quite due east and west.

In Figure 3.12, it is summer at the South Pole, with sunlight 24 h per day, and at the North Pole, there is no sunlight. In the Northern Hemisphere, it is winter because the same amount of incident radiation is spread over a larger area, as shown by lines equal distance apart, and that amount of energy is spread over a larger area in the winter.

3.4.2 SUN POSITION

Location on the Earth (Figure 3.13) is measured by longitude (lines from the North to the South Pole, 0 at Greenwich, England) and latitude (north latitude and south latitude from equator) in degrees (360), minutes (60), and seconds (60). Also, for global positioning systems (GPSs) and some maps, latitude and longitude are given in units of degrees and minutes (decimal), so there are no seconds. Position or location on the Earth can be found to within 2 to 3 m with a GPS, which are now inexpensive and are in mobile phones and tablets. The 360° of longitude are divided into 24 time zones, with universal mean time (UMT) the time at longitude 0, so you need to know how many time zones you differ from UMT to determine the local time. For central time in the United States, there is 6 h difference from UMT. Also, remember that during the summer some locations have daylight savings time.

The position of the Sun (Figure 3.14) can be calculated for any location and any time [1]. The position of the Sun is given by two angles, altitude and azimuth.

FIGURE 3.13 Longitude and latitude lines. Notice the lines for 23.5° for latitude N (Tropic of Cancer) and S (Tropic of Capricorn), where the Sun will be directly overhead at noon on June 21 and on December 21.

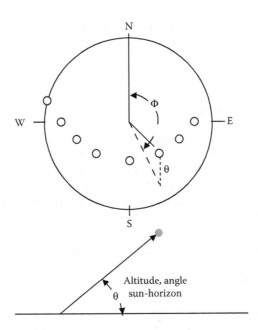

FIGURE 3.14 Position of the Sun given by altitude θ and azimuth Φ.

The altitude is the angle of the Sun above the horizon, and azimuth is the angle from north to the projection on the Earth of the line to the Sun. The position is symmetrical about solar noon (which is different from 12 noon). Previously, sun path diagrams were published for different latitudes and months and showed the sun angle. Now, websites are available to calculate the position of the Sun for any location (see the Links section). In some websites for the position of the Sun, the direction of zero azimuth can be chosen, generally north or south.

3.5 INSOLATION

Insolation is the solar radiation at the surface of the Earth and is given in units of energy/square meter (for some time period, which is really power) or power/square meter, which is generally for an average by day, month, or year. Global insolation is composed of two components, direct and diffuse. *Direct radiation* is the amount of radiation on a surface perpendicular to the Sun, and *diffuse radiation* is the indirect radiation from the sky and reflected radiation from other objects on the surface. The primary component of diffuse radiation is sunlight scattered in the atmosphere by air molecules, dust, water vapor, pollution, and the like. The predominant color on a clear day is blue, and on overcast days, the sky is gray. Insolation depends on length of day, clouds, haze and pollution, and elevation above sea level.

A class 1 monitoring station measures global and diffuse radiation. Many sites just measure the total insolation for a horizontal surface. The units and time period can be different; however, generally the units used now are kilowatt hours per square meter per day or watts per square meter averaged for a time period. Previously, the units were langleys (cal/cm^2) per day. Remember, the day includes night hours when there is no solar radiation. Another way of looking at insolation is to say that the sunshine can be averaged over the day and is equivalent to the peak value for 6 h. On clear days, that insolation is around 1.0 kW/m^2.

3.6 SOLAR RESOURCE

From the measurements of solar radiation, there are solar maps (Figure 3.15) available for the world [2,3], regions, and nations of the world and for regions or states within nations. As expected, the regions with the highest solar energy potential are the deserts on land and regions with few clouds over the oceans. Remember that elevation also is a factor. A good interactive map for the Western Hemisphere is available from 3Tier [4], which shows insolation in kilowatt hours per square meter per day. Overlays are solar, satellite, terrain, and hybrid, with tools for map size and selection, plus the additional feature of radiation type (direct normal, global horizontal, and diffuse).

Solar maps are also calculated for each day using data from satellites. These values are calculated using visible satellite images for each daylight hour. The approximate amount of energy reaching the ground is calculated for each image. This model takes into account the effect of clouds if there are clouds present. Every night, the images are integrated to give the estimated total amount of energy reaching the ground from the Sun. Maps and data sets (Table 3.2) are available for average day, month, and annual (22 years of data) for the world and regions [5]. Global solar energy data for 1,105 ground sites are also available.

A major source of information for solar insolation for the United States is the Renewable Resource Data Center (RReDC) [6] at the National Renewable Energy Laboratory (NREL). NREL provides interactive mapping tools, solar power prospector [7]. Solar resource radiation maps by month and for different applications are available from NREL. In some cases, information on solar insolation for concentrated solar power, tracking flat plate, and flat plate at different angles are available (Figure 3.16).

3.7 GREENHOUSE EFFECT

The atmosphere lets in solar radiation and reduces infrared radiation from the surface from going out, thereby maintaining an energy balance with a higher temperature (Figure 3.17). The greenhouse gases and their concentrations in the atmosphere are water vapor (around 1%), carbon dioxide (0.04%), methane (0.00018%), nitrous oxides, and other trace gases. The amount of effect per molecule and the residual time differ, for example, methane (CH_4) is around 20 times more effective than CO_2, but the residual time is a few years compared to hundreds of years for CO_2. Note for the students, remember that temperature and heat are different as heat is thermal energy and temperature is an indicator of potential flow for heat (always from high to low temperature).

The greenhouse effect is amply demonstrated on a sunny day by observing your car interior with the windows closed. The incident light passes through the windows

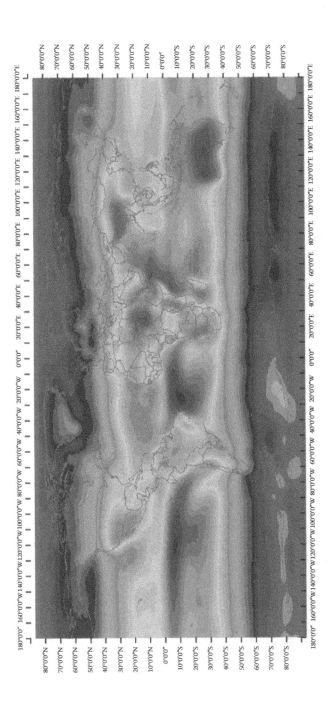

Yearly mean of irradiance in W/m²

0–10 10–20 20–30 30–40 40–50 50–60 60–70 70–80 80–90 90–100 100–110 110–120 120–130 130–140 140–150 150–160 160–170 170–180 180–190 190–200 200–210 210–220 220–230 230–240 240–250 250–260 260–270 270–280 280–290

Realized and produced by Michel Albuisson, Mireille Lefèvre, Lucien Wald.
Edited and produced by Thierry Ranchin. Date of production: 23 November 2006.
Centre for Energy and processes, Ecole des Mines de Paris / Armines / CNRS.
Copyright: Ecole des Mines de Paris / Armines 2006. All rights reserved.

FIGURE 3.15 Average insolation for 1990–2004 for the world. (Courtesy of Ecole des Mines de Paris/Armines 2006.)

TABLE 3.2

Average Insolation (kWh/m²/day) for Some Cities in Canada

City	Jan	Feb	Mar	Apr	May	Jun	Jul	Aug	Sep	Oct	Nov	Dec	Avg
Edmonton	1.45	2.36	3.41	4.25	4.91	5.42	5.55	4.76	3.52	2.18	1.43	1.21	3.37
Victoria	1.00	1.82	2.93	4.01	5.13	5.54	5.85	5.28	3.88	2.17	1.11	0.86	3.29
Winnipeg	1.21	2.08	3.27	4.55	5.54	5.8	5.85	4.84	3.32	2.21	1.33	1.02	3.41
St. Johns	1.56	2.27	3.48	4.19	4.76	5.05	5.05	4.54	3.53	2.29	1.43	1.27	3.28
Halifax	1.56	2.31	3.46	4.09	4.82	5.27	5.41	4.86	3.92	2.54	1.53	1.3	3.42
Toronto	1.44	2.27	3.19	4.13	5.15	5.83	5.67	4.82	3.66	2.47	1.48	1.20	3.44
Montreal	1.45	2.36	3.41	4.25	4.91	5.42	5.55	4.76	3.52	2.18	1.43	1.21	3.37
Regina	1.14	1.96	3.02	4.69	5.48	5.79	6.14	4.96	3.42	2.29	1.30	0.95	3.42

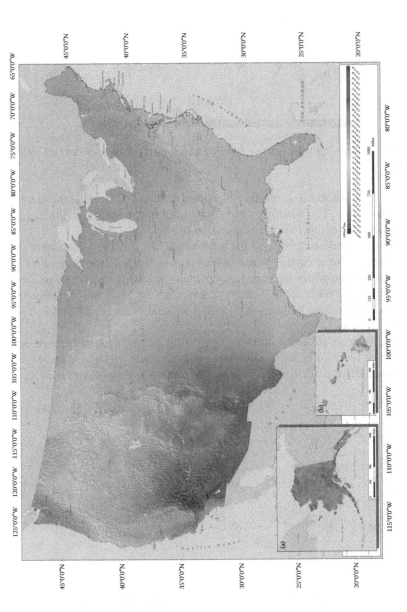

FIGURE 3.16 Solar map for the United States, flat plate tilted south at latitude. Insets: Annual average solar resource data is shown for a tilt = latitude collector. (a) The data for Alaska is a 40 km dataset produced by the Climatological Solar Radiation model (NREL, 2003). (b) The data for Hawaii and the 48 contiguous states is a 10 km, satellite modeled dataset (SUNY/NREL, 2007) representing data from 1998–2005. (From National Renewable Energy Laboratory [NREL], Dynamic Maps, GIS Data and Analysis Tools, Solar Maps, http://www.nrel.gov/gis/solar.html.)

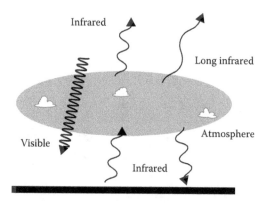

FIGURE 3.17 Greenhouse effect due to different transmission, absorption, and emission of radiation.

and is absorbed in the material inside, which then radiates (infrared) at the corresponding temperature. The windows are opaque to infrared radiation, and the interior becomes hotter until there is again an energy balance.

The amount of carbon dioxide in the atmosphere changes the temperature at which the energy balance occurs. Venus, our sister planet, is a drastic example of a dense CO_2 atmosphere where the surface temperature is 467°C, hot enough to melt lead. Even though the major component of the atmosphere of Mars is CO_2, temperatures are very cold because Mars has lost most of its atmosphere because it has less gravity due to less mass.

There is an increase in carbon dioxide (Figure 3.18) in the atmosphere due to the increased use of fossil fuels and agriculture, and most scientists say this results in global warming [8]. In 1850, the concentration was 285 parts per million (Figure 1.2)

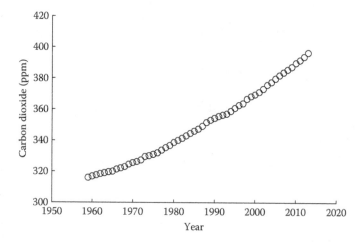

FIGURE 3.18 Annual mean atmospheric carbon dioxide concentration, parts per million. (Data from National Ocean and Atmospheric Administration [NOAA], Trends in Atmospheric Caron Dioxide, Mauna Loa data. http://www.esrl.noaa.gov/gmd/ccgg/trends/index.html.)

and note that the increase to 400 ppm in 2014 is having an impact on the climate of the Earth. To continue with business as usual of burning large amounts of fossil fuels means the climate of 2100 will be drastically different than today. Previously, industry and many politicians said the same thing about global warming as said about the ozone problem. Industry maintained that we cannot reduce the production of CO_2 because of economics and because the science for CO_2 and global warming is not completely certain. The former U.S. policy under the G. W. Bush administration was in sharp disagreement with that of the other industrialized countries, and the reasons were that it would cost too much and not enough provisions were made to curtail future emissions from developing countries. An interesting comment by Robert Romer [9] in 1976 was that "human activities do not now have a large effect on the global climate, however this calculation should not be considered as justification for complacency." So, within 30 years we are doing an uncontrolled experiment on the atmosphere of the Earth, and now most predictions on global warming and climate change (see Links section) are for problems (even catastrophe) by 2050–2100 unless carbon dioxide emissions for the world are reduced to 1990 levels.

REFERENCES

1. R. Foster, M. Ghassemi, and A. Cota. 2010. *Solar Energy, Renewable Energy and the Environment.* CRC Press, Boca Raton, FL, Chap. 2.
2. *SoDa: Service for Professionals in Solar Energy and Radiation.* 2006. http://www.soda-is.com/eng/map/index.html.
3. Solargix. http://solargis.info/doc/free-solar-radiation-maps-GHI.
4. 3Tier. http://maps.google.com/gallery/search?hl=en&q=renewable+energy.
5. NASA. Atmospheric Science Data Center: surface meteorology and solar energy. http://eosweb.larc.nasa.gov/cgi-bin/sse/sse.cgi?+s01.
6. NREL, RReDC. http://www.nrel.gov/rredc/solar_resource.html.
7. NREL. http://www.nrel.gov/gis/solar.html.
8. Intergovernmental Panel on Climate Change. 2014. *Fifth Assessment Report.* http://www.ipcc.ch.
9. R. H. Romer. 1976. *Energy: An Introduction to Physics.* Freeman, New York.

RECOMMENDED RESOURCES

Links

Sun Position and Sun Path

Solar position and intensity. http://www.nrel.gov/midc/solpos/solpos.html.

Sustainable by design. http://www.susdesign.com; shareware, also output tables, SunAngle, SunPosition, Sol path, interactive. http://www.susdesign.com/tools.php. Code available from NREL.

University of Western Australia. The Sun's path. http://engnet.anu.edu.au/DEpeople/Andres.Cuevas/Sun/SunPath/SunPath.html.

Global Warming and Climate Change

Carbon Trust, Reports. http://www.carbontrust.com/resources/reports.

Environmental Protection Agency. http://www.epa.gov/climatechange/index.html.

Global warming: early warning signs. http://www.climatehotmap.org.

Intergovernmental Panel on Climate Change. http://www.ipcc.ch.

National Climate Data Center, NOAA. http://www.ncdc.noaa.gov/ monitoring-references/faq/ global-warming.php.

Natural Resources Defense Council. Global warming solutions. http://www.nrdc.org/ globalWarming.

Union of Concerned Scientists. Global warming. http://www.ucsusa.org/global_warming.

U.S. Global Change Research Program. http://www.globalchange.gov.

GENERAL REFERENCES

D. R. Archer. 2011. *Global Warming: Understanding the Forecast*, 2nd Ed, Wiley, New York.

R. Ayers. 2001. How economists have misjudged global warming. *World Watch*, September/ October, 12.

T. Flannery. 2006. *The Weather Makers, How Man is Changing the Climate and What it Means for Life on Earth*. Atlantic Monthly Press, New York.

J. Hansen. 2009. *Storms of My Grandchildren. The Truth about the Coming Climate Catastrophe and Our Last Chance to Save Humanity*. Bloomsbury, New York.

T. R. Karl and K. E. Trenbrith. 1999. The human impact on climate. *Sci Am*, 281(6), 100.

E. Kolbert. 2006. *Field Notes from a Catastrophe, Man, Nature, and Climate Change*. Bloomsbury, New York.

National Academy of Sciences and Royal Society. 2014. *Climate Change, Evidence and Causes*. Download booklet at http://nas-sites.org/americasclimatechoices/ events/a-discussion-on-climate-change-evidence-and-causes/.

S. R. Weart. 2008. *The Discovery of Global Warming*. Harvard University Press, Boston, MA.

PROBLEMS

3.1. Take a penny or a ring. How far away from your eye would it have to be to have the same angular size as the full moon (i.e., to just block out the moon)? Note that the Moon and Sun have the same angular diameter because of difference in size and distance from the Earth. *Do not* try this with the Sun as looking directly at the Sun will damage your eyes.

3.2. Choose any day; go outside and estimate the angle (altitude) of the Sun at solar noon (high point) and then the angle (azimuth) at sunrise or sunset. Be sure to note date and location. *Do not look directly at the Sun.*

3.3. Choose any day; go outside and estimate for either sunrise or sunset the number of degrees from due east or due west. Be sure to note date and location (estimate latitude).

3.4. With the windows rolled up on your car for over an hour on a sunny day (outside temperature 15°C or higher; 15°C = 60°F), estimate the temperature inside.

3.5. Calculate the power of the Sun. Use $E = mc^2$ and the amount of mass converted to energy per second.

3.6. Find the altitude and azimuth for sunrise (and the same for sunset) and solar noon for your hometown on your birthday. Use Sustainable by Design website (http://www.susdesign.com).

3.7. Find the path of the Sun across the sky for June 1 and September 1 for your hometown. Use Sustainable by Design website (http://www.susdesign.com). What is the altitude at solar noon? What is the azimuth at sunrise and sunset?

3.8. OM (order of magnitude): Has the burning of fossil fuels (stored solar energy from the past) changed the energy balance of the Earth? Only use energy consumption for the world from fossil fuels for the past 10 years.

3.9. Find an annual solar insolation map for your country. What is the insolation for your hometown?

3.10. Go to the 3Tier website (http://maps.google.com/gallery/search?hl=en&q=renewable+energy) and find the annual solar insolation for Salem, Oregon, and Austin, Texas.

3.11. Which regions of the world have high solar insolation (Figure 3.15)? Why?

3.12. Estimate the average incident solar energy for a 2-m^2 flat-plate solar collector for your location. The collector is installed at an angle equal to the latitude, (a) for December and (b) for June.

3.13. Estimate how much more energy a gamma-ray photon has than a green (visible) photon.

3.14. An infrared wave has a wavelength of 0.001 m. What is the frequency? A red light has a wavelength of $7 * 10^{-7}$ m. What is the frequency?

3.15. Why is Venus so hot?

3.16. What is the range of predictions for the temperature increase for Earth by 2050? By 2100?

4 Heat Transfer and Storage

4.1 INTRODUCTION

Two primary uses of solar energy are providing heating and lighting for buildings. The building must be designed for the collection of solar energy, transfer of that energy, storage, and the release of the solar energy. The electromagnetic radiation is absorbed and becomes thermal energy (*heat*); therefore, we need to understand how heat is transferred. Heat is just internal kinetic energy of a material and is the random motion of atoms. Two important aspects for buildings using solar energy are reduction in the transfer of heat and sizing of sufficient thermal mass so there will not be large temperature variations from day to night or even over 3 to 5 days with no solar input to the building due to clouds. The reduction of heat transfer means that a solar-heated building must first be a well-insulated building, a heat trap.

Heat only flows in one direction, from hot objects to cold objects, high temperature to low temperature. *Remember that temperature and heat are not the same.* To move heat from a cold place to a hot place requires energy. For example, the air conditioner for your home uses a lot of energy. The government has promoted higher-efficiency appliances, which has saved a lot of energy. If there is no input of energy, then the heat transfer or flow is such that the objects come to the same temperature. In the winter, if there is no heat input for the building, the temperature inside will reach the outside value. In this case, the outside is so large that it can be considered a reservoir, and its temperature is not affected by the small amount of heat transferred from the house. The same is true in the summer; if no heat is removed from the building, then the inside temperature will reach the outside value, and again the outside is so large that it can be considered a reservoir; its temperature is not affected by the small amount of heat transferred into the house.

Heat can be transferred by conduction, convection, and radiation. A building, especially a building designed around the climate to use solar energy, must consider and control all three methods of heat transfer. You should be able to use spreadsheets to calculate the heat gain and loss. Water vapor and condensation are problems that have to be considered in buildings.

4.2 CONDUCTION

Conduction is the transfer of heat in a solid. If one side or end is at a high temperature and the other is at a low temperature, heat flows from the hot to the cold side (Figure 4.1). Conduction depends on the type of material. For example, metals are good conductors, and polystyrene (Styrofoam) is a poor conductor of heat. The heat flow or rate (amount/time) depends on the properties of the material, thickness of the material, and the difference in temperature of the two sides. Thermal conductance is the rate of heat flow through a unit area at the installed thickness and any given delta temperature (difference between two temperatures). Heat flow across slabs of

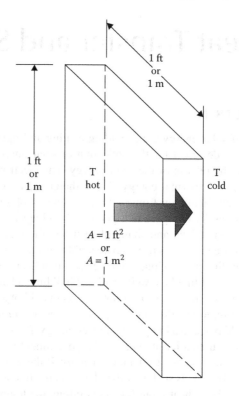

FIGURE 4.1 Conduction of heat from a hot side to a cold side.

material is measured experimentally and is defined in terms of thermal conductance (U) with units of watts per square meter per degree kelvin, W/(m^2 °K). The conversion between metric and English units is 1 W/(m^2 °K) = 0.1761 Btu/(ft^2 °F h) or 1 Btu/(ft^2 °F h) = 5.678 W/(m^2 °K).

Resistance (R) is a property of a material to retard the flow of heat and is the inverse of conductance. Good insulators have a high R value, which is low conductance. The total R value for a composite material is just the sum of the R values of the component parts.

$$U_t = \frac{1}{R_t} \tag{4.1}$$

$$\frac{1}{U_t} = \frac{1}{U_1} + \frac{1}{U_2} + \frac{1}{U_3} + \ldots + \frac{1}{U_n}$$

$$R_t = R_1 + R_2 + R_3 + \ldots + R_n \tag{4.2}$$

U values are measured experimentally for different materials, thicknesses, and unit area; of course, R values have inverse units ([m^2 °K]/W or [ft^2 °F h]/Btu). Note that for a delta temperature, use of degree kelvin or centigrade is the same.

A source for thermal conductivity values of some common materials and products is available from the Engineering Toolbox at http://www.engineeringtoolbox.com/thermal-conductivity-d_429.html.

The amount of heat loss or gain by conduction is given by

$$H_{CON} = U * A * \Delta T * h \qquad (4.3)$$

where:
A is the area
ΔT is the temperature difference
h is the number of hours

In the winter, you want to reduce the heat loss, and in the summer, you want to reduce the heat gain of the house.

Example 4.1

Calculate the heat transfer for a wall composed of the following materials with corresponding R value:

Component	R ([m²°K]/W)	R ([ft²°F h]/Btu)
Outside air film, 15-mph wind	0.03	0.17
Wood siding	0.14	0.81
Plywood sheeting, 0.5 in.	0.11	0.62
Fiberglass batt, 3 in.	2.29	13.00
Gypsum board, 0.5 in.	0.08	0.45
Inside air film	0.12	0.68
Total	2.77	15.73

Source: Equation 4.1, $U_t = 1/R_t = 1/2.77 = 0.36$ W/(m²°K).

The conduction heat loss for a 4.5-m² wall over an 8-h time period with the inside temperature at 21°C and the outside temperature at 5°C can be calculated from Equation 4.3.

$$H_{CON} = 0.36 \text{ (J/ s)/(m}^2°C) * 4.5 \text{ m}^2 * (21 - 5)°C * 8 \text{ h}$$

$$= 0.36 * 4.5 * 16 * 8 * 3{,}600 \text{ Wh}$$

$$= 746 \text{ kJ} = 710 \text{ Btu}$$

The answer has units and significant digits. Results cannot be more accurate than the one input data with least significant digits used in the calculation.

There is conduction through the framing (studs), windows, ceilings, and foundation, as all parts of the building allow heat to flow through it. The wood frame (studs) of a house conducts more heat than fiberglass batts, so Example 4.1 was not correct for a wall of a house. What would be the difference if aluminum studs were used in a building? In general, there should be more insulation in the ceilings. Superinsulated houses have high R values for ceilings, walls, and windows (double or triple pane).

There are a number of websites with information on buildings and heat flow through the exterior (building envelope). The Oak Ridge National Laboratory Building Envelopes Program [1] has information on insulation and radiation barriers. The insulation fact sheet has information for new (Figure 4.2) and existing houses. A map and table of recommended insulation R values are given for eight climate zones of the United States (Figure 4.3). A calculation for recommended insulation in the United States by zip code is also available, and interactive calculators are available, one of which is for whole-wall R value if you know the components of the wall.

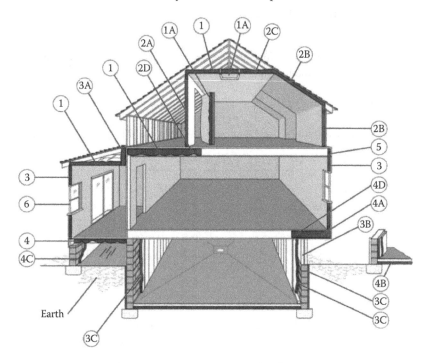

FIGURE 4.2 Locations where insulation is needed. Insulation fact sheet (insulating a new house) from Oak Ridge National Laboratory, http://www.ornl.gov/sci/roofs+walls/insulation/ins_05.html. (1) In unfinished attic spaces, insulate between and over the floor joists to seal off living spaces below. (Note: Well-insulated attics, crawl spaces, storage areas, and other enclosed cavities should be ventilated to prevent excess moisture buildup.) (1A) Attic access door. (2) In finished attic rooms with or without dormer, insulate. (2A) Between the studs of *knee* walls. (2B) Between the studs and rafters of exterior walls and roof. (2C) Ceilings with cold spaces above. (2D) Extend insulation into joist space to reduce airflows. (3) All exterior walls, including (3A), walls between living spaces and unheated garages, shed roofs, or storage areas; (3B) foundation walls above ground level; (3C) foundation walls in heated basements, full wall, either interior or exterior. (Note: For new construction, slab or grade insulation should be installed to the extent required by building codes or greater). (4) Floors above cold spaces, such as vented crawl spaces and unheated garages. Also insulate (4A), any portion of the floor in a room that is cantilevered beyond the exterior wall below; (4B) slab floors built directly on the ground; (4C) as an alternative to floor insulation, foundation walls of unvented crawl spaces; (4D) extend insulation into joist space to reduce airflows. (5) Band joists. (6) Replacement or storm windows and caulk and seal around all windows and doors.

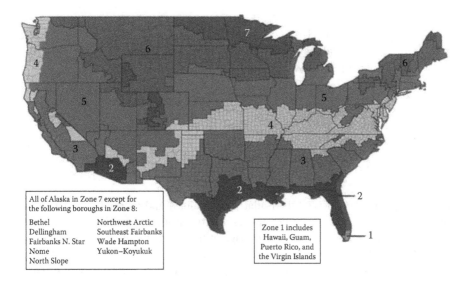

All of Alaska in Zone 7 except for
the following boroughs in Zone 8:

Bethel	Northwest Arctic
Dellingham	Southeast Fairbanks
Fairbanks N. Star	Wade Hampton
Nome	Yukon–Koyukuk
North Slope	

Zone 1 includes
Hawaii, Guam,
Puerto Rico, and
the Virgin Islands

FIGURE 4.3 Insulation recommendations (R values) for new wood-framed houses by climate zone. Zone 1 includes Hawaii, Guam, Puerto Rico, and the Virgin Islands. Zone 7 includes most of Alaska, and Zone 8 is the northern part. (Map and table, Oak Ridge National Lab, Building Envelopes Program.)

Zone	Heat System	Attic	Cathedral Ceiling	Wall Cavity	Wall Sheathing	Floor
1	All	30–49	22–38	13–15	None	13
2	Gas, oil, heat pump	30–60	22–38	13–15	None	13
	Electric					19–25
3	Gas, oil, heat pump	30–60	22–38	13–15	None	25
	Electric				2.5–5	
4	Gas, oil, heat pump	38–60	30–38	13–15	2.5–6	25–30
	Electric				5–6	
5	Gas, oil, heat pump	38–60	30–38	13–15	2.5–6	25–30
	Electric			30–60	13–21	5–6
6	All	49–60	30–60	13–21	5–6	25–30
7	All	49–60	30–60	13–21	5–6	25–30
8	All	49–60	30–60	13–21	5–6	25–30

Windows have higher conductivity, so double-pane windows or storm windows will reduce conduction losses (Table 4.1). Wood frame windows have better insulation than aluminum frames; however, aluminum is cheaper. Fiberglass frames are gaining in popularity with better insulation than wood and are becoming easier to manufacture. A comparison of many window frame types and glass styles is shown in U.S. Department of Energy website (http://energy.gov/energysaver/articles/window-types). Some aluminum windows use vinyl coverings or a thermal break to reduce conduction heat loss. Thermal shutters or shades can greatly reduce the

TABLE 4.1
U Values for Windows, 2.5 mm of Glass,
Vinyl or Wood Frame, 2.3 cm Air Space

No. of Panes	W/(m² °K)	Btu/(ft² °F h)
Single	4.7	0.84
Double	2.8	0.50
Double, low e	1.8	0.32

heat loss through windows [2]. Insulation (R values) and thermal conductivity of building materials and air spaces (air is a good insulator) can be obtained from the American Society of Heating, Refrigerating and Air-Conditioning Engineers (ASHRAE) 2001 *Fundamentals Handbook*. Its Chapter 25, "Thermal and Water Vapor Transmission Data," Table 4, "Typical Properties for Common Building Materials and Insulating Materials" provides thermal resistance per inch for various building materials. The handbook chapter can be purchased individually and downloaded at http://www.ashrae.org. R values and values for thermal conductivity for a number of different materials are available on a number of websites. A source for thermal conductivity values of some common materials and products is available from the Engineering Toolbox at http://www.engineeringtoolbox.com/thermal-conductivity-d_429.html.

4.3   CONVECTION

Convection is the transfer of heat by the movement of fluids—gases or liquids (Figure 4.4). Heated fluids can move by natural convection or by forced convection by pumps and fans. In natural convection or thermosiphoning, as a fluid is heated it expands and becomes less dense, thereby the hot air and the hot part of a liquid rise while the cooler part descends. So, a solar hot water system can be constructed that requires no pumps. Also, natural convection can move heat around a properly designed structure.

Convection works in conjunction with conduction. Heat from a warm or hot surface is conducted to the adjacent fluid, which is then carried away by convection. In forced convection, the quantity of heat moved depends on the amount of fluid moved (rate) and the heat capacity of the fluid. It takes a lot more air than water to move the same amount of heat.

Calculation of infiltration, convection heat loss or gain through open doors and cracks, is just an educated estimate. Even though conduction heat loss or gain through the exterior of the building is easier to calculate, reduction of infiltration (in or out through the exterior) is more important and will save energy. Infiltration barriers are now installed on the exterior of most new buildings.

$$H_{INF} = c * Q * L * \Delta T * h \qquad (4.4)$$

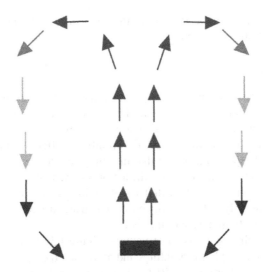

FIGURE 4.4   Convection currents from heat source.

where:
    c is the heat capacity of the fluid
    Q is the volume of air leakage per length of crack per hour
    L is length of crack
    ΔT is the temperature difference
    h is the number of hours

Example 4.2

A wooden, double-sash window with no weather stripping has the following values: Wind = 4.5 m/s, T inside = 22°C, T outside = 5°C, heat capacity of air = 1,297 J/(m³ °K), Q = 1.9 m³/(m h), L = 5 m, ΔT = 17°C, h = 8 h.

H_{INF} = 1,297 J/(m³ °K) * 1.9 m³/(m h) * 5 m * 17°C * 8 h = 1,700 kJ = 1,600 Btu

Now, you can see why weatherization programs are cost effective. Think how much energy and money Russia could save by installing weather-tight, double-pane windows on old buildings. It would be the same as discovering a giant oil field but with the most important benefit: no depletion.

You can check for infiltration in your home on a cold, windy day by placing your hand near suspected areas, such as edges of doors and windows, electrical outlets, ceiling lights, range hoods, and a clothes dryer vents. You will be surprised. In the home, 70% of infiltration is around the soleplate, wall electrical outlets, exterior windows, and heating and cooling duct system. The soleplate is the bottom member of the wall resting on the foundation or subfloor. If you have exterior electrical outlets, the inside wall may be cool where they are located.

Dead air spaces between window panes, between window and storm window, and in walls are good insulators. The problem is the maximum width that can be used before convection reduces the R value of the space. In other words, a large-width dead air space will have convection currents.

4.4 RADIATION

Radiation was discussed in Chapter 3. Remember that all objects emit electromagnetic radiation, and the amount and wavelengths depend on the temperature. Radiation barriers, such as aluminum foil on insulation, are now well-accepted building practices. You can be in a cooler surrounding and absorb infrared (IR) radiation directly from radiant heaters to keep warm. Notice that in a fireplace, the red coals emit a lot of energy (IR radiation) into the room. Do the astronauts wear aluminum underwear?

On commercial building roofs with flat roofs or low slopes, reflective coatings on the roof can significantly reduce energy costs. There are also reflective coatings that can be applied to the bottom side of the roof.

IR detectors can find heat leaks in structures (Figure 4.5). You can view IR photos of building at google IR photos buildings and at the photo exchange of the National Renewable Energy Laboratory (NREL) (http://www.nrel.gov/data/pix) (do a text search for *infrared*). Notice that IR photos show leaks in the winter as hot from the outside of the house and cold from the inside.

4.5 THERMAL MASS

Thermal mass in solar buildings reduces temperature variations between day and night, and if there is a large mass, it can reduce temperature variations for days such that no or less auxiliary heating or cooling is required. In the summer, the thermal mass can be cooled at night, and then it absorbs heat during the day, keeping the house cool. Thermal mass is also useful in ordinary buildings as it serves as a reservoir or sink for both heating and cooling.

Thermal mass provides a means of storing the solar energy that enters through the windows. The thermal mass absorbs the solar energy during the day and keeps the house from overheating. At night, the thermal mass releases the heat, keeping the house

FIGURE 4.5 IR image of heat loss from a house. (Courtesy of Sierra Pacific Innovations, Las Vegas, NV.)

warm. For thermal mass to be effective, air must circulate freely throughout the house to carry the heat from the thermal mass to the rest of the house. Natural convection will circulate the air; however, fans may be needed to assist heating in some rooms. As a general rule, more thermal mass is better. The most common materials are rock, stone, concrete, and water. In general, the denser a material is, the more thermal storage is available per mass. Remember that in cold climates water can freeze and present problems in terms of breaking containers. Solutions to keep water from freezing are using antifreeze or containers placed in an area where temperatures will never be below freezing. Also, water is corrosive, so ordinary barrels with water may be a problem after some time period. Finally, using water for storage presents a problem if the storage area is above ground level as a stronger structure is needed to support the weight.

4.5.1 THERMAL MASS PATTERNS

There are a number of methods or patterns for estimating the amount of thermal mass per window area and for location of the thermal mass. The rules of thumb for sizing thermal mass for passive solar homes are summarized in five patterns [3]:

Floor or wall in direct sunlight
Floor, wall, or ceiling in indirect sunlight
Floor, wall, or ceiling remote from sunlight
Mass wall (Trombe wall) or water wall in direct sunlight
Partial mass wall or water wall (containers) in direct sunlight

Then, the following interrelated factors determine the sizing characteristics:

Area of windows exposed to the Sun
Mass surface area
Mass thickness
Mass material type

4.5.2 SPECIFIC HEAT

The amount of heat a material can store is determined by the mass of material and its specific heat, a property measured experimentally (Table 4.2). Specific heat is the amount of energy (J) needed to raise 1 kg a degree centigrade or the amount of energy (Btu) needed to raise 1 lb a degree Fahrenheit. In fact, that was how the British thermal unit was defined, and the calorie is the amount of heat needed to raise the temperature of 1 g of water by 1°C. The amount of heat (thermal energy) stored in a given amount of material for a temperature difference is given by

$$H_{TE} = S * m * \Delta T \qquad (4.5)$$

where:
S is the specific heat
m is the mass
ΔT is the temperature difference

TABLE 4.2

Specific Heat and Heat Capacities of Common Materials

Material	Specific Heat (kJ/[kg °K])	Specific Heat (Btu/[lb °F])	Density (lb/ft³)	Heat Capacity (Btu/[ft³ °F])
Water	4.2	1.00	62.5	62.5
Air	1.0	0.24	0.075	0.018
Concrete	0.9	0.22	144	34.0
Brick[a]	0.9	0.22	123	24.6
Gypsum	1.1	0.26	78	20.3
Limestone	0.9	0.22	103	22.4
Wood	2.4	0.57	47	26.8
Rock	1.2	0.28		
Soil (dirt)	0.8	0.19		

Note: For specific heat and heat capacity of some common solids (kJ/kg °K) see the Engineering Toolbox website (http://www.engineeringtoolbox.com/specific-heat-solids-d_154.html).

[a] If magnesium is added to the brick, the heat capacity is larger.

Example 4.3

For heat stored in rocks, $m = 500$ kg of rock, $S = 840$ J/(kg °C), T final $= 50°C$, T initial $= 40°C$.

$$H_{TE} = 840 \text{ J/(kg °K)} * 500 \text{ kg} * 10 \text{ K} = 4.2 * 10^6 \text{ J}$$

This is the same amount of heat given up as mass cools from 50°C to 40°C.

If the heat is stored in a liquid, then the quantity is probably given in terms of volume, so you need to know the density to calculate the mass or know the heat capacity of the liquid, energy/(volume * degree).

$$\rho = m/V \tag{4.6}$$

where:
ρ is the density
V is the volume

Example 4.4

For heat stored in 4 m³ of water, $S = 4.19$ kJ/(kg °C), T final $= 26°C$, T initial $= 18°C$. 1 m³ water $= 1,000$ kg, so $m = 4,000$ kg.

$$H_{TE} = 4.19 \text{ kJ/(kg °C)} * 4,000 \text{ kg} * 8°C = 1.3 * 10^5 \text{ J} = 1.3 * 10^8 \text{ J} = 1.2 * 10^5 \text{ Btu}$$

4.6 SEASONAL HEATING OR COOLING

Once the U values of all exterior surfaces (envelope) have been calculated, the conduction heat loss is calculated using Equation 4.1 for each surface and then summed to estimate the total conduction heat loss. An important quantity is the hourly heat loss of the house at the outside temperature close to the lowest expected value, the design temperature. Design temperatures are listed for a number of U.S. cities. For example, the design temperature for Amarillo, Texas, is 8°F. A heating system needs to be able to deliver this amount of heat per hour during the coldest days. In general, building contractors use rules of thumb to size the heating system for the size and type of house, amount of insulation, and the design temperature of the area. Can the size of the heating system be reduced with solar heating and thermal mass?

The conduction heat loss for a heating season or months within the season is estimated using degree-days. The standard practice is to use 65°F for the inside temperature because most buildings do not require heat until the outdoor air temperature falls between 60°F and 65°F. A degree-day is then the difference between 65°F and the average temperature for the day. For example, if the average outside temperature is 50°F for 7 days, then the number of degree-days is 105. See the Links section for data for heating and cooling degree-days.

Example 4.5

Calculate the heat loss for a season for a wall for Amarillo, Texas.

$U_t = 0.064$ Btu/(ft^2 °F h), area = 50 ft^2, season degree-days = 3,985

Season degree-hours = 3,985°F day * 24 h/day = 95,640°F h

$H_{CON} = 0.064$ Btu/(ft^2 °F h) * 50 ft^2 * 95,640°F h = 3.1 * 10^5 Btu = 3.3 * 10^8 J

Generally, the calculation is done by month because the heat loss will be highest in December and January.

A similar estimation can be made for season cooling, which also uses 65°F as the base. Conduction heat gain can be calculated for the hottest day of the summer to estimate the design size of the cooling system. A conduction heat gain for a cooling season is estimated by the same procedure. Again, building contractors use rules of thumb for sizing cooling systems.

4.7 THERMAL COMFORT

As you well know, thermal comfort is subjective and differs from person to person. In the tropics, when the temperature dips to 30°C (86°F), most people feel cold. We want our houses to be comfortable with a feeling of heat in the winter and a feeling of cooling in the summer. We set the thermostat at 25°C (77°F) or lower in the summer and at 23°C (73°F) or higher in winter.

Thermal comfort depends on environmental and physiological factors:

Environmental: Air temperature (dry bulb), relative humidity, air speed, and radiation (mean radiant temperature, MRT)
Physiological: Metabolic rate and amount of clothing (insulation)

If you are active in the winter, then you feel comfortable at lower temperatures. Previously, we dressed for winter with long underwear, sweaters, and so on, but now we expect to wear summer clothes inside buildings during the winter.

The environmental factors are interrelated in our perception of thermal comfort. Air temperature affects the amount of heat lost or gained due to convection and affects evaporation of sweat.

MRT affects our perception of temperature because hot and cold objects emit or absorb radiant energy, which activates the same sensory organs as heat transferred by conduction or convection. The net exchange of radiant energy between two objects is proportional to their temperature difference multiplied by their ability to emit and absorb heat. MRT is the area-weighted mean temperature of all surrounding objects, which is positive when surrounding objects are warmer than the skin and negative when they are colder. People are highly responsive to changes in MRT as the human skin has extraordinarily high absorptivity and emissivity (0.99). This is why people have their thermostats set lower in summer than in the winter. In my office at the university, even though the thermostat is at the same level in the winter, on really cold days, I am cold because I am radiating heat to a brick outside wall and not getting radiation in return. The Efficient Windows Collaborative has information on the benefits of efficient windows and how they can reduce the radiation to enhance thermal comfort (http://www.efficientwindows.org).

Relative humidity is the amount of moisture vapor in a specific volume of air. For any dry-bulb temperature, there is only a certain amount of moisture vapor that can be absorbed in the air before it becomes saturated and precipitation occurs. Relative humidity affects the evaporation, and in hot, dry climates, sweat is readily evaporated. At relative humidity above 80%, sweat is produced, but most of it cannot evaporate as the air immediately surrounding the body becomes saturated. Humidity less than 20% can dry out mucous membranes and greatly increase susceptibility to infection. In the winter, humidity in buildings may need to be increased for thermal comfort. In hot, humid climates, the control of humidity is a major factor in cooling.

Air speed is also an important factor in thermal comfort. Stagnant air in artificially heated spaces often contributes to a feeling of stuffiness, and any air movement in cold environments is often considered drafty. We can accept higher air temperatures in the summer if there is air movement; therefore, ceiling fans are now installed in many houses and other buildings. If the air temperature is less than skin temperature, air movement increases convective losses substantially. In 30%–85% humidity, air movement increases the evaporation of sweat; however, air speed makes only minimal differences for relative humidity below 30%. In winter, we accept lower temperatures if there is no air movement, so the wind chill index is a combination of air temperature and wind speed as it affects people. For example, the wind chill index is −7°C when $T = 0$°C and the wind speed is 10 m/s. See the National Weather Service Windchill Chart website (http://www.nws.noaa.gov/om/winter/windchill.shtml).

A psychrometric chart (Figure 4.6) is used to determine the thermal comfort zone using local climatic data. That comfort zone can be enlarged (Figure 4.7) by changing air speed, humidity, and evaporative cooling in dry climates and of course by air conditioning (changing temperature).

Thermal comfort is highly subjective as is it depends not only on personal preference and acclimatization but also on the integration of internal and external temperature sensing. The overall sensation may be pleasing or displeasing depending on whether the overall effect is toward or away from the restoration of deep body temperature. A cold sensation will be pleasing when the body is overheated but unpleasant when the body core is already cold. At the same time, the temperature of the skin is by no means uniform. The wearing of clothes also has a marked effect on the level and distribution of skin temperature. For the purposes of building design, comfort will be defined as the absence of thermal stress on people.

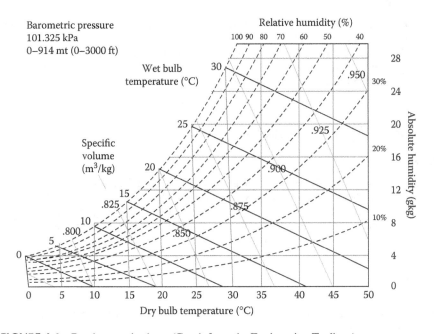

FIGURE 4.6 Psychrometric chart. (Graph from the Engineering Toolbox.)

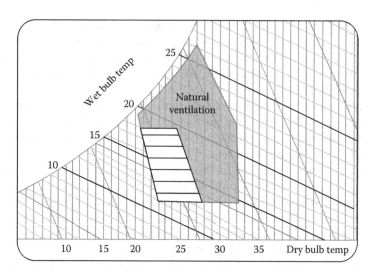

FIGURE 4.7 Thermal comfort zone (gray area) is enlarged by natural ventilation (big area). (Graph from the Engineering Toolbox.)

REFERENCES

1. Building Envelopes Program, Oak Ridge National Laboratory. http://www.ornl.gov/ sci/roofs+walls/index.html.
2. W. A. Shurcliff. 1980. *Thermal Shutters and Shades.* Brick House, Amherst, NH.
3. *The Thermal Mass Pattern Book, Guidelines for Sizing Heat Storage in Passive Solar Homes.* 1980. Total Environmental Action, Harrisville, NH.

RECOMMENDED RESOURCES

LINKS

Energy Efficiency and Renewable Energy, Buildings Technologies Office. Information resources. http://www1.eere.energy.gov/buildings/information_resources.html.
The Engineering Toolbox. Resources, tools, and basic information for engineering and design of technical applications. Great site for information. http://www.engineeringtoolbox .com/metabolism-activity-d_116.html.

CLIMATE DATA

National Oceanic and Atmospheric Administration, National Climatic Center. http://www .noaa.gov/climate.html.
World Climate. Heating and Cooling Degree Days. http://www.climate-charts.com.

HEATING AND COOLING DEGREE-DAYS

BizEE. Calculates degree-day (heating and cooling) data for the world by day, week, month, or average. Results downloaded as a spreadsheet. http://www.degreedays.net/.

The Engineering Toolbox. Design conditions in U.S. states and cities summer and winter. http://www.engineeringtoolbox.com/us-outdoor-design-temperature-humidity-d_296.html.

Environmental Change Institute. Data for United Kingdom. http://www.eci.ox.ac.uk/research/energy/degreedays.php#degreedays.

National Climatic Data Center, NOAA. Heating and cooling degree data. http://lwf.ncdc.noaa.gov/oa/documentlibrary/hcs/hcs.html.

National Weather Service. Degree-day monitoring and data. http://www.cpc.ncep.noaa.gov/products/monitoring_and_data/DD_monitoring_and_data.shtml.

Sustainable by Design. U.S. climate data. http://susdesign.com/usa_climate/index.php.

THERMAL COMFORT

Canadian Centre for Occupational Health and Safety. Thermal comfort for office work. http://www.ccohs.ca/oshanswers/phys_agents/thermal_comfort.html.

Health and Safety Executive. Thermal comfort. http://www.hse.gov.uk/temperature/thermal/index.htm.

INNOVA, AirTech Instruments. *Thermal Comfort.* http://www.lumasense.dk/Booklets.60.0.html.

Thermal comfort. http://personal.cityu.edu.hk/~bsapplec/thermal.htm.

PROBLEMS

4.1. Place your hand close (do not touch) to an incandescent lightbulb. What do you feel? How did that energy get from the lightbulb to your hand? Do the same thing with a fluorescent lightbulb or an LED lightbulb that emits around the same amount of light. What is the difference?

4.2. When there is significant temperature difference between outside and inside the house, place your hand on a window (inside and then go outside). Note the date and inside and outside temperatures. Does the window feel hotter or colder for each situation compared to the inside and outside temperatures?

4.3. Place a wooden pencil and a spoon in a cup of hot water. Feel both after a short time (10 s) and then after a longer time period (2 min). Their thermal conductivity differs quite a bit. Write down your observations. What if there is a metal casing on the pencil for the eraser? How does that feel compared to the wood?

4.4. Calculate the season heating loss for a single-pane picture window 1.2 m high by 3 m long. Use heating degree-days for Amarillo, Texas.

4.5. Calculate the season heating loss for a double-pane, low-e glass picture window 1.2 m high by 3 m long. Use heating degree-days for Amarillo, Texas.

4.6. Calculate the season cooling need for a single-pane picture window 1.2 m high by 3 m long. Use cooling degree-days for Amarillo, Texas. In other words, your air conditioner needs to remove this amount of heat from your house.

4.7. Calculate the season cooling need for a double-pane, low-e glass picture window 1.2 m high by 3 m long. Use cooling degree-days for Amarillo, Texas. In other words, your air conditioner needs to remove this amount of heat from your house.

4.8. Calculate the season heating loss for the south-facing windows of your house. Use heating degree-days for the city closest to you for which data are available.

4.9. Calculate the season cooling need for the south-facing windows of your house. Use heating degree-days for the city closest to you for which data are available.

4.10. Calculate the conduction heat loss for the following wall, 4 m long, 2.5 m tall. Calculate for 8 h, inside temperature at 20°C and the outside temperature at −5°C. Wall components from outside to inside are: brick; 2 cm Styrofoam board; 1.2 cm plywood; wood studs, 5 × 10 cm, on 50 cm centers with air space between studs; 1.2 cm gypsum board.

4.11. Calculate the conduction heat loss for the following wall: 15 ft long, 8 ft tall. Calculate for 8 h, inside temperature at 78°F and the outside temperature at 3°F. Wall components from outside to inside are brick; 1.2-cm (1/2-in.) plywood; wood studs, 5 × 10 cm (2 × 4), 45-cm (18-in.) centers with 9-cm (3.5-in.) fiberglass batt between the studs; 1.2-cm (1/2-in.) gypsum board. Remember that the air film adds to the insulation.

4.12. Estimate the infiltration loss at your home on a windy day (24 h). The outside temperature is at freezing. This will be a rough estimate. When doors are opened, they let in or out a lot of air.

4.13. Use Figure 4.2. When should insulation be placed for foundations?

4.14. Use Figure 4.3. What is the recommended R value for a wall cavity for a new wood frame house in your hometown? Include the town and state in your answer.

4.15. Use Figure 4.3. What is the recommended R value for the attic for a new wood frame house in your hometown? Include the town and state in your answer.

4.16. Calculate the heat stored in a concrete floor (10 cm thick, 3.5 m × 4.5 m), initial temperature 15°C, final temperature 27°C.

4.17. Calculate the heat stored in water (fifty 4-liter jugs), initial temperature 15°C, final temperature 27°C.

4.18. What is the design heating temperature for your home? Use the city closest to you for which there are data. Use your parents' home if needed.

4.19. What is the design cooling temperature for your home? Use the city closest to you for which there are data. Use your parents' home if needed.

4.20. At what temperature do you set your thermostat in summer? In winter?

4.21. Most ceiling fans will run in both directions. Why?

5 Solar Heating and Cooling

5.1 BUILDING

Heat is transferred by conduction, convection, and radiation, and the ratio of heat transfer by the three modes can vary widely depending on amount of insulation, number and types of windows, radiation barrier, and infiltration. Air infiltration through open windows, doors, and vents and through cracks in the envelope of the house or around windows and doors can account for 20% to 55% of the heat loss. Even the damper in the fireplace must be closed when it is not in use.

In general, older homes have large U values due to single-pane windows and little insulation. Because many older homes are horrible in terms of heat loss or gain, the first step in reducing energy cost is always insulation and reducing infiltration. Weatherization programs for low-income groups are more economical than money for assistance in paying utility bills when energy costs are high. And weatherization continues to save energy for a long period of time. The primary aspects of weatherization are addition of storm windows, caulking cracks, and adding insulation, especially for the ceiling.

Superinsulated houses [1] have R-40 insulation in the ceilings, 15-cm or more stud walls with R-22 batts, and 5 to 10 cm of Styrofoam insulation around the exterior foundation. All windows would be double or triple pane in very cold climates.

For new houses, states are now implementing building codes for residences that emphasize energy conservation (reduction in heat transfer). The other aspects promoted are energy efficiency in appliances, lighting (incandescent lights are being phase out), daylighting, furnaces, and air conditioning.

5.1.1 AIR QUALITY

The reduction in infiltration can mean less fresh air. Houses with less than 0.5 air changes per hour can have excessive levels of carbon dioxide, moisture, other air pollutants, and even radon from soils and groundwater in some areas. In super-insulated houses, an air-to-air heat exchanger may be necessary as the heat exchangers remove the heat from the exhaust air and transfer it to fresh intake air.

5.1.2 AIR AND VAPOR BARRIERS

The activities of the family can produce 2–3 gal of water vapor per day, and the teen-ager who takes a 20-min shower adds to that total. Most of the moisture is carried by infiltration. Water vapor migrates or diffuses from areas of greater to lower vapor pressure, which can be prevented with a vapor barrier. Vapor barriers are large, thin sheets of transparent polyethylene around the inside of the building envelope that limit the movement of moisture.

During the winter, moisture in the warm air can condense in wall and ceiling cavities where it meets a cold surface. If enough vapor condenses, it can saturate the insulation, reducing its R value, and may eventually cause rot and decay in wood materials. The following guidelines will help control condensation:

- Use materials in the outer skin that are five times as permeable as the inner skin.
- Seal all cracks and joints.
- The air barrier should be as tight as possible.

The insulation outside the vapor barrier should be twice as much as inside. In high-moisture areas, the vapor barrier should be on the warm side of the insulation.

Condensation on double-pane windows may indicate that you have problems with inadequate vapor barriers.

Air barriers are high-density polyethylene fiber films placed around the outside of the house to reduce infiltration. Air barriers allow moisture vapor to pass. If a vapor barrier is sealed carefully, it can also serve as an air barrier.

5.1.3 WIND AND VEGETATION

Wind can drastically change the heat loss or gain for a house. The conduction heat losses through the surfaces of a house increase with wind speed, as the R value of the air film is reduced from 0.68 to 0.17 in a 7-m/s wind. This reduction is most important for windows and doors. In addition, winds increase air infiltration through any crack or opening, and it is easily detected inside the house in cold weather just by feeling around doors and windows.

A belt of evergreen trees sheltering the house from prevailing winter winds can reduce heat loss by as much as 30%. The trees should eventually reach the height of the house, and the distance from the house should be less than five times the building height. Deciduous trees on the east and west can reduce heat gain in the summer by as much as 25%. The local agricultural extension service can recommend trees and shrubs suited to your climate. In addition, vegetation produces oxygen and absorbs carbon dioxide.

Baffles and shutters can also reduce heat transfer due to wind. Entrances should not be on the side of the house in prevailing winter winds. Windows should also be kept to a minimum on the north due to winter winds and kept to a minimum on the west due to afternoon heating in the summer, unless the windows are shaded.

Information on wind speed and direction (wind rose diagrams) can be obtained from local, regional, or national weather service agencies. Local terrain will affect both wind speed and direction, so if you are new to an area, check with residents who have been in the area for some time.

5.2 PASSIVE

In a passive system, the solar collector and storage are an integral part of the house, and there is no auxiliary assistance for the transfer of heat within the house. You can have both passive heating and cooling; however, heating is the most common. *The most significant factors are the size and placement of windows and the sizing of the thermal mass.* Other factors are orientation, color, and shape of the house; color inside the house; plus vegetation for shading and wind control. The materials of the house act as the thermal mass; however, for conventional homes there is not enough mass for a passive system. So, for a retrofit or remodeling for passive solar, mass will have to be added.

5.3 WINDOWS AND GLAZING

Fenestration is the openings in the house, which refer to windows, skylights, and doors. In general, we think of windows as having glass panes; however, there can be other materials, so the covering that transmits the solar energy is called *glazing*. Remember that solar radiation is composed of the direct and diffuse (radiation from the sky and reflection from other bodies, respectively). What you want is to let the direct sunlight in the house in the winter and to shade that direct sunlight in the summer. You want the diffuse sunlight for daylighting in both winter and summer.

The incoming solar electromagnetic (EM) radiation is composed primarily of three ranges, infrared from 25 to 7.6 * 10^{-7} m, visible (sometimes also referred to as *light*) from 7.6 to 4.0 * 10^{-7} m, and ultraviolet (UV) from 4.0 to 3.0 * 10^{-7} m. The characteristics of the glazing are given by the division of the incoming EM radiation into the following: transmission, reflection, absorption, and the solar heat gain coefficient (SHGC). The absorbed component is reradiated in the infrared range, so in reality the incoming EM radiation is composed of transmitted and reflected radiation (Figure 5.1). The transmittance depends on the angle of incidence, being fairly constant from normal to 50° and then decreases as more EM radiation is reflected (Table 5.1). So, vertical windows, when the Sun is high in the sky, transmit less direct sunlight due to less area perpendicular to the Sun (cosine factor) and more reflection. However, the best strategy is to shade south-facing (north-facing in the Southern Hemisphere) windows in summer so that little to no direct sunlight enters the house. Also, west windows let in too much sunlight (solar energy) in the summer during late afternoon.

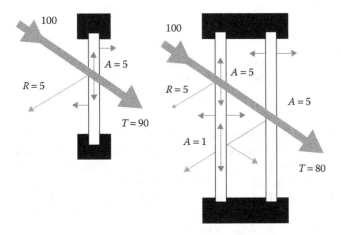

FIGURE 5.1 Incoming solar radiation for single- and double-pane windows. The absorbed radiation will not be equal for inside and outside due to the temperature differences.

TABLE 5.1

Transmittance of Glass versus Angle of Incidence

Angle of Incidence	Single Glaze	Double Glaze
0	0.90	0.81
20	0.90	0.81
40	0.89	0.80
60	0.82	0.71
70	0.77	0.59
80	0.44	0.29
90	0.00	0.00

TABLE 5.2

Transmittance of Different Glass Glazings

Glazing	Solar (%)	Visible (%)	R ([m² °K]/W)	R ([ft² °F h]/Btu)
Single clear	90	80	0.15	0.87
Double				
Clear	80	65	0.36	2.04
Coated/tinted				
Clear low-e	52	74	0.55	3.13
Green Solex	41	68	0.36	2.04
Bronze tinted	41	48	0.36	2.04
Reflective				
Gray	6	8	0.40	2.27
Gold	9	18	0.52	2.94
Silver-gray	23	29	0.38	2.17
Silver-blue	12	18	0.40	2.27
Triple				
Clear	62	75	0.45	2.56
Clear low-e	48	70	1.2	7.0

Since the glazing has lower R values than insulated walls, you do not want all the walls of your house to be glass, although some houses with almost all glass have been built. Also, a structure with a glass envelope would overheat in the summer as too much solar radiation would be admitted. Therefore, large buildings with glass envelopes have highly reflective glass (Table 5.2) to reduce direct sunlight. There was a case when a skyscraper with reflective glass increased the heat load significantly for an older adjacent building.

Low-emissivity (low-e) glass has a thin film on the inside of the third surface to reduce the transmission of the long-wavelength infrared radiation. This reduces heat loss in the winter from inside the house to the colder space outside and reduces some of the heat gain in the summer as it blocks the infrared portion of the solar radiation. Remember that the U value for windows includes everything, frame and glazing. For solar heating, you want a glazing that has high solar heat gain. The reason to go to double pane is lower U (higher R) values since the dead air (or even noble gas) increases the insulation value. The U.S. Department of Energy (DOE), Energy Efficiency and Renewable Energy (EERE), Building Technologies program, and Efficient Windows Collaborative have information for new and replacement window types [2,3] with U values, transmission, and solar heat gain.

5.3.1 OTHER GLAZINGS

Fiberglass-reinforced plastics, polycarbonate (Lexan), and polyvinyl chlorides (PVCs) are available for windows, passive solar systems, daylighting, greenhouses, and so on. They can be obtained in single sheets or as component systems of two sheets plus insulation for higher R values. Previous disadvantages were weathering and coloring with age due to UV radiation. Sun-Lite has a Tedlar film on the exterior surface to reduce UV damage.

Material	Transmission (Depends on Thickness)
Sun-Lite	80%–90%
Lexan	64%–84%

5.3.2 SOLAR HEATING

The amount of solar energy (insolation) that comes into a building through the windows (size, orientation, glazing, and shading) depends on time of year, location, and percentage sunshine (clouds and haze). The solar heating can be estimated from tables of clear-day insolation by latitude and percentage sunshine by month (Figure 5.2 and Table 5.3). Average insolation values by month that take into

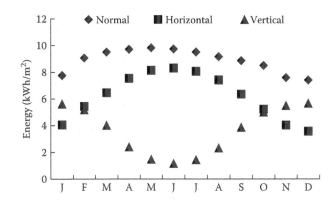

FIGURE 5.2 Clear-day insolation, average for month, for latitude 32° N.

TABLE 5.3

Clear-Day Insolation (kWh/m²) for 32° N and Estimated Percentage Sunshine by Month for Amarillo, Texas, Area

Month	Normal	Horizontal	Vertical	% Sun
Jan	7.8	4.1	5.6	70
Feb	9.1	5.4	5.2	70
Mar	9.5	6.5	4.0	70
Apr	9.7	7.5	2.4	70
May	9.8	8.1	1.5	70
Jun	9.7	8.3	1.2	80
Jul	9.5	8.1	1.4	80
Aug	9.2	7.4	2.3	80
Sep	8.9	6.4	3.9	75
Oct	8.5	5.2	5.0	75
Nov	7.6	4.0	5.5	75
Dec	7.4	3.6	5.7	70

Note: Insolation is for 21st day of month.

account cloud cover and different angles of surfaces are available on the Internet. The data for vertical insolation indicate that in most locations, vertical windows on the south can be net heat gainers, depending on climate and R value of the windows. Surprisingly, even single-pane windows on the south can be net energy gainers in temperate climates. Also, during the summer, vertical windows let in less solar energy due to the angle of the Sun (cosine factor). There is more reflection due to the larger angle of incidence.

Example 5.1

Calculate the amount of solar heat that comes through a south-facing window (single pane) for January (70% sunshine).

Vertical window, 1.2 by 2.5m, single pane, transmission = 90%

Area = 3 m², insolation for January is 6 kWh/m² per clear day

Energy hitting window per day = 3 m² * 6 kWh/m² = 18 kWh/day

Energy transmitted = 0.9 * 18 kWh/day = 16 kWh/day = 55,000 Btu/day

Energy for month = 16 kWh/day * 31 days * 0.70 (% sunshine)

= 350 kWh = 1.2 * 10⁶ Btu

Maps of solar insolation for the United States by month are available from the National Renewable Energy Laboratory (NREL) for different types of collectors and orientation (http://rredc.nrel.gov/solar/old_data/nsrdb/1961-1990/redbook/atlas//). The two-axis

TABLE 5.4
Average-Day Solar Insolation (kWh/m²)
for Amarillo, Texas, from NREL Solar Maps

Month	Normal	Horizontal	Vertical
Jan	6.5	2.5	4.5
Feb	6.5	3.5	3.5
Mar	7.5	4.5	3.5
Apr	9.0	6.5	3.5
May	9.0	6.5	2.5
Jun	9.0	7.5	2.5
Jul	9.0	6.5	2.5
Aug	9.0	6.5	2.5
Sep	7.5	5.5	3.5
Oct	7.5	4.5	4.5
Nov	6.5	3.5	4.5
Dec	5.5	2.5	4.5

concentrating collector gives the normal to the Sun. The maps also take into account the percentage sunshine (Table 5.4), so the average values of energy per day do not have to be adjusted.

Example 5.2

Calculate the amount of solar heat that comes through a south-facing window (double pane) for January for Amarillo, Texas.

Vertical window, 1.5 by 2 m, area = 3 m², double pane, transmission = 80%

Average insolation for January for vertical window is

$4.5\,\text{kWh}/\text{m}^2$ per day (takes into account cloudy weather)

Energy hitting window per day = $4.5\,\text{kWh/m}^2 * 3\,\text{m}^2 = 13.5\,\text{kWh/day}$

Energy transmitted per month = $0.80 * 13.5\,\text{kWh/day} * 31\,\text{days}$

$$= 335\,\text{kWh} = 1.1 * 10^6\,\text{Btu}$$

The double-pane window will transmit less solar heat into the house; however, it will reduce the heat loss. Therefore, in colder climates, double-pane windows are better.

During the winter, vertical windows on the south (Northern Hemisphere) capture most of the solar energy available (Figure 5.3) because of the position of the Sun. On December 21 (35° N, 102° W), there are 9 h of sunshine with around 6 h for solar heating (Figure 5.3). At 0900 and 1500, the altitude is 15° with azimuths of 120° and 217°, respectively. The maximum elevation is 32°, so during that 6 h, the vertical windows let in most of the sunlight. On July 21, sunrise is at 0549 and sunset is at 1958. That means, there are 14 h of sunshine, with the Sun due east at 0900 and

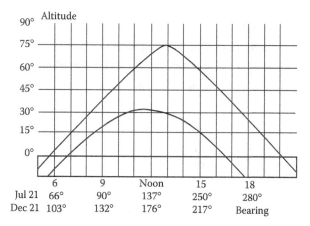

FIGURE 5.3 Path of the Sun for December 21 and July 21, location 35° N, 102° W. Solar noon occurs at different times primarily because of daylight savings time.

due west at 1650. There are 4 h when the elevation angle is above 60°, so more solar radiation is reflected by the window. However, it is still better to have the south-facing windows shaded during the summer months. The sun path across the sky can be plotted for any latitude and longitude for any day (see Chapter 3).

Sustainable by Design has a number of shareware design tools [4]: sun angle tools, window tools, panel shading, and U.S. climate data. Instead of using solar maps and climate data for calculating heat gain, the window heat gain tool calculates solar heat gain through vertical windows in temperature latitudes by month and time of day. For cities in the United States, the tool uses climate data and percentage sunshine. General input data include window type, orientation, and ground reflectance (default = 0.2).

5.3.3 Shading

The south-facing windows need to let sunlight in during the heating season and be shaded during the cooling season so that little direct sunlight enters the house. The simplest shading devices are exterior to the house, such as overhangs and awnings. Deciduous trees that provide shade on the east and west sides of the house are also effective. Even though the Sun is the highest and the longest day is on June 21, shading needs to be provided through August (Figure 5.4). There can be a problem with awnings as they require manual control, and if you live in a windy area they tend to get torn.

FIGURE 5.4 Overhang gives shade in summer and permits direct solar radiation into the structure in winter.

South-facing windows transmit more energy during the winter than the summer due to the solar position of the Sun, even though there are more hours of sunshine during the summer. The primary factor is the angle, such that less area is perpendicular to the Sun, and the other factor is that there is more reflection for angles above 60°.

Solar rights means the property owner should have access to sunlight and the right to install a solar system. Communities are considering building codes, solar easements, and solar rights. Also solar access is where for a certain amount of sunlight should be available for building facades and ground areas [5,6]. Sites are available for determining the sun angle for shading for buildings [7]. For the Southern Hemisphere, the sun angles are symmetric, so shading needs to be on the north. For the tropics, the days are all about the same, 12 h of sunshine. However, now you need shading for windows on both the south and north sides of the house.

5.4 PASSIVE HEATING AND COOLING

5.4.1 DIRECT GAIN

Direct gain occurs when the sunlight comes through windows, skylights, or clerestories (glazings can be glass or plastic) and is absorbed in the room (Figures 5.5 and 5.6). In general, larger-area skylights are needed for solar heating; however, they are heat gainers in the summer since the Sun is more directly overhead. One solution to reduce heat gain for skylights would be to install movable shades. Clerestories are a way of getting direct solar energy into rooms that are not on the south. As noted, there has to be enough thermal mass to reduce the temperature variations between sunny days and cold nights in the winter. Thermal mass also helps for passive cooling.

5.4.2 INDIRECT GAIN

Indirect gain occurs when the sunlight is absorbed and then transferred to the rest of the house by natural or forced convection, conduction, and radiation. Examples of indirect gain occur in a sunroom (solarium), attached greenhouse, Trombe wall, water wall, and solar pond. If you live in areas with freezing temperatures, then water storage may be a problem.

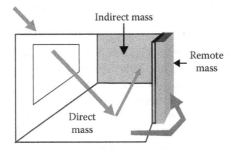

FIGURE 5.5 Direct gain with solar heat stored in direct mass (direct sunlight), indirect mass (reflected sunlight), or remote mass through convection.

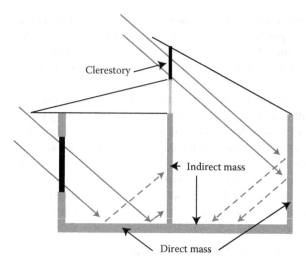

FIGURE 5.6 Direct gain through south-facing windows and clerestory. Clerestory gives direct gain into rooms that are not on the south side of the house. Overhangs shade windows in the summer.

FIGURE 5.7 Attached greenhouse at Amarillo Children's Home that was constructed by volunteers at AEI workshop.

Attached greenhouses (Figure 5.7), sunrooms, and solariums present a way to increase space and provide heating. Many are additions to an existing house; however, some sunrooms are an integral part of the house.

Sunrooms and attached greenhouses can overheat, even in winter, so heat has to be transferred to the rest of the house or enough thermal mass needs to be available to reduce temperature variations. They may need auxiliary power for fans to circulate air to the rest of the house. Also, sunrooms and attached greenhouses need ventilation during the summer, especially if there is glazing on the roof. This can be accomplished through windows that open or ventilation fans. A patio on the north side of the house (location 35° N, 102° W), open screen on sides with a fiberglass roof, was too hot during the summer. If the primary use is for plants, then you may need auxiliary heating in the winter, especially in cold climates. Thermal shades or blankets are one possibility for reducing heat loss at night for sunrooms and attached greenhouses. Again, thermal mass will reduce the temperature variations.

Sunrooms, solariums, attached greenhouses, and large skylights from kits to design information are available from a number of manufacturers. Photos of different types, layouts, and prices are available.

A Trombe wall consists of a masonry wall 20 to 40 cm (8 to 16 in.) thick coated with a dark, heat-absorbing material and covered with a single or double layer of glazing (Figure 5.8). The glazing is around 5 cm from the masonry wall, so the hot air will rise and go into the room during the day. Most of the solar energy is absorbed by the dark surface, stored in the wall, and conducted through the masonry such that peak temperatures on the inner surface of the room occur later in the day (heat travels through the concrete wall about 2.5 cm/h [1 in./h]). For a 20-cm-thick Trombe wall, the heat takes about 8 h to reach the interior of the building. This means that rooms remain comfortable through the day and receive slow, even heating after the Sun sets. Rooms heated by a Trombe wall often feel more comfortable than those heated by forced-air furnaces because warm surface radiates infrared radiation, even with lower air temperatures in the room.

If the wall is too thick, the heat takes too long to reach the room, plus there is more cost for the concrete. If the wall is too thin, it gets too hot, and the heat reaches the room from the wall during the day when you also have heating from convection. Dampers must be placed on the openings, so there is no reverse thermosiphoning at night.

The surface of the wall is painted with a flat black paint for absorption. Applying a selective surface to a Trombe wall improves its performance by reducing the amount of infrared energy radiated back through the glazing. A selective surface absorbs almost all the radiation in the visible range and emits little in the infrared range. A selective surface on a sheet of metal foil is glued to the outside surface of the wall.

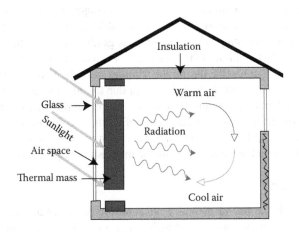

FIGURE 5.8 Indirect gain, Trombe wall.

FIGURE 5.9 Visitor center has windows and clerestory for direct gain and daylighting and Trombe wall for indirect gain. The Trombe wall is the lower two rows of glass. That glazing is a single-covering, high-transmittance, patterned glass. Notice that the windows can open for ventilation. (Data from National Renewable Energy Laboratory [NREL], http://images.nrel.gov.)

Windows for direct gain and daylighting can be combined with a Trombe wall, which absorbs and stores heat for evening use. Overhangs for the Trombe wall prevent the wall from getting too hot during the summer. A Trombe wall [8] provides most of the heating for the Zion National Park Visitor Center building (Figure 5.9). Winter surface temperatures on the inside of the Trombe wall can often reach 38°C (100°F).

Since water has a higher heat capacity than masonry, water walls can be used. These can be tubes, barrels, or even milk jugs. There are two considerations for barrels, freezing and rusting; both can result in leaks, but antifreeze and anticorrosion substances can be used for these problems. Steve Baer (Zomeworks) built a home that had indirect gain, which consisted of insulated cover, glazing, and then barrels, which were painted black. The insulated cover, which was moved manually, was on the ground during the day with a sheet of aluminum foil to reflect more sunlight onto the barrels.

Harold Hay designed a house (Skytherm) in California that had a water pond on the roof with a sliding insulating cover. The water bags absorbed heat during the day and then the heat conducted through the concrete roof and radiated into the house at night. The system worked in reverse during the summer, with the pond covered during the day and uncovered at night. It cooled at night due to radiation to the night sky. This system was not duplicated on any scale for the following reasons: weight on the roof requiring a stronger structure, manual control of insulated cover, primary suitability for mild climates with no freezing, and condensation on the concrete during the summer. An additional problem in California could be the seismic response to earthquakes.

5.4.3 COOLING

Passive cooling is the use of natural ventilation to replace hot air with cooler air by convection. At higher elevations, most nights are cool, so if you have thermal mass for passive solar heating, you cool off that mass at night, close the space during the day, and then the cooler thermal mass absorbs the heat from the air, slowing increases in temperature. Thermal chimneys are another way of getting rid of excess heat in hotter climates, as hot air in the upper part of the structure is replaced by cooler air from the outside. Soffit vents with ridgeline vents can keep attics from excessive temperatures. Low windows on the south and high windows on the north, both of which can be opened, can also create convective currents to get rid of the warmer air, especially in regions with lower night time temperatures in the summers, for example, the high plains region of the United States.

5.5 ACTIVE HEATING

In most active systems, the system (collector, storage) is separate from the building, so the system can be added to existing structures. An active heating system (Figure 5.10) requires power for pumps or fans for moving the fluid (air, water, or even silicone) and power for the controller for turning on pumps, valves, and so on. The main components of the system are the collector, storage, and the controller (Figure 5.11). The components of the system are as follows:

Collector: Insulated box, glazing (one or more), and absorber
Heat transfer: Air, liquid, and other
Controls
Storage
Distribution: Pumps, valves, fans, and dampers
Controller (thermostat): Sensors for temperatures to turn distribution on and off
Auxiliary heat: May be tied into present heating system or may be two distri-
    bution systems

With a solar hot water system, the space heating can be provided by radiant water heaters or by air heating, similar to a furnace, where the air passes over a heat exchanger. The heat is provided to the space when needed; however, the cost is too high to provide all the heating by solar energy (Figure 5.12), so an auxiliary heating system works in conjunction with the solar heating system. This means that during the fall and spring, the solar system will provide all the heat, and during the coldest

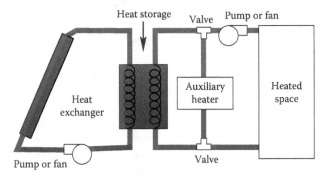

FIGURE 5.10 Active solar heating system. Notice that solar collector loop is closed so it can have antifreeze or other fluid for freezing climates.

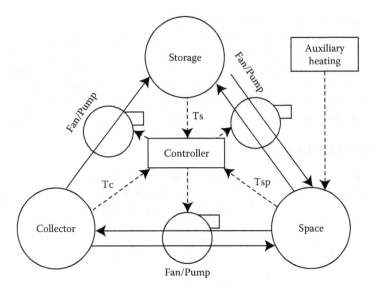

FIGURE 5.11   Diagram of an active solar heating system.

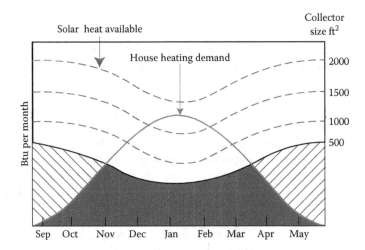

FIGURE 5.12   Size of collector and solar heat available versus house heating demand.

months, the solar heating system will be sized to provide around 60% of the heat. During late spring, early fall, and the summer, the excess heat could provide assistance with domestic hot water.

Again, freezing of water is a problem in moderate and cold climates. There can be an open-loop solar heating system with a drain back at freezing temperatures, a closed loop on the solar side with antifreeze or silicone fluid, or a closed loop or open loop with a two-tank system with heat exchangers.

A solar hot air system must move a larger volume of air than water to transport the same amount of heat. The operation modes depend on the relative temperatures

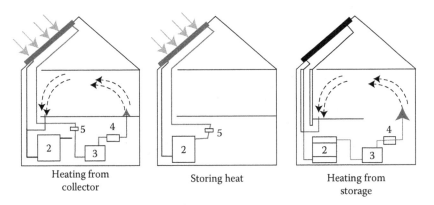

Heating from
collector

Storing heat

Heating from
storage

FIGURE 5.13   Active solar heating system using air for heat transport.

of the collector, storage, and the space (Figure 5.13). The rock storage bin for an air system is sized for 2 to 3 days of heating. At present in the United States, there are no commercial active solar hot air systems for residences, although some special systems have been constructed for commercial buildings.

Most of the systems for active solar heating use flat-plate collectors, so they need to be installed at an angle to the horizontal that gives the best performance. For solar heating, the collectors should be placed at an angle around the latitude minus 15°; for year-round use, they should be at an angle close to the latitude; and for summer cooling, they should be at latitude plus 15°. NREL has a site that calculates the solar insolation for different types of collectors and for the flat-plate collectors at angles of latitude and angles of latitude plus or minus 15° (http://rredc.nrel.gov/solar/old_data/nsrdb/1961-1990/redbook/atlas/).

5.5.1   FLAT-PLATE COLLECTORS

Flat-plate collectors are the most common solar collectors for space heating and domestic hot water. The different components (Figure 5.14) can be fabricated from a variety of materials. In cold climates, two or three glazings are needed, and

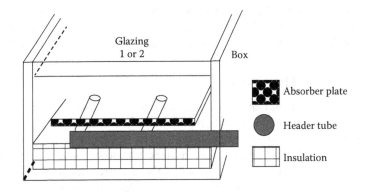

Glazing
1 or 2

Box

Absorber plate

Header tube

Insulation

FIGURE 5.14   Components of flat-plate collector.

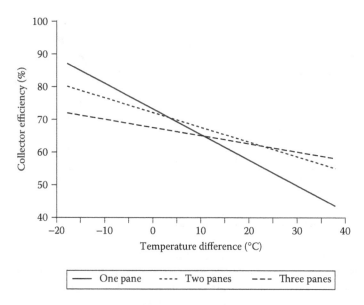

FIGURE 5.15   Daily collector efficiency for temperature difference between absorber plate and outdoor air. Notice efficiency is higher when temperatures are the same.

more insulation of the box is needed to reduce heat loss. The efficiency changes with number of glazings and the temperature differential between the outside and the absorber plate (Figure 5.15). In mild climates (no freezing) or for extending the swimming pool season, all that is needed is an absorber and a way to transfer the water. Absorbers can be copper, aluminum, or iron, with copper the most common for plates, tubes, and header pipes. Glazings can be plastic or glass with the box being aluminum, metal, or vinyl. Remember that different materials have different coefficients of expansion with temperature. If too dissimilar, joints and connections tend to come loose. Higher efficiencies can be obtained with evacuated glass tubes for the collector, with the inner tube having a selective coating (Figure 5.16). Glass glazings need to be able to withstand hail.

The components of a flat-plate air collector do not include tubes for absorber plate or header pipes (Figure 5.17). However, the air must come in contact with all parts of the back of the absorber plate (no stagnation areas), and the airflow needs to be slower.

5.5.2   DOMESTIC HOT WATER

Generic types of systems that use flat-plate collectors to heat water are thermosiphon, closed-loop freeze-resistant, drain back, drain down, and air-to-liquid systems. The thermosiphon (passive) system does not use pumps as the tank is generally insulated and is above the collector; again, there is a check valve to stop reverse circulation [9]. However, placing a hot water storage tank in the attic presents problems, primarily due to weight and the possibility of leaks. For colder climates, a heat exchanger is necessary to keep the collector fluid separate from the water system.

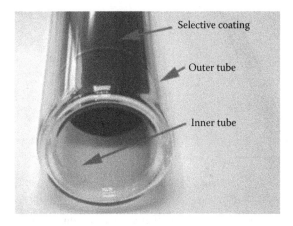

FIGURE 5.16   Evacuated tube collector.

FIGURE 5.17   Cross section of vertical hot air collector. Actual collector is in the background. From outside, fiberglass glazing, dead air space, absorber plate (selective coating), air channel, insulation with aluminum foil, plywood, and metal frame. Notice that glazing on the actual collector has become more translucent with UV aging. Light transmission is still about the same. Ken Starcher is holding the cross section.

In a direct circulation system, pumps circulate the household water through the collector. This works well in mid-temperature climates where it rarely freezes. In indirect circulation systems, pumps circulate a nonfreezing heat transfer fluid through the collector and a heat exchanger in the hot water tank (Figure 5.18).

Other types of systems are an integral collector-storage system (batch or bread box) and evacuated tube solar collectors. The batch hot water system has one or more black tanks or tubes in a collector box with a glazing, which preheats the water in conjunction with an ordinary hot water system; in some cases, it is the only source of

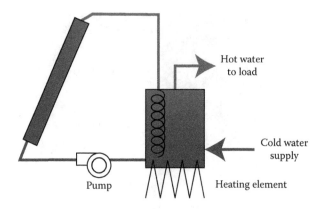

FIGURE 5.18   Solar hot water system. A differential temperature controller turns the pump on and off.

FIGURE 5.19   Copper absorber fin with copper tube.

hot water. The bread box is a simple system to build; however, this system loses too much heat in cold climates.

   There are commercial solar hot water systems available, and with different incentive programs, there is renewed emphasis on installing these systems. In Israel and Japan, solar hot water heaters are common. Components (Figure 5.19) can be purchased for the handy person to fabricate a personal system [12], and there are numerous sites with design information.

FIGURE 5.20 Solar hot water system for heating a swimming pool. Fixed tilt angle of array is 27° for summer sun at 35° N. Eight panels, thermoplastic, each 1.2 by 3 m (4 by 10 ft).

5.5.3 SWIMMING POOLS

Solar hot water systems can also be used for swimming pools, and in many climates, they do not require any glazing (Figure 5.20) because they are used during the summer and used to extend the swimming season; therefore, they are less expensive. Unglazed solar collectors are generally made of thermoplastic rubber (flexible rubber mat) or polypropylene plastic treated with an UV light inhibitor to extend the life of the panels. These solar panels may be either semirigid or have individual pipes. In areas with high evaporation rates, swimming pools have to be heated. For example, in the High Plains of Texas, the evaporation rate is over 2 m/year due to wind and low humidity, so city swimming pools are heated with natural gas during the summer.

A free software program, Energy Smart Pools, is available from Reduce Swimming Pool Energy Costs [10]. The program lets owners analyze the current energy consumption of their pool and to project potential savings by energy management, pool cover, or solar heating.

5.6 ACTIVE COOLING

Active solar systems can cool a house during the summer; however, higher temperatures are needed to drive the absorption cooling cycle, so in general evacuated tubes or concentrating collectors are used but are more expensive. An absorption cooling unit uses two working fluids: an absorbent such as water and a refrigerant such as ammonia. The principle is the same as for refrigerators powered by burning natural gas. However, a flat-plate collector, with two or even three glazings, has to be very efficient. Approximately 50% of installed commercial air conditioning in Japan uses solar collectors.

5.7 DAYLIGHTING

Daylighting is the use of natural light for building spaces and is now recognized as an important part of residential, commercial, and industrial building design. Daylighting reduces the need for electric lights, and during the summer, it also saves on the heat load, thereby reducing the need for air conditioning. Indirect lighting can be spread throughout the structure using skylights, shaded windows, atriums, light pipes, reflecting shelves, fiber optics, and so on. Effective design strategies assist in optimizing the use of natural light. The strategies include the following: optimize the amount of daylight, preserve visual comfort, avoid direct sunlight, and preserve thermal comfort. A disadvantage of daylighting is the variability due to clouds and reflection from terrain and other buildings. Proper design of glazed areas has to take into account heat loss in the winter and heat gain in the summer. There are many images of daylighting available on Google Images. A number of websites explain, discuss, and give examples. For commercial and industrial buildings, consult http:// www.wbdg.org/resources/daylighting.php.

5.8 HYBRID AND OTHER

The Thompson system for a home consists of an open water system piped to the top of a corrugated metal roof. During sunny days in the winter, the water trickled down the south side of the roof, and then, the heated water was used to heat the space at night. During summer, the water trickled down the north side of the roof at night, and then, the cooler water was used to cool the space during the day. Of course, there are some problems that have to be eliminated, dirt and dust, too much evaporation in dry climates, and efficiency would be low in cold climates.

Roof spray for low, flat roofs of commercial buildings is effective for cooling in many areas. One problem is hard water clogging up the nozzles. Some people have advocated that 5 cm of water on flat roofs would greatly reduce cooling costs.

5.9 DRYING AGRICULTURAL PRODUCTS, LUMBER

In the developing world, agriculture products were and are dried by spreading them in the Sun. Before cars were common in China, during harvest the asphalt highways were reduced to one lane due to grain being spread on the roads. In isolated areas, the Sun is the only source for drying low-cost agriculture products such as fruit and vegetables for cash crops and to feed the village during the year.

The benefits of dry food are preservation to eliminate or reduce spoilage from bacteria, yeast, and mold; dry food takes up less room; and it is lightweight and easier to transport. The disadvantages of traditional outdoor drying are relatively low drying temperatures and the high dependence on ambient temperature. Slower drying time causes greater spoilage, and health issues arise from dust and insects (ants and flies). Finally, birds, squirrels, rodents, and lizards eat the food.

Drying of agriculture products by different methods (Table 5.5) indicated that solar drying is the best option for small-scale farmers. The solar thermal collector is around 60% efficient. The benefits of the solar dryers are as follows:

TABLE 5.5
Advantages and Disadvantages of Agriculture Product Drying Methods

Type	Advantage	Disadvantage
Open air	No cost, no fuel	Dust, animals, ants, rain
		Labor to turn product over
		Collect at night?
		Longer time to dry
		Decreased effectiveness with high humidity
Firewood	Faster than open air	Cost of dryer box, fuel
		Forest degradation
		Smoke in product
		Labor for operation and maintenance
Electric	Drying well controlled	High cost, high operation cost
	Excellent quality	Electricity not available
Solar	Fairly low cost	New technology
	No fuel cost	Requires proper operation
	Faster (three to four times) than open air	
	Protected drying environment	
	Built locally	

Dries food between 50°C and 80°C, which is a good range for drying and pasteurization
Reduces moisture content to 10%–20%
Faster drying time

The two methods of solar drying are direct (Figure 5.21) and indirect (Figures 5.22 and 5.23). The benefits of the indirect solar dryer are as follows:

No direct UV rays (UV light degrades vitamins and color)
No need for sulfur, blanching
Better control of drying process; regulation of temperature by passive or active venting

In either method, almost all the components can be obtained locally and built locally (Figure 5.23).

Drying lumber can be a complex operation to reach the desired quality: wood without cracks. Lumber can be air dried, but in many locations the humidity is too high, so a kiln is used. There are a number of designs for solar kilns [12,13], which in general have a sloping roof with a glazing and a big door for access. Slope of the roof depends on latitude and ease of construction. A solar kiln should be relatively inexpensive and have simple operation.

Solar kilns designed, constructed, and tested at Virginia Tech [14] will dry a load of lumber in around 1 month in the medium sunny climate around Blacksburg,

FIGURE 5.21   Direct solar dryer for fruit and vegetables, Kabul, Afghanistan. Solar dryer designed by Robert Foster. (Courtesy of Robert Foster.)

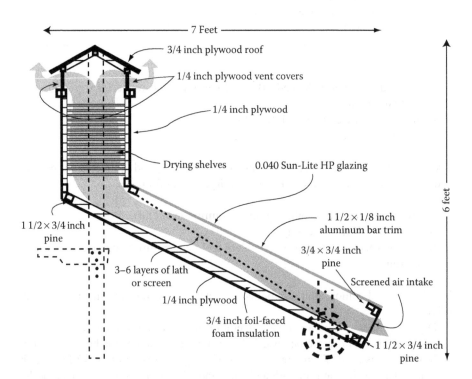

FIGURE 5.22   Diagram of indirect solar food dryer, based on design, Appalachian State University, 1997.

FIGURE 5.23 Indirect dryers built at local shop, Kabul, Afghanistan. In 2009, there were 130 built. Left to right: Khatera Obaid, Robert Foster, and carpenter Mohammad Hakeen Popa. (Courtesy of Mohammad Salim.)

Virginia. The kilns have a sloping roof of clear, greenhouse-rated, corrugated polyethylene, four insulated walls, and an insulated floor (Figure 5.24).

5.10 SOLAR COOKERS

Many people in underdeveloped countries use biomass (wood, charcoal, and dung) for cooking; with the resulting problems of deforestation and desertification, children and women spend an inordinate amount of time collecting wood, children have less time for school, and health problems from smoke and fumes in confined spaces occur. Solar cookers have now become part of the solution with over 2 million in use worldwide.

Of course, solar cookers depend on a climate with enough sunshine, that is, climates that are dry and sunny for at least 6 months of the year. These regions are generally within 40° of the equator. Solar cookers will not work during cloudy and inclement weather, and because most solar energy occurs between 1000 and 1400, the solar cooker can be used for two meals per day, noon and evening. The principles are simple: collection of radiant energy to heat a cooking vessel and retention of heat. Solar cookers can be constructed primarily from locally available materials, and many are portable.

FIGURE 5.24 Solar kiln for drying wood—(a) front view and (b) back view shows load of lumber with the baffle up. (Courtesy of Wood Science and Forest Products, Virginia Tech University.)

Solar cookers require a change in method of cooking due to moderate temperatures and time for cooking. Temperatures are within the range for cooking food, 82°C to 91°C, which help preserve nutrients and kill bacteria and viruses (when heated to 71°C). However, as noted this may take a period of time, up to 1 or more hours, to fully cook the food [15]. Because of the moderate temperatures, the food can be placed in the cooker and left unattended. Dark, shallow, thin, metal pots with tight-fitting lids to hold in the heat and moisture work the best.

The three main types of cookers are box, panel, and curved concentrator, although there are lots of variations on these, and even large-scale solar cookers have been developed for institutions [16]. For example, a dark pot inside a transparent plastic bag or a large inverted glass bowl can serve as a solar cooker. Box cookers (Figure 5.25), also referred to as *bread box cookers* in the past, reach moderate to high temperatures and can accommodate multiple pots. One or more surfaces can reflect additional sunlight into the box cooker to increase the temperature. Panel cookers (Figure 5.26) have panels that reflect radiation onto the cooking pot, which can be inside a plastic bag or under a glass bowl. The advantage of the panel cooker is that it can be built in an hour, and the cost is low. Curved concentrating or parabolic cookers (Figure 5.27) reach higher temperatures, which means they cook faster; however, they require frequent adjustment and supervision. Numerous plans and kits are available for solar cookers [17].

FIGURE 5.25 Box solar cooker. (Courtesy of Solar Cookers International, http://www. solarcookers.org.)

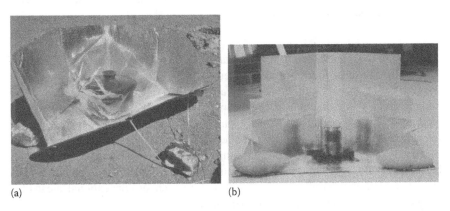

(a)   (b)

FIGURE 5.26 Panel solar cookers: (a) pot in plastic bag and rope tied to rock is to keep cooker in place in wind (Courtesy of Solar Cookers International, http://www.solarcookers. org.) and (b) student project for solar class at West Texas A&M University.

5.11 WATER PURIFICATION

Unsafe water is a major health problem as over $1 * 10^9$ people do not have potable water. Preventable waterborne diseases account for approximately 80% of illnesses and deaths in the developing world. Microbes in water that cause diseases are killed

FIGURE 5.27 Each family at the Solar Energy Research Institute, Lhasa, Tibet, has a concentrating solar cooker to heat its teakettle.

by exposure to water heated to 65°C for around 20 min (pasteurization). At around 70°C, milk and other foods are pasteurized.

Solar cookers can pasteurize water for a family at a rate of about 1 L/h. Water can be brought to a boil. One reason that people are told to boil water is that thermometers are not readily available in many places, and the boiling action serves as the temperature indicator. Water can be boiled in all three types of cookers; however, the time is much shorter for a concentrating collector. To make potable water, it only has to be pasteurized, not sterilized. A reusable water pasteurization indication [18] can be used to determine when water heated by solar or conventional means has reached appropriate temperatures to make it safe.

Solar stills (Figure 5.28) are another way of producing potable water, this time by distillation [19]; this is used for undrinkable (brackish, contaminated) water. This method also eliminates microbes. As the water evaporates (Figure 5.28), it condenses on the glass surface of the collector and drains down to a collecting reservoir. Water is added once per day, generally manually, and excess water drains out the overflow port. This also reduces salt buildup, a ratio of around 3/1 for input versus collected potable water. Of course, production depends on solar input, and with a good resource, production rates will vary from 2 L/m² per day in the winter to 6 L/m² per day in the summer. A solar still (Figure 5.29) (standard glass patio door,

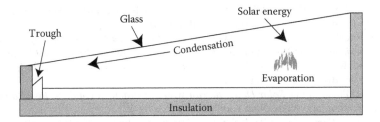

FIGURE 5.28   Diagram of solar still.

FIGURE 5.29   Solar still. (Courtesy of El Paso Solar Energy Association.)

1.1 by 2.5 m) produced around 10 L per day in the El Paso, Texas, region [20]. Cost of materials for this size solar still is around $300. Again, there are designs on the Internet for building solar stills.

REFERENCES

1. W. A. Shurcliff. 1988. *Superinsulated Houses and Air-to-Air Heat Exchangers.* Brick House, Amherst, NH.
2. Efficient Windows Collaborative. http://www.efficientwindows.org/index.php.
3. J. Carmody and K. Haglund. 2012. *Measure Guideline: Energy-Efficient Window Performance and Selection.* http://apps1.eere.energy.gov/buildings/publications/pdfs/building_america/measure_guide_windows.pdf.
4. Sustainable by Design. http://www.susdesign.com/tools.php.
5. D. Erley and M. Jaffee. 1997. *Site Planning for Solar Access: A Guidebook for Residential Developers and Site Planners*, Google eBook. Education Resources Information Center. http://eric.ed.gov/?id=ED193070.
6. D. White. 2008. *Site Design Strategies for Solar Access*, The Rocky Mountain Land Use Institute, http://www.law.du.edu/images/uploads/rmlui/rmlui-sustainable-siteDesign-StrategiesSolarAccess.pdf.

7. U.S. DOE, EERE, Building energy software tools (ShadowFX, Shading II). http://apps1.eere.energy.gov/buildings/tools_directory/.

8. P. Torcellini and S. Pless. 2004. *Trombe Walls in Low-Energy Buildings: Practical Experiences*, NREL/CP-550-36277. http://www.nrel.gov/docs/fy04osti/36277.pdf.

9. D. K. Reif. 1983. *Passive Solar Water Heaters: How to Design and Build a Batch System*. Brick House, Amherst, NH.

10. R. H. Plante. 1983. *Solar Domestic Hot Water: a Practical Guide to Installation and Understanding*. Wiley, New York.

11. RSPEC. *Energy Smart Pools Software*. http://www.rlmartin.com/rspec/software.htm.

12. Build It Solar. Site provides various links. http://www.builditsolar.com/Projects/WoodDrying/wood_kiln.htm.

13. Solar Kiln Info. http://www.solarkilninfo.com.

14. Virginia Tech, Department of Sustainable Biomaterials. Wood Drying. Virginia Tech Solar Kiln. http://www.woodscience.vt.edu/for-the-community/extension/solar-kiln/index.html.

15. Solar cooking: Frequently-asked questions. http://solarcooking.wikia.com/wiki/Solar_Cooking: Frequently-asked_questions.

16. Solar Cookers International. http://www.solarcookers.org.

17. Solar cooker plans. http://solarcooking.wikia.com/wiki/Category:Solar_cooker_plans.

18. Water pasteurization indication. http://solarcooking.wikia.com/wiki/Water_Pasteurization_Indicator.

19. Solar still basics. http://www.solaqua.com/solstilbas.html.

20. Solar water purification project. http://www.epsea.org/stills.html.

RECOMMENDED RESOURCES

LINKS

Solar radiation data manual for flat-plate and concentrating collectors. http://rredc.nrel.gov/solar/pubs/redbook.

U.S. Solar Radiation Resource Maps. http://rredc.nrel.gov/solar/old_data/nsrdb/1961-1990/redbook/atlas/.

PROJECTS, TECHNICAL INFORMATION, AND PHOTOS

Buena Vista. http://www.sunroom.com.

Kalwall. http://www.kalwall.com/main.htm.

National Renewable Energy Laboratory. 2004. *Trombe Walls in Low-Energy Buildings: Practical Experiences*. http://www.nrel.gov/docs/fy04osti/36277.pdf.

Sun-Lite HP. http://www.solar-components.com/sun.htm#sgmat.

Sunroom-sunroom.com home page. http://www.sunroom-sunroom.com.

SOLAR HOT WATER

Energy Efficiency and Renewable Energy, DOE. http://energy.gov/energysaver/energy-saver.

Florida Solar Energy Center. http://www.fsec.ucf.edu/en/consumer/solar_hot_water.

Integration of Renewable Energy on Farms. Solar thermal-water heating. http://www.farm-energy.ca/IReF/index.php?page=water-heating.

Reduce swimming pool energy costs. http://www.rlmartin.com/rspec/index.html.

U.S. DOE, EERE. http://energy.gov/eere/energybasics/articles/solar-water-heater-basics.

SOLAR AIR CONDITIONING

Austin solar uses parabolic troughs. http://www.austinsolarac.com/product.htm.

DAYLIGHTING

Daylighting guidance. http://www.ecw.org/daylighting.
National Institute of Building Sciences. http://www.wbdg.org/resources/daylighting.php.

SOLAR DRYING

Solar food drying. http://solarcooking.wikia.com/wiki/Solar_food_drying.
W. Weiss, J. Buchinger. Solar drying. http://www.aee-intec.at/0uploads/dateien553.pdf.

PROBLEMS

For Amarillo, Texas, heating is needed from mid-October through mid-May. There are 887 heating degree-days for January.

5.1. Calculate for January the season heating gain for a single-pane window (south facing), 1.2 m by 2.5 m, for Amarillo, Texas. Note, you can use wind heat gain tool, http://www.susdesign.com/windowheatgain/index.php.

5.2. Calculate, for January the season heating gain for a double-pane, low-e window (south facing), 1.2 m by 2.5 m, for Amarillo, Texas.

5.3. Calculate the season heating gain for the south-facing windows at your house.

5.4. Is the single-pane window in problem 1 a net heat gainer for January? You will have to do the heat loss calculation for January.

5.5. Calculate for January the solar heat gain for the window in problem 5.2. Calculate the heat lost. What is the difference between heat gain and heat loss?

5.6. Would you buy triple-pane, low-e windows for a home in the area of Amarillo, Texas. Justify your answer (yes or no will not suffice).

5.7. List two advantages and two disadvantages for a passive system.

5.8. What are the advantages of Sun-Lite as a glazing material? Go to their website for information (http://www.solar-components.com/sun.htm).

5.9. What are the R values for a wall composed of Kalwall. Go to their website for information (http://www.solar-components.com/panels.htm).

5.10. What are the main components of an active solar heating system?

5.11. What are the main components of a flat-plate collector?

5.12. Does anyone in your neighborhood in your hometown have a domestic solar hot water system? Briefly estimate size and performance.

5.13. List two benefits of daylighting? Do you want direct sunlight in the area you are daylighting? Give reason for your answer.

5.14. List two advantages and two disadvantages for solar cookers.

5.15. List two advantages and two disadvantages for solar stills for water purification.

6 Photovoltaics

6.1 INTRODUCTION

A photovoltaic (PV) cell converts sunlight directly into electricity. A number of materials are photoelectric: Light is absorbed, and an electron acquires kinetic energy to move it to another energy level within the material. Today, the primary materials for PV cells are semiconductors, although researchers are trying other materials, even organic polymers. A PV cell is used in things from small items such as calculators and watches to large installations for electric utilities. The cost of PV systems has decreased dramatically so grid-connected PV has increased, and there are a number of applications in which they are the least cost, especially for stand-alone systems some distance from the utility grid, mobile jobs such as construction signs, and even for low power (50–200 W) close to the grid. A PV system has the following advantages and disadvantages:

Advantages	Disadvantages
High reliability (30 years)	Higher initial costs
Low operating costs	Variability
Modularity	Storage increases costs
Low construction costs	Lack of infrastructure in remote areas
Silent operation	

PV has two other major advantages: PV and wind do not require water for the production of electricity, in contrast to conventional thermal steam plants, even those powered by nuclear reactors, and there is some load matching, for areas that need air conditioning and good solar resource, so peak loads for utilities can be reduced. There is still the problem of load for late afternoon and early evening.

6.2 PHYSICS BASICS

Charge is an inherent property of particles, and charged particles interact by electromagnetic (EM) interaction, which is strong in terms of force. Charge comes in units, positive or negative, equal to the value of the charge on an electron. Atoms are neutral due to an equal number of positive charges from protons in the nucleus and negative charges, electrons, that surround the nucleus. For our purposes in describing the electrical properties of materials, the nucleus stays in the same place because it has almost all the mass, and only the electrons move.

Most properties (mechanical, electrical, thermal, chemical, biological) of materials can be explained by their electron structure and, in general, by the outer electron structure of the atoms or molecules. Only the basic information on electricity is covered as beginning physics textbooks cover electricity and magnetism in more detail.

Charge Q or q is measured in coulombs (C).

One electronic charge $= 1.6 * 10^{-19}$ C, positive or negative. Notice that 1 C is a large number of unit charges.

Electric fields **E** are created by charged particles, and if a charged particle is placed in an external electric field (due to other charges), there is a force on it, which will make it move. Notice that *bold* indicates a vector; it has both magnitude and direction. A scalar only has magnitude (e.g., temperature).

$$\mathbf{F} = q * \mathbf{E} \tag{6.1}$$

Then, energy or work (see Equation 2.1) is available as the charged particle is moved through some distance. The electric potential V is the energy/charge.

$$V = W/q, \text{volt } (V) \tag{6.2}$$

Then, the equation can be rearranged, Energy $= V * q$. An electron volt (eV) is the energy one electron would acquire from moving through a potential of 1 V, so 1 eV $= 1.6 * 10^{-19}$ joules.

The electrical properties of materials are divided into three general classes: conductors, semiconductors, and insulators. Metals (e.g., copper and gold) are good conductors, and wood and glass are good insulators. However, if there is enough excess charge of one type, then all materials will break down and conduct electricity. You do not want to be the lightning rod in a thunderstorm.

Current is the flow of charge moving past a point in a given amount of time:

$$\mathbf{I} = \frac{dq}{dt}, \text{ampere } (A) \tag{6.3}$$

So a flow of one Coulomb per second is one ampere. Direct current (DC) is the flow of charge in only one direction (convention is from positive to negative), and alternating current (AC) is the flow of charge in both directions, variable in terms of cycles or frequency (number/time). In the United States, the frequency of the electric grid is 60 Hz, and in Europe, it is 50 Hz (hertz = number of cycles/s). For conductors and semiconductors, if a voltage is applied across the material, there will be a current. However, there will be a resistance to the flow of that current. Resistance is measured experimentally, and it will have a low value for metals and a large value for insulators.

$$R = \frac{V}{I}, \text{ohms } (\Omega) \tag{6.4}$$

Ohm's law is the linear relationship of voltage and current for metals, $V = I * R$. Then, the power, which is energy/time, can be obtained from the voltage and current.

$$\text{Power} = V * I, \text{ watt (W)} \tag{6.5}$$

Example 6.1

An element has a resistance, $R = 2\ \Omega$, and a voltage, $V = 12$ V, applied across it. What is the current and power?

$$I = \frac{V}{R} = \frac{12\ V}{2\ \Omega} = 6\ A$$

$$P = V * I = 12\ V * 6\ A = 72\ W$$

Notice: For problems that have units, answers have to have units.

6.3 ENERGY BANDS

In materials, the energy states for electrons are close together, so they form bands that explain the three main types of material: Conductors have free electrons that can move easily under an applied electric field, semiconductors have a few electrons that can move, and insulators have no free electrons (Figure 6.1). In a conductor, the conduction band is partially filled with electrons, so there are energy states available for free electrons; therefore, metals are good conductors.

In a semiconductor, at room temperatures some electrons have enough energy to get into the valence band, leaving a hole in the conduction band. The band gap is small, on the order of an electron volt. Another way the electron can obtain enough energy to move from the conduction band to the valence band is by the absorption of a photon (EM radiation with enough energy).

In an insulator, such as glass, there are not any free electrons as all electron states are filled in the conduction band, and the band gap energy to the valence band is large, so there are no electrons in that band.

In a semiconductor, the current is explained by the movement of electrons (negative) and holes (positive). The movement of electrons in the conduction band fills an

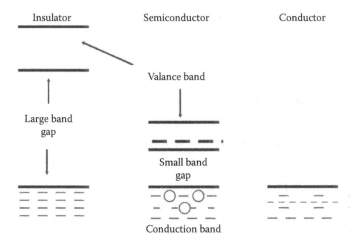

FIGURE 6.1 Electron energy states form bands in materials.

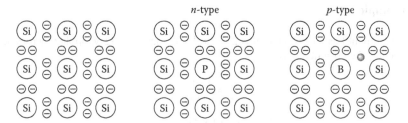

FIGURE 6.2 Semiconductors doped with phosphorus and boron. Electrons are –, and holes are ○, which are treated as a positive charge with a mass essentially the same as the electron mass.

energy state, which then leaves an empty state, so it is easier to describe that current by the movement of holes, which have an equivalent positive charge. The number of electrons in the valence band can be increased by doping the silicon with phosphorus, which has an extra electron for bonding (*n*, negative-type semiconductor), or the number of holes in the conduction band can be increased by doping the silicon with boron, which is deficient by one electron for bonding (*p*, positive-type semiconductor) (Figure 6.2). Either one increases the electrical conduction. The *n* and *p* materials are fabricated in layers and islands to create diodes, transistors, and integrated circuits (i.e., solid-state electronics).

EM radiation is described as waves with the electric and magnetic fields varying in time and wave travels at the speed of light. The other way to describe phenomena is by particles: quantization. So, EM radiation occurs in units, called *photons*, which have no mass and travel at the speed of light. The energy of a photon is proportional to the frequency (Equation 3.2), so X-ray photons have a lot more energy than visible photons, and blue photons have more energy than red photons.

Photons are produced and absorbed in materials but always in those units of energy. So, when a semiconductor has a certain band gap, photons with less energy than the band-gap energy cannot be absorbed. When photons with energy equal to or greater than the band gap are absorbed, an electron is shifted to the valence band, which leaves a hole in the conduction band. Excess energy becomes heat. Many times, physicists like to have numbers without powers of ten, so electron volts (eV) are used in describing semiconductors.

6.4 PHOTOVOLTAIC BASICS

The solar cell (Figure 6.3) is fabricated by having an *n* and *p* layer, which is a junction. At the junction, the excess electrons in the *n*-type material flow to the *p* type, and the holes thereby vacated during this process flow to the *n* type. The junction acts as a battery, creating an electric field at the surface where they meet. This field causes the electrons to move from the semiconductor out toward the surface and make them available for the electrical circuit. At the same time, the holes move in the opposite direction, toward the positive surface, where they await incoming electrons. So, when photons are absorbed by the PV cell, they create more electrons and holes, which become available for a current for an external load [1].

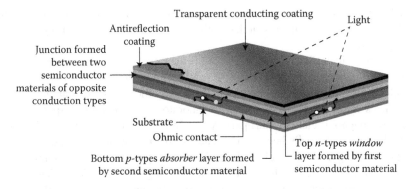

FIGURE 6.3 PV cell formed by *n* and *p* layers. (Data from U.S. DOE, EERE.)

To improve efficiency, the materials should be modified to have band-gap energies for photons in the visible range, where the most energy is available from the sunlight. The spectrum from infrared to ultraviolet (UV) covers a range of about 0.5 to about 2.9 eV. For example, red light has energy of about 1.7 eV, and blue light has energy of about 2.7 eV. Effective PV semiconductors have band-gap energies ranging from 1.0 to 1.7 eV. Band gaps in semiconductors are in the 1–3 eV range. For example, the band-gap energy of silicon is 1.1 eV.

The solar cell (Figure 6.4) is the basic building block of a PV system. Individual cells can vary in size from about 1 to 10 cm. However, one cell only produces 1–2 W, which is not enough power for most applications. So, cells are combined into a module, modules into a panel, and panels into arrays for large power applications. Thin-film materials like amorphous silicon and cadmium telluride can be made

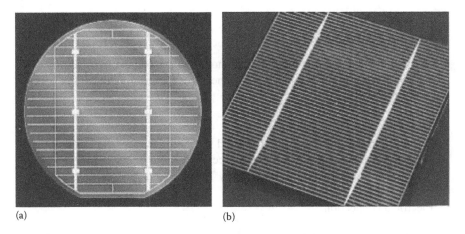

(a)   (b)

FIGURE 6.4 Examples of PV cells: (a) older cell type, wafer cut from ingot of silicon; (b) polycrystalline cell, wafer cut from cast square ingot of silicon. (From National Center for Photovoltaics, National Renewable Energy Laboratory [NREL], http://www.nrel.gov/ncpv.)

directly into modules. Another manufacturing process is drawing of ribbons of semi-conductor material. Finally, research on other materials and manufacturing processes is progressing.

The conversion efficiency of the PV cell is the ratio of electrical energy produced over the insolation on the cell. Today, PV devices convert 7%–20% of light energy into electric energy; however, multiple layers and other improvements have raised efficiencies into the 35%–45% range for laboratory experiments (http://www.nrel.gov/ncpv/images/efficiency_chart.jpg). About 55% of the energy of sunlight cannot be used by most PV cells because this energy either is below the band gap or carries excess energy, which results in heat. Another way to improve efficiency is using multiple *p/n* junctions, which have efficiencies as high as 45%. Cell efficiencies for silicon decrease as temperatures increase, and higher temperatures also threaten the long-term stability of the material. Therefore, PV cells must be kept cool, especially for concentrating collectors.

Cell types are as follows:

Single crystal: It is the most expensive production method; it is reliable; and its module efficiencies average 10%–12%.
Semicrystalline: Production costs are lower; efficiencies are lower; and module efficiencies are 10%–11%. However, cell performance may degrade over time.
Polycrystalline thin films: These are less efficient due to boundaries between crystals.
Amorphous: Material is vaporized and deposited on glass or stainless steel; production costs are lower; and module efficiencies are 7%–8%. It degrades over time.
Thin film: It is created by deposition or thin ribbon (Evergeen Solar, 10 cm wide, which are then cut into wafers).

6.5 PERFORMANCE

Cells are combined into modules, and the modules have outputs from 10 to 300 W. The amorphous and thin films are cut to panel size. A typical module or panel consists of the following:

Transparent top surface: Usually glass
Encapsulant: Usually thin sheets of ethyl vinyl acetate that hold together the top surface, solar cells, and rear surface
Rear layer: Thin polymer sheet, typically Tedlar, to seal module
Frame: Typically aluminum
Electrical connection

The performance of a module is given by the current-volt characteristics (Table 6.1). The open circuit, no current, gives the maximum volts, and a short circuit, no voltage, gives the maximum current. Remember power is volt * current, and the maximum

---

TABLE 6.1

Example Specifications for a PV Module, 53 W_P

Operating Point

P	53 W_P
VMP (at peak power)	17.2 V
IMP (at peak power)	3.08 A
V_{OC} (open circuit)	21.5 V
I_{SC} (short circuit)	3.5 A
Standard test conditions (STC)	1,000 W/m², 25°C

Note: NOCT would be around 50°C with ambient temperature at 25°C.

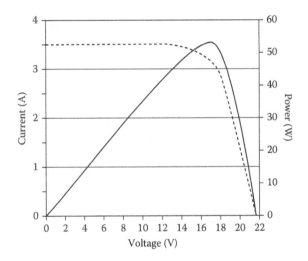

FIGURE 6.5 Typical PV cell I-V curve (power indicated by solid line). Notice that maximum power occurs at the knee in the I-V curve.

efficiency is at the knee of the curve (Figure 6.5). The power curves will change with temperature, and maximum power output will decrease.

The performance ratings of modules include the following:

Peak watt (W_P): Maximum power under laboratory conditions of high light level and low temperature.

Normal operating cell temperature (NOCT): Measures nominal operating cell temperature after module comes into equilibrium with a specified ambient temperature. Results in a power output lower than W_P.

AMPM standard: Measures performance for a day, not peak value, based on standard solar global-average day in terms of light, ambient temperature, and mass of air (primarily elevation). Solar insolation on a sunny day is greater at higher elevations.

Some PV cell specifications provide power curves at 1 and 0.5 sun (incident energy of 1,000 and 500 W/m²), which represents full-sun and cloudy day performance.
   Four factors affect array performance:

I-V operating point (load matching for maximum power)
Solar intensity (Figure 6.6)
Operating temperature (Figure 6.7)
Sun angle

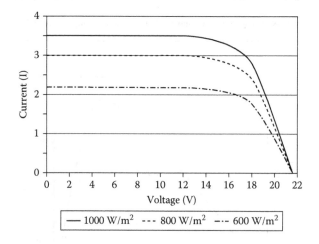

FIGURE 6.6   PV cell performance as a function of incident radiation.

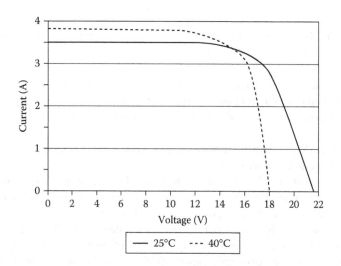

FIGURE 6.7   Performance of a PV cell due to temperature.

TABLE 6.2
Average PV System Component Efficiencies

PV array	80%–85%
Inverter	80%–90%
Wire	98%–99%
Disconnects, fuses	98%–99%
Total grid tied (AC)	60%–75%
Batteries, round trip	65%–75%
Total off grid (AC)	40%–56%
Total off grid (DC)	49%–62%

A PV system provides all the power requirements of an application. The system includes one or more PV modules, power conditioning or controller, wires and other electrical connectors, and the load. Batteries for backup power or float are an option. The simplest PV system consists of flat-plate modules, panels, or arrays in a fixed position. The advantages are no moving parts and lightweight structure. Of course, there is less conversion of light to electricity as the angle of the Sun to the fixed modules changes during the day and by day of the year (season).

The performance of a PV cell/module can be described in terms of its energy conversion efficiency, the percentage of incident solar energy (input) that the cell converts to electricity (output) under standard conditions of 1,000 W/m^2, 25°C, one air mass (sea level). The average energy conversion efficiencies for cells/modules [2] were crystalline silicon (single crystal), 15%–20%; crystalline silicon (cast), 13%–16%; crystalline silicon (ribbon), 13%–14%; thin film (amorphous silicon, CdTe, CuInGaSe, organic), 7%–9%; and concentrator PV systems, 30%–45%. Research and development are focused on improving cell/module efficiencies and reduction in manufacturing costs. Cell efficiencies of 46% have been reported for multijunction semiconductors in the laboratory.

The overall performance and efficiency is the conversion of sunlight to the end product: electricity, water pumped, potable water produced, and so on. Generally, the end product is electricity, DC (direct current) for stand-alone systems or AC (alternating current) for village power or grid connected. In any case, there are efficiencies (Table 6.2) for the balance of the system (BOS): charge controllers, storage (generally batteries), inverters and converters, pumps, and so on. In the final analysis, the important factors are energy produced and the cost of that energy for the PV system and the competitive cost of energy from other sources, fossil fuels, and even other renewable energy sources.

6.6 DESIGN CONSIDERATIONS

Primary considerations for design of a PV system are the load, solar resource (percentage of hours of sunshine) by month or by season of load, and storage (how much), and for stand-alone or village power, considerations are also for load growth over the next 10 years [3,4].

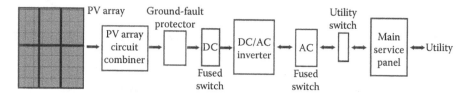

FIGURE 6.8   Typical PV system connected to the utility grid.

Stand-alone or isolated systems generally have a charge controller and batteries; however, some systems, such as water pumping, may only have a controller. The size of PV cell and batteries depends on load and how much storage is needed.

For grid connection [5], the PV system is a parallel source to the grid, and an inverter changes the DC to AC at the proper frequency (Figure 6.8). All the energy produced by the PV system may be used on site, and depending on the size of the PV system, energy may be fed back to the grid. If there is a fault in the utility grid, the inverter keeps the PV system disconnected from the grid for safety reasons. If the end user wants power if the grid is down, then battery storage is added, and more stringent controls are needed for disconnecting and reconnecting to the utility grid and the possibility of only powering critical loads.

Then, there are PV power plants for which the PV system is another generator on the grid. These systems range from 300 kW to megawatts. PV panels are grouped into an array and a number of arrays form a power plant.

6.6.1   Sizing

The solar resource will vary widely depending on the location; however, most temperate and tropical locations have adequate resources. Locations where there is continuous cloud cover for weeks present challenges, so PV systems would have to be larger. Power on overcast days is around 10% of that on sunny days with clear sky. Concentrating solar energy systems need direct sunlight, so they are located in excellent solar resource areas.

Maps and tables are available (see Chapter 2) for average monthly solar insolation (kWh/m²) for many regions of the world. Insolation data that are closer to the project site should be used and should take the conservative estimate for the amount of solar resource. The PV system is designed for the seasonal solar resource, seasonal load (or month of highest demand), or yearly average.

The surface of the module or array will receive both direct and diffuse radiation; on clear days most energy is direct, while on cloudy days most energy is diffuse. The amount of radiation received (surface perpendicular to the Sun) on a clear day varies by time of day due to difference of path length through the Earth's atmosphere (Figure 6.9). For practical considerations, an average day is used (power/area or kWh/area). Since the standard power used for sunlight is 1,000 W/m², then over 1 h that would be 1 kWh/m², sometimes referred to as *peak sun hour* or *sun hour*. So,

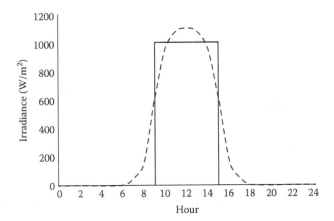

FIGURE 6.9 Irradiance (dashed line) and corresponding or equivalent sun hours (solid line) for average day.

on a clear day there would be 6 sun hours (dashed line in Figure 6.9). In temperature zones, the number of peak or sun hours will vary by season; of course, for fixed arrays, less energy is absorbed due to the angle factor (cosine factor) between array and sun position.

For a fixed-tilt array, placement of the array with the azimuth south (north in the Southern Hemisphere) and tilt angle at the latitude will give the best performance. However, the tilt angle is not too critical as tilt angle within 10° of the latitude will give about the same results—just the peak values occur at different times of the year. One way to increase energy output is by tracking or even by changing tilt angle for a fixed array by season.

6.6.2 Tracking

There are one- and two-axis tracking systems. Common PV systems have a flat plate and fixed angle (tilt), so one way to improve the performance of flat-plate collectors is by tracking the Sun (Figure 6.10), especially for PV panels since they are expensive. Methods for improving performance by tracking are as follows:

A manual change of panel tilt from summer to winter is used.

One-axis tracking: The axis of rotation can be either the north–south or east–west line.

Two-axis tracking: One possibility is to have passive east–west tracking and change the tilt angle by month manually. Otherwise, it requires an active tracking mechanism, with inherent problems of power and moving mechanical parts. One method of control is to track the Sun, and a newer method of control is to use a geographic position system.

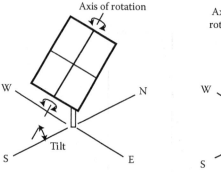
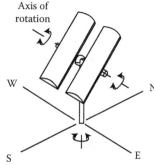

(a)   One-axis tracking flat-plate collector
      with axis oriented north-south

(b)   Two-axis tracking concentrator
      Axis of rotation

FIGURE 6.10 Flat-plate collectors with (a) one- and (b) two-axis tracking. (Data from National Center for Photovoltaics, National Renewable Energy Laboratory [NREL], http://www.nrel.gov/ncpv.)

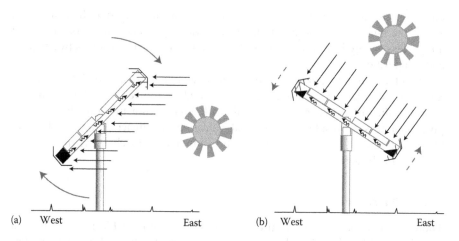

(a)   West   East

(b)   West   East

FIGURE 6.11 (a,b) Passive tracking for flat-plate collectors. (Data from Zomeworks, http://www.zomeworks.com.)

Passive trackers use two canisters where the solar direct radiation increases the vapor pressure, driving liquid from one side to another (Figure 6.11) to keep panels oriented toward the Sun. Passive tracking systems do not require extra energy or motors and gears (which require more maintenance). However, in windy areas, passive trackers may not work well as the wind force is often larger than the passive tracking force. Tracking collectors produce 25%–45% more power in the summer and pump up to 50% more water than a fixed flat-plate, PV water-pumping system.

6.6.3   ESTIMATION OF ENERGY PRODUCTION

Generally, the energy production for the PV system is estimated by month from average day insolation for a flat plate. Some databases [6] have values ($kWh/m^2/day$) for different tilt angles (0, latitude, latitude $\pm 15°$, and $90°$), while the National Renewable Energy Laboratory (NREL) has maps by month for a flat plate tilted at latitude angle (http://www.nrel.gov/gis/solar.htm or http://rredc.nrel.gov/solar/old_data/nsrdb/1961-1990/redbook/sum2/).

For a fixed array, the energy day can be estimated from average day and area.

$$E_D = E_S * E_C * IN_D * A \qquad (6.6)$$

where:
E_S is the system efficiency (includes derating for operating temperature)
E_C is the cell efficiency, which depends on cell type
IN_D is the average insolation at the tilt angle (month or annual average)
A is the array area

Another method is to use the average-day insolation (map value, tilt at latitude) and the power rating of the module.

$$E_D = E_S * RP * HS \qquad (6.7)$$

where:
RP is the rated power
HS is the average number of sun hours

Example 6.2

Estimate the energy output for a 0.5-kW PV system for Amarillo, Texas, for the month of January. The system is as follows: BP solar, crystalline silicon, 225-W module, 1.65 by 1 m, area = 1.65 m², array of two modules tilted at latitude. See California Energy Commission [5] for Amarillo data: January average day = 4.9 kWh/m²/day, $E_S = 70\%$, $E_C = 17\%$. Use Equation 6.6.

$$\text{Energy} = 0.7 * 0.17 * 4.9 \text{ kWh} / (m^2 day) * 3.3m^2 * 31 \text{ days} = 60 \text{ kWh}$$

Cell efficiency = 17%, and system efficiency is 70% (better output because of colder weather), so estimated energy is 60 kWh.

If you need more energy, use more modules, modules with higher rating, or a tracking system.

Example 6.3

As a rough estimate, two modules = 450 W, average of 5 peak hours/day, $E_S = 70\%$. Use Equation 6.7.

$$E_D = 0.7 * 450W * 5h = 1.5kWh/day$$

$$\text{Energy/month } E = 1.5 * 30 = 45kWh/month$$

$$\text{Energy/year } E = 1.5 * 365 = 550kWh/year$$

The PVWatts calculator [7] creates hourly performance simulations that provide estimated monthly and annual energy production and power (kilowatts) for grid-connected systems for locations throughout the world. Users can select a location and choose to use default values or their own system parameters for size, electric cost, array type, tilt angle, and azimuth angle.

The PVWatts calculator determines the solar radiation incident on the PV array and the PV cell temperature for each hour of the year using typical meteorological year weather data for the selected location. The DC energy for each hour is calculated from the PV system DC rating and the incident solar radiation and then corrected for the PV cell temperature. The AC energy for each hour is calculated by multiplying the DC energy by the overall DC-to-AC derate factor and adjusted for inverter efficiency as a function of load. Hourly values of AC energy are then summed to calculate monthly and annual AC energy production.

Example 6.4

For this example, use PVWatts and a PV system for Canyon, Texas: 2 kW, 30° fixed tilt, south facing. It displaces electricity at $0.84/kWh. The output for an average year, energy = 3,027 kWh, value of electricity = $256. *Caution*: Actual values may vary from the long-term average by ±30% for monthly values and ±10% for yearly values. The energy production values in the table are valid only for crystalline silicon PV systems.

Find the output from PVWatts program.

Station Identification	
Cell ID	0208377
State	Texas
Latitude	35.0° N
Longitude	101.9° W
PV System Specifications	
DC rating	2.00 kW
DC to AC derate factor	0.770
AC rating	1.54 kW
Array type	Fixed tilt
Array tilt	30.0°
Array azimuth	180.0°
Energy Specifications	
Cost of electricity	8.4 ¢/kWh

	Results		
Month	Solar Radiation (kWh/m²/day)	AC Energy (kWh)	Energy Value ($)
1	4.83	230	19.43
2	4.92	209	17.65
3	6.04	275	23.23
4	6.56	283	23.91
5	6.35	276	23.31
6	6.60	268	22.64
7	6.69	280	23.65
8	6.25	264	22.30
9	6.05	253	21.37
10	5.92	263	22.22
11	4.85	214	18.08
12	4.41	210	17.74
Year	5.79	3027	255.69

A geographical information system for estimating PV production (PVGIS) is available for Europe, Africa, and Southwest Asia [8]. It is an interactive-map program that calculates solar resource (monthly and average daily radiation) and estimation of PV performance (grid-connected, stand-alone, fixed tilt, and tracking options). Performance for crystalline silicon, CdInSe, CdTe, and others is available. The calculator can suggest the optimum inclination/orientation of the PV modules to obtain maximum yearly production.

6.7   INSTALLED CAPACITY AND PRODUCTION

The installation of PV systems has increased dramatically (Figure 6.12), and the world installed capacity was over 180 GW at the end of 2014 [9–11]. In 2013, the amount of PV installed (38.7 GW) surpassed the wind installed (35.5 GW) in that year, a major milestone. Increase was primarily driven by two factors: policy driven (feed in tariffs, renewable portfolio standards, and tax credits) and lower costs for modules due to increased production capacity.

Installation during the 1990s was primarily in the United States and Japan, then Germany surpassed Japan in 2005; Spain became number two in 2008; and then China started installing large numbers in 2010 to become number two by 2014 (Figure 6.13). However, Europe was still the leader in terms of cumulative installed capacity, with 89 GW as of 2014 [12].

The increased installations in the United States [13] are now primarily grid connected (Figure 6.14). In 2013, the amount installed was 0.9 GW for residential, 1.0 GW for non-residential (commercial), and 2.7 GW for utility installations. Note that some reports divide grid connected into the following sectors: residential, commercial, industrial, and ground mounted (utility). Due to the cost of PV, up to the year 2000 most installations were off grid for the world, even for developed countries as shown by

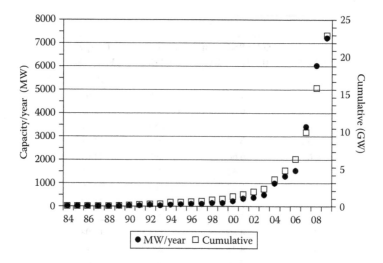

FIGURE 6.12   World PV production per year (MW) and cumulative installation (GW).

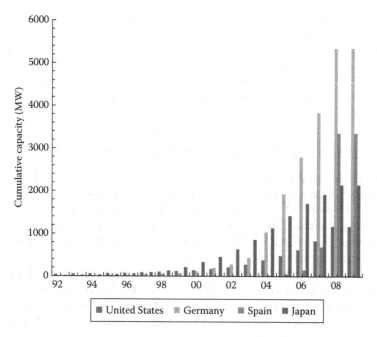

FIGURE 6.13   PV installed capacity (MW) by country.

installations per year in the United States (Figure 6.15). There is still a large market for off-grid PV in developing countries and another future market will be village power (mini grids). It is estimated that China installed 500 MW of off-grid PV in 2013.

Production of PV cells and modules was primarily in the United States, Japan, and Europe in the 1990s; however, the tigers of Asia have surged in production (Figure 6.16).

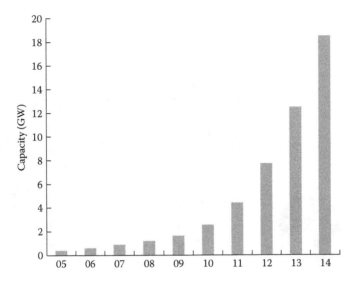

FIGURE 6.14   Yearly PV installed capacity in the United States.

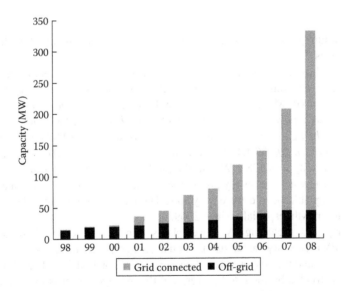

FIGURE 6.15   Yearly PV installed capacity from 2000–2008 in the United States.

In 2009, China led the world with 2.4 GW of production, and that increased to 25.6 GW in 2013, an astounding 64% of world production versus 1% in 2000.

Shipments of PV cells and modules are divided into three main categories: (1) crystalline silicon cells made from single-crystal or polycrystalline silicon, based on processes such as ingots, cast, and ribbon; (2) thin-film cells made from layers of semiconductor material, such as amorphous silicon (a-Si), cadmium telluride, or

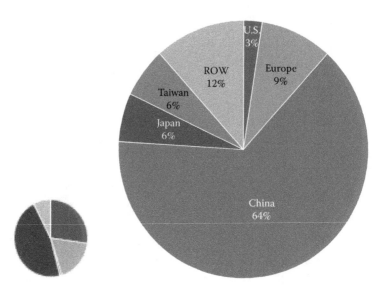

FIGURE 6.16   World PV production (MW) by country: left, in 2000 (world total = 277 MW); right, in 2013 (world total 39,800 MW). Ratio of the areas is over 1,000. ROW = rest of the world.

copper indium gallium selenide; and (3) concentrator systems, which include reflectors or lenses.

To show how changes in national and even state policies affect events, contrast my future predictions in 2003 for the online course *Solar Energy* with what has actually occurred. My comments were as follows: *Production of PV in the world has increased dramatically in the past few years. United States is over 80 MW with most of that being exported. Cumulative sales surpassed 400 MW and the forecast is expected to reach 10,000 MW by the year 2030.* Notice that in 2013 the actual world installed capacity was 138 GW, and the production of PV was around 38,000 MW (Figure 6.12), which means my predictions based on current information at that time by knowledgeable people were very low. The trend was up, but nobody expected such rapid growth, especially since 2009.

Power plants, including PV power plants, can be characterized by rated power (W_P for PV), capacity factor (CF), annual energy production, and, of course, the installed cost and the cost or value of energy ($/kWh or $/MWh). The CF is the average power divided by the rated power, and the average power is usually calculated from energy/time, which is usually for a year.

$$CF = P_{avg}/P_{rated} = \left(\text{Annual energy/8760} \right)/P_{rated}$$

6.8   DISTRIBUTED SYSTEMS

Distributed systems are installations on the retail side of the electric meter for residential and non-residential systems. In Japan, 85% of the installed capacity is residential systems. Residential installations accounted for 22% of the market in

Europe in 2013, and in the Czech Republic, the Netherlands, Belgium, and Denmark, residential installations account for over 50% of installed capacity. Residential and non-residential systems accounted for 41% of the market in the United States in 2013, of which 95% had net metering.

A major advantage is the world rooftops provide area for installation of PV, with capacity of 5 to 20 kW for residential buildings and 100 kW or more for non-residential buildings. Google rooftop PV and click on images for photos. Another prospect for buildings is to have PV as an integral part of the infrastructure [14], building-integrated photovoltaic (BIPV) systems (Figure 6.17). PV could be part of the wall or roof, which would reduce the cost of being draped over the facade or installing on top of the roof. Google building-integrated PV and click on images for photos. In 2014, the market share was around one percent, but the market should be larger in the future.

FIGURE 6.17 Solar panels on the facade of the Convention Center, Austin, Texas. Panels are angled around 30° to the south, with strips of translucent blue glass behind and between panels. Specifications: 4.8 kW, 188 panels, 60 W, amorphous silicon.

An important issue in the growth of distributed generation is fair compensation for utilities for backup power. Distributed generation with net metering or feed-in tariffs reduces the income for utilities; however, distributed generation can enhance or improve the distribution system. Demand side management and smart grids will alleviate part of the problem of distributed generation for utilities. Finally, cheap storage would drastically change the impact of renewables on electric generation and the utility industry.

6.9   COMMUNITY SOLAR

The definition of community solar varies and there will be overlap with distributed solar, residential, non-residential, and even utility installations. One definition would be solar projects that incorporate local financial participation and control. Another definition would be projects completely owned by villages, towns, communities, and cooperative groups. *Community solar* is a mechanism for renters, property owners with inadequate solar access, groups, and/or communities to participate in renewable energy. Community solar is a way to finance the system, and it provides local economic benefit. In a shared installation, the electricity generated is credited to subscribers of the installation. There are over 50 shared renewable projects in the United States.

A *Guide to Community Solar* provides definitions, project models (two examples for each model), and other information and resources for community solar. Project models are as follows: (1) *Utility sponsored (green power)*: Utility owns or operates a project that is open to voluntary rate payer participation or the opportunity to buy shares in a solar farm. (2) *Special purpose entity*: Investors form a business enterprise to develop project. (3) *Non-profit*: Donors contribute to a community installation owned by a charitable non-profit corporation.

States are setting policy for community solar, for example, the Community Solar Gardens Act, in Colorado, 2010 has the following provisions.

Energy must be sold directly to an investor-owned utility.
Utility pays retail plus renewable energy credits.
System size limited to 2 MW.
Six MW total limit on the program for first 3 years.
There must be at least 10 subscribers.
Subscribers must be located in same county or city as the solar garden. Subscribers whose county has a population less than 20,000 may subscribe in a neighboring county.
Subscribers may buy up to 120% of their own power use worth of solar power.
Either a for-profit or a nonprofit entity may own and administer the solar garden.

Forty-five kW of PV was installed on the rooftops of six community-owned buildings for the Nimbin Community Solar Farm in Australia. The project generates around 61,000 kWh/year and created three permanent jobs. At the other end of the scale, the Westmill Solar Cooperative, UK, has a 5 MW community solar project. The offering sold out in 6 weeks and has 1,650 subscribers, with over half living within 40 km of the project.

The Clean Energy Collective has a list of their community solar projects with information on size and photos (http://www.easycleanenergy.com/aboutUS.aspx). Their model is to have larger projects to obtain economy of scale and that the customers of the utility own part of the solar farm and they are credited on their energy bill. At this point in time (2014), most of their projects are in Colorado. A different aspect is the University of Utah program for community solar offers discounted solar panels and installations to university faculty, students, staff, alumni and campus guests. Participants in the U Community Solar program can give associated renewable energy credits generated by their solar panels to the university, thereby it assists the university to achieve its carbon-reduction goals.

6.10   APPLICATIONS

Applications can be divided into the following types:

Grid connected
 Residential, non-residential (commercial and industrial), and utility
Village power
 Maybe PV alone or hybrid PV with diesel, wind, and other sources in parallel
Stand-alone
 Lights, radio, TV, refrigeration, water pumping, water desalinization, and
  water purification
 PV for schools, clinics, local government offices, and battery-charging
  stations

6.10.1   GRID CONNECTED

In the developed world, as the capacity, size, and number of systems have increased, the major PV type of installation is now grid connected. The advantages of the grid systems are fairly simple connection though an inverter, no storage need, and peak shaving as the PV system produces power primarily when the loads are higher. Today, most PV capacity is installed for grid connection in the developed world, with system size of kilowatts for residential, hundreds of kilowatts for non-residential, and megawatts for utility power.

• *Residential*: Size and number of PV systems depend primarily on incentives at the national and even state and local levels. In the United States, the average size for residential systems grew from 2 kW in 2000 to 4.9 kW in 2008. The average size varies by state, depending on available incentives, interconnection standards, net metering regulations, solar resources, retail electricity rates, and other factors. In the United States by 2013, there were 420,000 (grid connected) residential installations, which accounted for 19% of the capacity.

  Japan was a leader in PV production and had a gigawatt of PV installed by 2004, mainly residential [3, see Section 8.3], and by 2013, around 7 GW was in residential. Japan has more homes with PV systems than any other country in the world, with around 1.5 million installations on residential roofs, around 85% of total PV installed. Subsidies for residential systems were phased out by 2006 as the price of PV systems decreased. The government reinstated the program in 2009 but was again phased out in March 2014, which will reduce the number of new installations

* *Non-residential*: The average size in the United States is around 110 kW. Systems of megawatt size have been installed as large companies are investing in PV systems, for example, Apple has two, 20 MW each, solar installations and a 10 MW fuel cell installation at their data center in North Carolina.

    In Japan, the prime minister's residence, Japanese Parliament, and many government buildings have 30- to 50-kW PV systems mounted on the rooftops. This is in contrast to the United States where President Reagan ordered the solar hot water system (installed by President Carter in 1979) removed from the White House. Then President Obama ordered that solar panels be installed on the White House, and 6.3 kW of PV was installed in 2014.

* *Utility scale*: There were 40 plants in the world [16] with capacity of 100 MW_P or greater (2013), with large plants being in the United States, China, India, Germany, France, and Ukraine. In 2012–2013, ten plants (2.57 GW_P) were constructed to supply electricity to the California market, and two plants, which came online November 2014 and January 2015, each have a capacity of 550 MW_P (Figure 6.18). Another five plants (total around 1.8 GW) are under construction in California and Nevada. Contrast this with information in the first edition of the book to see the dramatic change in capacity and size of utility scale PV. Note that the largest PV plant was 60 MW. *Seventy percent of utility-scale PV plants in Europe are located in Spain (>3.2 GW_P), followed by Germany (~700 MW_P) and Italy (70 MW_P) [14]. Spain has 27 of the top 50 PV plants, ranging from 13 to the largest PV plant in the world (2009) at 60 MW, the Olmedilla Photovoltaic Park (aerial images available on Google). There are four PV power plants of*

FIGURE 6.18 Topaz solar farms, 550 MW, in California. Site area 25 km². (Courtesy of BHE Renewables.)

50 MW and greater (two in Spain and two in Germany), and there is a 46-MW$_P$ (expansion to 62 MW$_P$ under construction) in Portugal. The largest PV plant outside Europe and the United States is a 24-MW$_P$ plant in Korea. As of 2009, the largest utility PV plant in the United States was 25 MW$_P$, FPL's DeSota Next Solar Energy Center in Florida. Previously, it was the 14-MW$_P$ PV system at Nellis Air Force Base in Nevada.

6.10.2 VILLAGE AND HYBRID POWER

Around $1.4 * 10^9$ people do not have electric service as they are too distant from transmission lines. Extension of the grid is too expensive for most rural areas, and if extended, it has poor cost recovery, which means it is heavily subsidized. Thus, the extension of the grid is too expensive for most rural areas. These people depend on wood, biomass, or dung for cooking and heating purposes; mainly, these materials are collected and the fires tended by women and girls.

For remote villages and rural industry that have electricity, the standard is diesel generators. Diesel generators are inexpensive to install; however, they are expensive to operate and maintain, and major maintenance is needed from every 2,000 to 20,000 h, depending on the size of the diesel genset. Most small village systems that have diesel power only have electricity in the evening.

Renewable (single or hybrid) village power systems (Figure 6.19) are a minigrid powered by PV, wind, mini- and microhydro, or bioenergy with battery or conventional diesel/gas generators to supply reliable energy (limited). Village power systems can range from small (<100 kWh/day, ~15 kW$_p$) to large (tens of MWh/day, hundreds of kW$_p$).

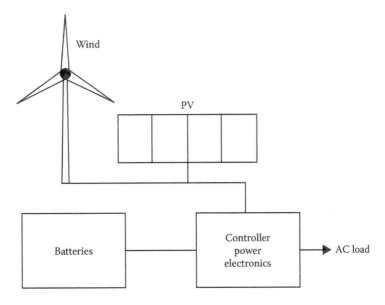

FIGURE 6.19   Diagram of a renewable village power system.

For present village systems powered by diesel generators, the electrical energy is subsidized from the state to national level, and it is difficult to find the actual cost. In general, the costs are \$0.30 to \$0.60/kWh, or even higher for some islands and remote areas. This means that renewable hybrid systems for village power are cost effective today. The problem is threefold: ensuring the installation of enough village power systems within a region or state that are robust (high reliability) and modular, reducing the costs by economics of scale, and having the infrastructure for long-term operation and maintenance.

The advantages of village power systems using renewable energy are as follows:

Provide AC or DC power for remote areas. For a system of any size, AC is the standard
Provide electricity for productive uses
Modular
No or low fuel costs
Lowest life-cycle cost of electricity
May be owned and operated by local cooperative

The disadvantages are as follows:

High initial capital cost compared to diesel generators
More complex—sophisticated controllers, power conditioning, batteries
High growth in demand means there is not unlimited usage (within a short time, there is the need for load management, load limitation)
Few suppliers, few systems installed; need high-volume production
Infrastructure—who maintains (trained personnel), how much do consumers pay per kilowatt hour?

Institutional issues are more important than the technical issues, especially for demonstration projects, as many demonstration projects can become political. Institutional issues are as follows:

Planning, which includes locals
Cost, subsidies (who, how much), repairs (maintenance fee or paid at occurrence)
Ownership
Operations and maintenance, training of operators
Financing; world (multilaterals), other nation aid agencies (lateral), and non-government organizations; national, state, local, and private organizations
Tariff design, metering, ability and willingness to pay
Load growth, education of users
Quality of service
Economic development versus social services (schools, clinics, local government offices)
Cultural response
Cooperation—local, state, national, electric utilities, financing

FIGURE 6.20 A 50-kW PV array with controller and a large battery bank were added to the diesel station in the village of Campinas, Amazonas, Brazil. (From National Center for Photovoltaics, National Renewable Energy Laboratory [NREL], www.nrel.gov/ncpv.)

The types of village renewable systems are the following:

> Fuel-saver system, addition of renewable energy system to existing diesel power plant (Figure 6.20)
> Low penetration
> High penetration, need to control system
> Renewable energy source, single or combination, with battery storage
>> *Hybrid system 1*: Renewable source (single or combination), diesel/gas generators
>> *Hybrid system 2*: Renewable source (single or combination), battery bank, diesel/gas generators

China is the world leader in village power systems [17]. The SDDX program for renewable village power and single-household systems (SHSs) for western provinces of China installed 721 PV/wind systems (15,540 kW), 146 minihydro (113,765 kW), and 15,458 SHSs (1,102 kW).

A major acquisition for any village is television, whether at school, a community center, or even individual households. As an example, a PV system was installed in a village in China, even though the grid was only 10 km away, and the village chief became an entrepreneur as he became the local cable provider.

6.10.3 Stand-Alone

In the past, the two main applications for PV were (1) small systems, single modules (50–100 W) for stand-alone electricity for remote areas and (2) water pumping for livestock and residences. These are still important applications with the addition of remote and mobile applications for signals and data transmission. More details on stand-alone PV systems are given in Foster et al. [3, see Section 8.5–8.13]. Many PV

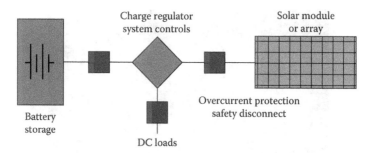

FIGURE 6.21   Diagram of simple PV system for lights, radio, and television.

FIGURE 6.22   PV systems (50 W) for two fluorescent lights, village of Cacimbas in the state of Ceara, Brazil. Notice the tilt angle of the collectors. (From National Center for Photovoltaics, National Renewable Energy Laboratory [NREL], http://www.nrel.gov/ncpv.)

systems had a single module for lights and a radio, with later installations having larger modules to have enough power for a TV (even color). The system consists of a module, controller, and battery (Figures 6.21 and 6.22).

In remote locations, PV is cost competitive with batteries and even with small diesel generators. Notice that for small power uses, PV is even cost competitive with connection to an adjacent utility power line as PV is used for school crossing lights, flashing stop signs, and so on.

Water pumping for livestock, residences, and villages is important in all regions of the world. Now, tens of thousands of PV water-pumping systems are installed throughout the world. The advantages are the same as for other PV systems: reliability, durability, and no fuel cost. The disadvantage is the higher initial cost. Any water-pumping system requires maintenance, no matter the power source.

The standard for water pumping is diesel engines. Depending on the hydraulic power and total power (dynamic head * volume, m–m³/day) needed, there is a range of pumping systems, from hand pumps to diesel (Figure 6.23). Of course, the PV and wind systems depend on the availability of the resource as a good resource makes them more economical. The solar resource should be >3 kWh/m² per day and the wind resource should be an

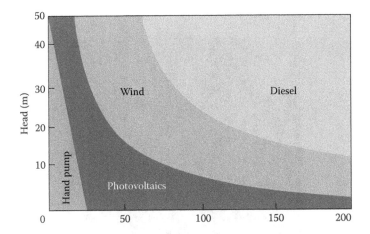

FIGURE 6.23 Competitive ranges for water-pumping systems in terms of hydraulic power (m^3/day).

average of 3–4 m/s for the month of highest demand. For small systems, PV is the most economical; for midsize systems, wind is more economical; and for very large systems, diesel is still more economical. PV should be considered when the hydraulic power is from 200 to 1,500 m–m^3/day. One of the good things about water-pumping systems is that storage tanks are fairly cheap, so wind and solar systems can supply water for a number of days. Now, there are some wind/PV hybrid water-pumping systems.

PV systems are now becoming the choice for pumping small amounts of water and even for village systems (Figure 6.24). The farm windmill will only pump around 10–20 L/min and will not pump the large volumes needed for a village. PV water-pumping systems can use a direct connection through a controller from the modules or an array to the pump and no batteries. PV water pumping for livestock has replaced many farm windmills in the United States (Figure 6.25), especially as many farm windmills are 30 or more years old, and the trade-off is the difference among repair, a new farm windmill, or a new PV system. PV water pumping is even the economical choice for replacing electric pumps at the end of long utility lines.

UV radiation and ozone can be used for water purification, which can be driven by electricity produced by renewable energy: PV, wind, and PV/wind hybrid [18,19]. Light-emitting diodes, which have long life and low energy requirements, now provide a good source for UV radiation powered by renewable energy [20]. An UV water purifier powered by different configurations of renewable energy was tested [21]: two by PV (100 W), two by wind (500 W), and one by hybrid wind/PV. The PV-only system was more efficient and cost effective than the wind-only system. However, the wind/PV system was more reliable in terms of power. The system purified 16,000 L/day, which is enough potable water for around 4,000 people at an estimated equipment cost of around $5,000. A hybrid system in Afghanistan [21] powered by a 1-kW wind turbine, 280 W of PV, a small battery bank, and an inverter provides around 160 W to generate 7.5 L/h of ozone. The ozone is used to treat 500 L of water at a time (batch process), and most communities using the system treat about 2,000 to 4,000 L/day.

FIGURE 6.24 PV water-pumping system for village in India. (From National Center for Photovoltaics, National Renewable Energy Laboratory [NREL], http://www.nrel.gov/ncpv.)

FIGURE 6.25 PV with passive tracker for pumping water for livestock. (From National Center for Photovoltaics, National Renewable Energy Laboratory [NREL], http://www.nrel.gov/ncpv.)

6.11 COMMENTS

An intriguing development is transparent PV collectors that generate electricity from UV radiation and allow visible light to pass through. A company makes glass containing dyes that concentrate and deflect solar energy to PV cells, so frames of windows and skylights could become producers of electricity that are integrated into a building.

The growth of installed PV has been dramatic and will continue to grow due to national, state, and local policies, especially in reaction to global warming, and due to the decrease in price. This means that the costs of utility-scale PV systems will become competitive with other sources: fossil fuels, nuclear, and other renewables. Micro inverters with individual panels, more efficient PV materials, flexible PV membranes [23], integrated PV with buildings (e.g., PV shingles, rooftops, and facades), and other technical advances mean PV has a bright future.

REFERENCES

1. U.S. DOE, Office of Energy Efficiency and Renewable Energy. Crystalline Silicon Photovoltaic Cell Basics. http://energy.gov/eere/energybasics/articles/crystalline-silicon-photovolatic-cell-basics.
2. Energy Informative. *The Homeowner's Guide to Solar Panels.* Which solar panel type is best? http://energyinformative.org/best-solar-panel-monocrystalline-polycrystalline-thin-film/.
3. R. Foster, M. Ghassemi, and A. Cota. 2010. Photovoltaic system sizing and design. In: *Solar Energy, Renewable Energy and the Environment.* CRC Press, New York, Chap. 7.
4. Sandia National Laboratory. 1995. *Stand Alone Photovoltaic Systems: A Handbook of Recommended Design Practices.* https://www.wbdg.org/ccb/DOE/TECH/sand87_7023.pdf.
5. California Energy Commission. 2001. *A Guide to Photovoltaic (PV) System Design and Installation.* http://www.energy.ca.gov/reports/2001-09-04_500-01-020.PDF.
6. *Solar Radiation Data Manual for Flat Plate and Concentrating Collectors.* http://rredc.nrel.gov/solar/pubs/redbook/.
7. National Renewable Energy Laboratory. PV Watts. http://www.nrel.gov/rredc/pvwatts.
8. Photovoltaic Geographical Information System (PVGIS), Joint Research Commission. http://re.jrc.ec.europa.eu/pvgis/index.htm.
9. EC, Joint Research Centre, Institute for Energy. 2013. *Arnulf Jäger-Waldau, PV Status Report.* http://iet.jrc.ec.europa.eu/remea/pv-status-report-2013.
10. *Global Market Outlook for Photovoltaics 2014–2018.* http://www.epia.org/fileadmin/user_upload/Publications/EPIA_Global_Market_Outlook_for_Photovoltaics_2014-2018_-_Medium_Res.pdf.
11. L. R. Brown. 2009. *Plan B 4.0, Mobilizing to Save Civilization.* Earth Policy Institute, http://www.earth-policy.org. Great source of information for data, download report, and a large number of spreadsheets (population, health, and society; natural systems; climate, energy, and transportation; food and agriculture; economy and policy).
12. SEIA. *Solar Industry Data.* http://www.seia.org/research-resources/solar-industry-data.
13. L. Sherwood. July 2014. *U.S. Solar Market Trends 2013.* Interstate Renewable Energy Council. http://www.irecusa.org/publications.
14. P. Eiffert and G. J. Kiss. 2000. *Building-Integrated Photovoltaic Designs for Commercial and Institutional Structures: A Sourcebook for Architects.* http://www.nrel.gov/docs/fy00osti/25272.pdf.
15. U.S. DOE, EERE. 2011. *A Guide to Community Solar: Utility, Private, and Non-Profit Project Development.* http://www.nrel.gov/docs/fy11osti/49930.pdf.

16. *Large-Scale Photovoltaic Power Plants*, ranking 1–50. http://www.pvresources.com/PVPowerPlants/Top50.aspx.
17. C. Dou. 2008. *Capacity Building for Rapid Commercialization of Renewable Energy in China*. CRP/97/Geq. UNDP/GEF, Beijing, China.
18. A. Agraw. 2003. *Renewable Energy in Water and Wastewater Treatment*. NREL/SR-500–30383. http://www.nrel.gov/docs/fy03osti/30383.pdf.
19. Himalayan Light Foundation. *Solar Powered Ozone Water Treatment System—SPOWTS*. http://www.hlf.org.np/programdetail.php?id=7.
20. A. H. Crawford, M. A. Banas, M. P. Ross, D. S. Ruby, J. S. Nelson, R. Boucher, and A. A. Allerman (Eds.). 2005. *Final LDRD Report: Ultraviolet Water Purification Systems for Rural Environments and Mobile Applications*, Sandia 2005–7245. http://prod.sandia.gov/techlib/access-control.cgi/2005/057245.pdf.
21. B. D. Vick, R. N. Clark, J. Ling, and S. Ling. 2003. Remote solar, wind, and hybrid solar/wind energy systems for purifying water. *J. Solar Energy Eng.*, 125, 107.
22. *Bergey Windpower Case Study, Ozone Based Water Treatment System*. http://bergey.com/parwan-district-afghanistan.
23. A. C. Mayer, S. R. Scully, B. E. Hardin, M. W. Rowell, and M. D. McGehee, 2007. Polymer-based solar cells. *Mater. Today*, 10, 28–33.

RECOMMENDED RESOURCES

LINKS

Distributed Generation. http://www.distributed-generation.com.
Enerank. Energy ranking. http://www.enerank.com/index.php.
Institute for Self-Reliance. *The Political and Technical Advantages of Distributed Generation.* http://www.ilsr.org/political-and-technical-advantages-distributed-generation/.
International Energy Agency, PV Power System Programme. http://www.iea-pvps.org. A good source for national data for 24 countries.
National Center for Photovoltaics. NREL. http://www.nrel.gov/ncpv.
Photos of PV and village systems available from NREL. Photographic Information eXchange. http://www.nrel.gov/data/pix/.
Photovoltaics. Solar Energy Technologies Program, EERE, DOE. http://www1.eere.energy.gov/solar/photovoltaics_program.html.
PV Resources. http://www.pvresources.com.
Sandia Labs. http://www.sandia.gov/pv/main.html.
Solar Electric Power Association. http://www.solarelectricpower.org.
Solar Energy Industries Association. http://www.seia.org.
http://www.pvresources.com/BIPV.aspx.
Solar Gardens Community Power. http://www.solargardens.org/directory/.

PROBLEMS

6.1. List two advantages and two disadvantages of PV systems.
6.2. The peak of solar intensity is in the yellow region (see Figure 3.5). What is the energy of photons of yellow light in electron volts if the yellow light has a frequency of $5 * 10^{14}$ Hz and a wavelength of $6 * 10^{-7}$ m?
6.3. Pick any manufacturer and one of their solar modules. Find specifications for rated power, voltage, and current at rated power. Be sure to state manufacturer and module number.

6.4. BP solar, SX 3200, 200 W_P. If voltage at $P_{max} = 24.5$ V, what is the current? Will that be larger or smaller than the short-circuit current?

6.5. An array (fixed tilt at latitude) of six modules, each 250 W_P, is connected to the grid through an inverter. What is the average power from the system? Show steps or reasoning.

6.6. An array (fixed tilt at latitude) of six modules, each 250 W_P, is connected to the grid through an inverter. Estimate annual energy output for Amarillo, Texas.

6.7. One of the largest PV plant in the world, 550 MW, is in California. There are 1,700 annual average sun hours. Estimate the annual energy output.

6.8. You want an average of 600 kWh/month for a home near Denver, Colorado. How big (power) a PV system should you buy? How much area is needed for the array? Use crystalline silicon modules.

6.9. You want an average of 600 kWh/month for a home near Denver, Colorado. How big (power) a PV system should you buy? How much area is needed for the array? Use amorphous silicon modules.

6.10. How much more energy can be obtained from a flat-plate PV system with one-axis tracking?

6.11. Use the PVWatts calculator. What is the estimated annual energy for a 4-kW PV system (fixed tilt at latitude $-15°$) for Phoenix, Arizona?

6.12. Use the PVGIS for Casablanca in northern Africa. What is the estimated annual energy for a 4-kW PV system (fixed tilt at latitude)?

6.13. What is the average-day kilowatt hour per square meter for Roswell, New Mexico, for a PV system at fixed tilt at latitude $-15°$? For a 50-MW utility plant and crystalline silicon modules, what is the approximate area of the array?

6.14. Estimate the CF for a 50-MW PV utility plant located in area with an average of 4.7 sun hours/day for a fixed-plate system tilted at latitude.

6.15. Why have grid-connected systems taken much of the market in the world?

6.16. In the United States, what is the approximate size of the PV market (MW/yr) for the following: residential, industrial, and utility scale? Estimate the size of the market for utility scale for 2020 and 2030.

6.17. Find a solar community project near you. Describe project and state name and location. What is the capacity, estimated annual kWh?

6.18. Go to the Internet. Find a building-integrated PV project. Describe project and state location. What is the capacity, estimated kWh?

6.19. What would you consider as the three major institutional issues for village power? Give reasons for your answers.

6.20. From the Web, find any example or case history of PV water pumping. Then, write down system parameters and output (if given).

6.21. To consider PV water pumping as an option, what is the approximate range of hydraulic power?

6.22. Contrast water purification by UV radiation or ozone using PV power. Which treatment would you use? Why?

6.23. Find any commercial water purification unit that uses renewable energy for power. List specifications for power and average production.

7 Concentrating Solar Power

7.1 INTRODUCTION

The most common collectors are flat-plate collectors for space and water heating. It is possible to get higher temperatures for process heat and for solar cooling with well-insulated flat-plate collectors with two to three glazings; however, most higher-temperature collectors use some form of concentration for thermal generation of electricity, cooling (absorption cooling), or process heat. Concentrating collectors can also reduce the amount of photovoltaic (PV) cell area for producing electricity. Concentrating solar power (CSP) requires direct radiation, so these collectors need to be in areas with a good-to-excellent solar resource (2,000–3,000 kWh/m²/year or an average of 5.5–8 sun hours/day. See solar maps in Chapter 3 for those regions of the world with that solar resource, which are primarily deserts and arid regions. There are resource maps for CSP for the Southwestern United States that include locations of power plants, plus state maps in that region (http://www.nrel.gov/csp/maps.html).

Even though it is not a conventional CSP, solar ponds can be used to generate electricity in the same manner. The collector is the pond, and the salt gradient allows for stratification of the thermal energy, which can be used to drive the turbine to generate electricity.

CSP systems can provide firm or peak power or even base load capacity due to thermal storage or fuel backup; they produce the greatest amount of power during the afternoon when electricity demand is high. The main types of solar thermal generation of electricity are power tower, line or linear focus, and dish/engine.

Power tower (point focus)
  Heliostat (parabolic shape or smaller individual mirrors to approximate parabolic shape)
  Two-axis tracking
  Focal length about hundreds of meters
  Concentrating ratio about 800
Line or linear focus
  Parabolic two-dimensional (2D) shape or linear Fresnel
  One-axis tracking, east to west
  Focal length about 3 m
  Concentrating ratio about 30–40
Dish (point focus)
  Parabolic three-dimensional (3D) shape
  Two-axis tracking
  Focal length about 1–4 m
  Concentrating ratio about 3,000

The power tower uses tracking mirrors (heliostats) to focus sunlight onto a boiler located on a tower (power tower); line focus uses parabolic troughs or Fresnel lenses to focus on a linear collector tube (receiver); the dish/engine uses a parabolic dish and a receiver at the focal point, and a number of Fresnel lenses can be used to focus sunlight on PV cells. The CSP collectors can use one- and two-axis tracking (Section 6.6.2). In both systems, there are operation and maintenance problems with the control and operation of motors and gears and the match between materials for thermal expansion at rotating joints. Heliostats are two-axis reflectors; most parabolic troughs use one-axis tracking. Some systems use tracking systems and concentrators to focus sunlight onto small PV cells, either line focus or point focus with individual units.

CSP projects are listed by country, project name, technology, and status [1]. Also, overviews are available for four technologies: power tower, parabolic trough, linear Fresnel reflectors, and dish/engine. Sandia Laboratories is the main national laboratory for CSP projects [2]; it includes images. CSP projects (≥ 1 MW) are included in the list of major solar projects in the United States [3], which includes under construction and under development.

For CSP and large-scale PV projects, the land needed is around 4 ha/MW, or another way to state the same information, around 25 MW/km^2, with most of the land for the solar field. Remember you have to have space between the collectors or heliostats so they do not shade one another. The land area for a wind project is larger than for CSP project; however, actual land taken out of production is less since the capture area is vertical rather than horizontal. However, CSP projects are located in arid regions with excellent solar insolation, so the land generally has low productive value for other uses.

7.2 POWER TOWER

Scientists in Russia first proposed the power tower concept in the mid-1950s, and after the oil crisis of 1973, a number of experimental systems were built in various countries [4]. A power tower system uses a large field of heliostats to focus and concentrate the sunlight onto a receiver on the top of a tower (Figure 7.1). The concentrated beam has very high energy; a bird flying into the beam would be vaporized. A heat transfer fluid is used to generate steam for a conventional electric generator. The heat transfer fluid could be water or steam or molten salt because of its superior heat transfer and energy storage capabilities. Thermal storage allows the system to continue to dispatch electricity during cloudy weather or at night, and with the addition of fossil fuel, the system can provide firm or base power.

The annual capacity factor of most solar technologies without storage is around 25%; however, with the addition of thermal storage, power towers can have annual capacity factors of up to 60% and as high as 80% in the summer. There are enough heliostats to provide sufficient energy to power the turbine and provide the extra energy for storage. At night or during extended cloudy periods, the turbine is powered with the stored thermal energy (Figure 7.2). In Figure 7.2, sunrise was around 0730, and the intensity of sunlight rose rapidly then dropped off before sunset after 1630. Notice that power output was started at 1100, when it was needed by the utility.

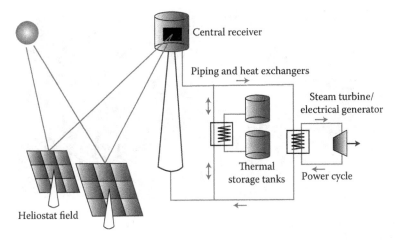

FIGURE 7.1 Diagram of power tower for generating electricity. (Courtesy of Sandia Labs, Albuquerque, NM.)

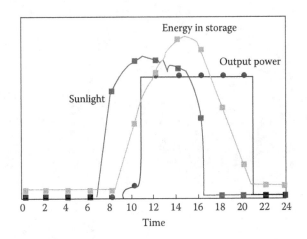

FIGURE 7.2 CSP input, storage, and output power for a winter day. (Courtesy of Sandia Labs, Albuquerque, NM.)

In 1976, an experimental power tower was constructed at Sandia National Laboratories with a field of over 200 heliostats. It is now the central receiver test facility (Figure 7.3) with 5 MW of thermal power and a peak flux of 260 W/cm². Then, Solar One, a 10-MW demonstration project, was constructed near Barstow, California, that had 1,818 heliostats, each 40 m². During operation from 1982 to 1988, it produced around 38 million kWh. In 1995, Solar Two (Figure 7.4) was a retrofit to Solar One to demonstrate the advantage of molten salt over water or steam for heat transfer and storage. A second ring of 108 larger 95-m² heliostats was added, for a total of 1,926 heliostats with an area of 82,750 m². Solar Two was decommissioned in 1999 due to economics and in 2001 was converted by the University of

FIGURE 7.3   Sandia central receiver thermal test facility. (Courtesy of Sandia Labs, Albuquerque, NM.)

FIGURE 7.4   Solar Two, Barstow, California. Notice receiver, which is white hot, and the oil storage tanks at the bottom. (Courtesy of Sandia Labs, Albuquerque, NM.)

California, Davis, into an air Cherenkov telescope, measuring gamma rays hitting the atmosphere.

There was renewed interest in power towers with the construction of 11- and 20-MW plants in Spain (Figure 7.5) from 2007 to 2009. There was an additional 1.5 MW plant installed in Germany in that period. For the 11-MW plant, there are 624 heliostats, each 120 m², to focus sunlight onto the 115 m tall central receiver. There is 1 h of storage of pressurized steam, 50 bar at 285°C. For the 20-MW plant, there are 1,255 heliostats, each 120 m², covering an area of 95 ha (235 acres), and

FIGURE 7.5 Abengoa power towers, Seville, Spain. Top is 11 MW, bottom is 20 MW. Notice that there are also PV panels. (Courtesy of Abengoa.)

the central receiver is on a 160-m tower. Unless noted otherwise for the generation of electricity, project capacity is in MW electrical.

Now there are 12 power towers with 517 MW of capacity and another 7 are under construction with 541 MW of capacity (Table 7.1). Out of the total of 1058 MW, 657 MW is in the United States, primarily to serve the California market. The number of heliostats ranges from hundreds to hundreds of thousands, depending on size, 1–140 m^2, and on capacity of the project, and the area for the heliostats ranges from 1 to over 250 ha. Tower heights are from 50 to 200 m. One project in the United States was selling electricity for \$0.135/kWh on a long-term contract. In general, land utilization is around 0.25 MW per hectare, primarily covered by the solar field as the heliostats need to be spaced such that they do not shade one another. In 2014, the largest project, 392 MW, had three towers (Figure 7.6) with 173,500 heliostats. Besides the problem of bird morality, the glare from the project [5] interferes with pilot's ability to scan the sky in certain directions during late morning and early afternoon.

Currently, 40% of California's oil production uses steam injection for enhanced oil recovery and the percentage will increase. Thus, a unique power tower application is the generation of steam for enhanced oil recovery, Coalinga, California. Natural gas is used to generate the steam, and now a demonstration project of 29 MW$_t$ provides steam, up to 12 h on sunny days, displacing natural gas.

The following are some design consideration for power towers:

Place another set of mirrors at the top of the power tower to reflect the sunlight onto a boiler on the ground.

Heat the air at the focal point to 1,000°C (solar flower or Tulip, power tower) and then the hot air drives a turbine to generate electricity, with the addition

TABLE 7.1
Power Tower Projects in the World

Project	Country	MW	Heliostats Area Ha	#	Ap Area M²	Land Area Ha	Storage hr	Date op
Planta Solar 10	Spain	11	7.5	624	120	55	1	2007
Jülich	Germany	1.5	1.8	2,153	8.2	17	1.5	2008
Planta Solar 20	Spain	20	15	1,255	120	80	1	2009
Sierra Sun Tower	United States	5	2.7	24,360	1.1		None	2009
Gemasolar	Spain	19.9	30.4	2,650	120	195	15	2011
Acme Solar Tower	India	2.5	1.6	14,280	1.1	5.0	None	2011
Lake Cargeligo	Australia	3	0.62	620	9.8			2011
Dahan	China	1	1.0	100	100	40.3	1	2012
Greenway Mersin	Turkey	1.4					4	2012
Ivanpah	United States	392	260.0	173,500	15	6	None	2013
Dehli	China	10		217,440	2	330	2.5	2013
Khi Solar One	South Africa	50	57.7	4,120	140		2.0	2014
Subtotal			517					
Under Construction								
Crescent Dunes	United States	110	107.1	17,170	62.4	647	10	2015
Jemalong	Australia	1.1	1.5	3,500	4.3	10	3	2015
Rice	United States	150	107.1	17,170	62.4	570	12?	2016
Ashalim	Israel	121						2017
Cerro Dominador 1	Chile	110	148.4	10,600	140	700	17.5	2018
Cerro Dominador 2	Chile	110	148.4	10,600	140	700	17.5	2018
Supcon	China	50	43.5	217,440	2	330	2.5	
Subtotal			541					
Total			1058					
Dehli first phase								
Coalinga	United States	29		3,822		100	None	2011

FIGURE 7.6 Ivanpah solar power tower system, 392 MW; bottom photo shows glare.

of using the exhaust for process heat. No water is needed for cooling in contrast to the conventional solar thermal plants driving a conventional steam turbine. A demonstration plant (http://aora-solar.com/active-sites/) in Spain has a power tower (tulip) of 110 kW combined with a 110 kW natural gas turbine and the exhaust heat of 170 kW$_t$ for desalination, producing 3 m^3/day of fresh water.

Place the heliostats in a large abandoned pit mine, that way they can be closer together as you are using another dimension.

Build a very tall tower (chimney effect), have a large covering on the ground (greenhouse effect), and then use the rising air to power a turbine, an innovative wind turbine (applet, http://demonstrations.wolfram.com/SolarUpdraftTower/).

7.3 LINE OR LINEAR

The line focus collectors use parabolic troughs, reflecting mirrors, or linear Fresnel reflectors. Generally, the line focus system is oriented in the north–south direction, with one axis tracking in the east–west direction (Figure 7.7). Focus is by a parabolic trough (Figure 7.8) or linear Fresnel reflectors (Figure 7.9). A Fresnel lens takes the continuous surface of a standard lens into a set of surfaces of the same

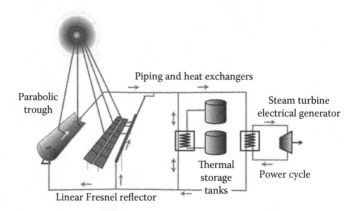

FIGURE 7.7 Diagram of line focus, parabolic trough or Fresnel lens, for generating electricity. (Courtesy of Sandia Labs, Albuquerque, NM.)

FIGURE 7.8 Parabolic troughs at Solar Energy Generating Systems, 345 MW. (Courtesy of Sandia Labs, Albuquerque, NM.)

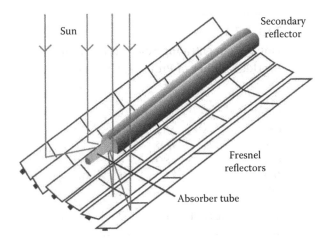

FIGURE 7.9 Diagram of linear Fresnel reflector system. Notice secondary reflector on top of receiver tube.

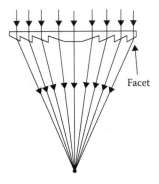

FIGURE 7.10 Diagram of Fresnel lens.

curvature (Figure 7.10), with discontinuities between them, thus reducing thickness and weight. For solar applications, they can be formed from plastic or glass. The typical heat transfer fluid is a synthetic oil heated to around 390°C, and electricity is generated in a conventional Rankine cycle steam turbine. The oil can also be used for heat storage.

There were around 4,200 MW of parabolic troughs (Table 7.2) installed in the world through 2014, with another 850 MW under construction. Spain has the largest capacity and number of projects, mostly 50 MW-plants, followed by the United States. The early systems (installed in 1985–1991 and still operating) were the Solar Energy Generating Systems (SEGS) in Southern California (Table 7.3) with another 857 MW installed in the United States since then (Table 7.4). Many of the energy production values for CSP projects are estimates, rather than one or more years of actual data.

TABLE 7.2

Linear Concentrating Solar Power in the World: Number of Projects, Capacity

Country	Parabolic		Fresnel	
	#	MW	#	MW
US	8	1456		
Spain	29	2175		
ROW	16	584		
Total	47	4215	8	189
UC	9	849	2	53

Note: ROW, rest of the world; UC, under construction.

A typical parabolic trough project in Spain has the following characteristics:

Rated power	50 MW
Solar resource	2,136 kWh/m²/year
# collector assemblies	624
Solar field aperture	51 ha
Land area	200 ha (0.25 MW/ha)
Energy production	175 GWh/year
Storage	7.5 h (28,500 tons of molten salt, 1,010 MWh)
Power purchase/tariff rate	€ 0.27/kWh
Power purchase/tariff period	25 years
Power block	Steam Rankine
Solar-electric efficiency	16%

There were two linear Fresnel reflector plants in operation in 2009: the Kimberlina 5-MW plant (Figure 7.11) near Bakersfield, California, and a 1.4-MW plant near Calasparra, Spain. Through 2014, there were 8 Fresnel projects with a capacity of 191 MW and two projects under construction (Table 7.2). The steel-backed mirrors are located near the ground and rotate individually while focusing on the fixed receiver tube (Figure 7.12), in contrast to the parabolic troughs where the receiver tube moves with the trough. The Kimberlina plant has the following configuration:

26,000-m² solar field aperture area
Three 385-m lines
Ten 2-m wide mirrors
Fixed 18-m high receiver
Water transfer fluid
40-bar power cycle pressure

TABLE 7.3

Parabolic Trough Systems Installed in the United States, 1985–2007

Plant	Location	Year	MW$_e$	Temperature Out (°C)	Area (m²)	Efficiency (%)	Power Cycle	Dispatch
SEGS I	Daggett, CA	1985	13.8	307	82,960	31.5	40 bar, steam	3-h TES
SEGS II	Daggett, CA	1986	30	316	190,338	29.4	40 bar, steam	Gas boiler
SEGS III	Kramer Junction, CA	1987	30	349	230,300	30.6	40 bar, steam	Gas boiler
SEGS IV	Kramer Junction, CA	1987	30	349	230,300	30.6	40 bar, steam	Gas boiler
SEGS V	Kramer Junction, CA	1988	30	349	250,500	30.6	40 bar, steam	Gas boiler
SEGS VII	Kramer Junction, CA	1989	30	390	194,280	37.5	100 bar, reheat	Gas boiler
SEGS VI	Kramer Junction, CA	1989	30	390	188,000	37.5	100 bar, reheat	Gas boiler
SEGS VIII	Harper Lake, CA	1990	80	390	464,340	37.6	100 bar, reheat	HTF heater
SEGS IX	Harper Lake, CA	1991	80	390	483,960	37.6	100 bar, reheat	HTF heater
APS Saguaro	Tucson, AZ	2006	1	300	10,340	20.7	ORC	None
Nevada Solar One	Boulder City, NV	2007	64	390	357,200	37.6	100 bar, reheat	None

Note: SEGS, Solar Energy Generating Systems; HTF, heat transfer fluid; ORC, organic Rankine cycle; TES, thermal energy storage.

TABLE 7.4

Parabolic Troughs Installed in the United States, 2009–2014

Project	MW	Solar Ap Ha	Land Area Ha	Storage hr	Date op	Energy GWh/Year
Holaniku at Keahole Point	2		1.2	2	2009	4
Martin Next Generation	75	46.5	200		2010	155
Solana	280	220	780	6	2013	944
Genesis	250		789	None	2014	580
Mojave	250		714	None	2014	600
Total	857					

FIGURE 7.11   Linear Fresnel reflector, Kimberlina plant, 5 MW. (Courtesy of Ausra.)

Of course, the same system could be used for auxiliary steam for existing power plants or for process heat. These systems are referred to as *integrated solar combined cycle* (ISCC). For example, in Australia there are two projects to provide steam to coal plants, the Kogan Creek Solar Boost (44 MW Fresnel system to 740 MW coal plant) and Liddell (9 MW_t, 3 MW_e Fresnel system to 2,000 MW coal). The City of Medicine Hat, Canada, project is a 1.1 MW parabolic trough system to provide steam to a 33 MW natural gas plant. There were two parabolic trough plants, one in the United States and the other in Oman, that generated steam for enhanced oil recovery. There is consideration for CSP to provide thermal energy for a geothermal plant that has a declining output. Note there is a difference between project size as electric, MW_e or MW_P, and thermal energy, MW_t, as the ratio is around 1/3 due to heat loss and the efficiency of the steam turbines. Most CSP project report size as electrical MW, but if they only provide thermal energy they generally rate them as MW_t. There are some installations of micro-CSP (0.25–25 MW), for example, the 500 kW project, Holaniku at Keahole Point has parabolic troughs (http://www.nrel.gov/csp/solarpaces/project_detail.cfm/projectID=71).

FIGURE 7.12   Close-up of mirrors (each 2 m wide) and receiver tube, which is 18 m above ground, Kimberlina plant. (Courtesy of Sandia Labs, Albuquerque, NM.)

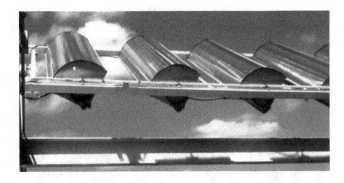

FIGURE 7.13   Linear Fresnel lens, PV strip (4 cm wide) at focus, two-axis tracking, system (around 400 W per unit), 300-kW array, at 3M Research Center in Austin, Texas (1990). Notice fins on the back for heat dissipation.

Some systems use tracking systems and concentrators to focus sunlight onto a line of PV cells (Figure 7.13), thereby reducing the PV cell area. Since a PV system decreases in efficiency with higher temperature, the extra heat needs to be used or dissipated. The same type of units provided 24 kW electricity for the Dallas-Fort Worth airport and 140 kW of thermal energy for its Hyatt Hotel.

7.4 DISH/ENGINE

A dish/engine system uses a mirrored parabolic dish for focus on a thermal receiver, which transfers heat to the engine/generator. The parabolic shape is generally made from individual mirrors or reflecting membranes. The most common type of heat engine is the Stirling engine, which uses the fluid heated by the receiver to move pistons and create mechanical power to run a generator or alternator. Stirling engines [6] depend on expansion and compression at different temperatures of a gaseous working fluid; the external heat can be from any source. Remember that conventional thermal generation of electricity requires cooling, generally by water; however, the Stirling engine does not require much water.

Sandia Labs has worked on experimental dish/engines for a number of years and has a test bed of six systems (Figure 7.14) with a power output of 150 kW. A later configuration, 25 kW, is round (Figure 7.15) and reached an efficiency of 31%, sunlight to electricity. A unique experimental system was built at Crosbyton, Texas, where the collector (spherical dish with mirrors) was fixed, and the line focus was movable.

Stirling engines have a long history; however, they have not attained large commercial production. One project in Utah at the Tooele Army Depot was to have 430 dishes, rated power of 1.5 MW. Infina (Figure 7.16) installed a number of dishes at Tooele, but had financial problems and was acquired by Qnergy, an Israel company. With the decrease in price of PV, it will be difficult for the dish/engine to acquire much of the CSP market.

Industry uses around a third of the total energy and over 70% is for process heat and 57% of that falls with the temperature range 70°C to 400°C, which could be produced by concentrating solar systems. The applications for process heat are numerous: food processing, water distillation, paper and pulp, enhanced oil recovery, cooling, and others.

FIGURE 7.14 Test field for dish/engine at Sandia National Laboratories. (Courtesy of Sandia Labs, Albuquerque, NM.)

FIGURE 7.15   Suncatcher at Sandia test field. (Courtesy of Sandia Labs, Albuquerque, NM.)

FIGURE 7.16   Dish/Sterling engine, 3.5 kW, 11 m diameter, at Infinia Corporation, Kennewick, Washington.

FIGURE 7.17 Solar Total Energy System, 144 dishes, 7 m diameter, at textile plant near Shenandoah, Georgia. The dishes provided thermal energy for process heat and air conditioning at a nearby textile mill. (From National Center for Photovoltaics, NREL, http://www.nrel. gov/ncpv.)

A demonstration project used dishes for higher temperatures for process heat and air conditioning (Figure 7.17). Another demonstration dish system was a combined heat system, 400 kW of electricity and 150 tons of chilled water, that was installed in 1983; however, it only operated for 4 years.

7.5 POINT FOCUS

Other CSP systems with two tracking systems use individual Fresnel lenses to reduce the amount of PV material (Figure 7.18). The largest project was 30 MW from 504 units (60 kW) in Alamosa, Colorado (again go to Internet for images, google Amonix). Another similar system is the compound parabolic reflector (Figure 7.19).

In 1995, West Texas Utilities built a solar energy demonstration project at the end of a feeder line near Fort Davis, Texas. The project consisted of the following systems:

PV, 83-kW systems, line focus, Fresnel lenses, two axis (similar to Figure 7.13)
PV, 20 kW, flat with multiple Fresnel lenses, two axis (similar to Figure 7.18)
Dish/Stirling engine, 7 kW, multiple mirrors, stretched membrane, two axis
PV array, 100 kW, flat-plate panels, one axis

The dish was installed but Cummins never delivered the Stirling engine and within a few months dropped that business. Because of high operation and maintenance costs, the rest of the project was dismantled by 2000.

7.6 SOLAR POND

A solar pond is not really a concentrating solar system; however, the production of electricity is similar to other renewable energy systems that use a Rankine turbine. The major differences are that the collector and storage are together, and

FIGURE 7.18 Fresnel lenses, point focus, 53 kW, two-axis tracking, 22 m wide by 15 m high collector area. (Courtesy of Amonix, Seal Beach, CA.)

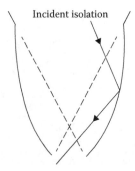

Incident isolation

FIGURE 7.19 Compound parabolic reflector.

there is a small temperature differential between the stratified layers. A solar pond consists of a salinity gradient to trap the heat (Figure 7.20), which can then be used for the generation of electricity, process heat, and freshwater. The temperature of the bottom layer is around 90°C, while the temperature of the top layer is around ambient temperature, 30°C. A 250,000-m² solar pond for the generation of electricity (5 MW) was constructed on the Dead Sea in Israel and operated until 1988 [7].

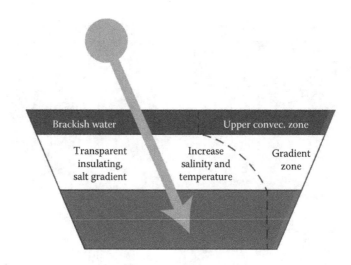

FIGURE 7.20   Diagram of salt gradient solar pond.

FIGURE 7.21   El Paso solar pond; notice wave suppression devices.

The El Paso solar pond (Figure 7.21) was a research and demonstration project of the University of Texas at El Paso started in 1984. In addition to the R&D aspect, process heat was provided for Bruce Foods, a canning company located adjacent to the solar pond. In 1986, the project was expanded to produce electricity (Figure 7.22). The solar pond, with 300-kW thermal capacity, had a Rankine cycle engine with a 100-kW generator that was grid connected. The surface area of the pond was 3,350 m², and it was 3 m deep. During one cold period, the top of the pond

FIGURE 7.22 El Paso solar pond with Rankine-cycle engine (100 kW) for electric generation in the foreground.

froze; however, it still produced some power. As expected, maintaining a proper gradient with little or no mixing between layers and evaporation are major problems. Wave suppression devices on the top of the pond were installed to reduce mixing. A membrane failure shut the plant down for three years, 1992–1995, and then it was closed in 2003 due to economics.

7.7 COMMENTS

There are enough operating projects using CSP, so now it is a question of economics and competition from other energy sources, both fossil fuels and other renewables. Again if life-cycle costs are used and the externalities of fossil fuels are included, then CSP is an option in regions of excellent solar resource. The Technology Roadmap [8] indicates that CSP could become cost competitive in intermediate and bulk loads by 2020 and base loads by 2030. There are a couple of problems for CSP: generation of electricity by steam need for water, generally a scare commodity in regions with excellent solar resources, and those regions are generally distance from major loads.

Technology and applications can and will change: production of hydrogen, assisted liquid fuel production, and storage materials. A research group at the Massachusetts Institute of Technology has developed a material to produce steam at ambient temperature. The porous, insulating material of graphite flakes and carbon foam floats on water and can convert 85% of incoming solar energy into steam. A group at Rice University used nanoparticles suspended in water to convert solar energy to steam. Another application would be to use solar energy or waste heat from a power plant to produce electricity by the thermoelectric effect (Seebeck effect), which does not require water.

REFERENCES

1. *Concentrating Solar Power Projects*. http://www.nrel.gov/csp/solarpaces.
2. Sandia National Laboratories. http://energy.sandia.gov/energy/renewable-energy/solar-energy/csp-2/.
3. Solar Energy Industries Association. 2015. *Major Solar Projects in the United States*. http://www.seia.org/sites/default/files/Full%20Public%20MPL%207-7-2015.pdf.
4. Sandia National Laboratories. 1995. *Today's Solar Power Towers*. SAND 91-2018. http://www.osti.gov/bridge/purl.cover.jsp?purl=/580539-nWYYRI/webviewable/.
5. *Sandia Report Presents Analysis of Glare Impacts of Ivanpah Solar Power Site*. 2014. http://energy.sandia.gov/sandia-report-presents-analysis-of-glare-impacts-of-ivanpah-solar-power-site/.
6. G. Walker, G. Reader, O. R. Fauvel, and E. R. Bingham. 1994. *The Stirling Alternative: Power Systems, Refrigerants and Heat Pumps*. CRC Press, New York.
7. C. Nielson, A. Akbarzedeh, J. Andrews, H. R. L. Becerra, and P. Golding. 2005. The history of solar pond science and technology. *Proceedings of the International Solar Energy Conference*, Orlando, FL.
8. International Energy Agency, 2010. *Technology Roadmap: Concentrating Solar Power*. http://www.iea.org/publications/freepublications/publication/csp_roadmap.pdf.

RECOMMENDED RESOURCES

CSP world map. http://www.csp-world.com/cspworldmap.

Interactive Maps

List of solar tower thermal power and solar furnances. https://www.google.com/maps/d/viewer?mid=zNzgsv-AKLRc.kBNgbWZ2QUFM&ie=UTF8&t=h&oe=UTF8&msa=0.

U.S. DOE. Concentrating solar power facilities and CSP energy potential gradient. http://energy.gov/articles/top-10-things-you-didnt-know-about-concentrating-solar-power.

Links

Abengoa Solar. http://www.abengoasolar.com/web/en/.

Concentrating solar power, energy from mirrors. 2001. http://www.nrel.gov/docs/fy01osti/28751.pdf.

CSP World. http://www.csp-world.com.

International Energy Agency. Technology roadmap, concentration solar power. http://www.iea.org/publications/freepublications/publication/csp_roadmap.pdf.

Kimberlina solar thermal energy plant. http://www.areva.com/EN/solar-164/areva-solar.

National Renewable Energy Laboratory. http://www.nrel.gov/csp/.

Pdfs on technology. http://www.solarpaces.org/library.

Solar ponds. http://www.solarponds.com.

Solar Power and Chemical Energy Systems. http://www.solarpaces.org.

Thermo Electric Gencell. http://thermoelectric-generator.com.

TroughNet. Parabolic trough solar power network. http://www.nrel.gov/csp/troughnet/.

More Technical Information

R. Foster, M. Ghassemi, and A. Cota. 2010. *Solar Energy: Renewable Energy and the Environment*. CRC Press, New York.

W. B. Stine and M. Geyer. 2001. *Power from the Sun*. Power from the SUN.net. http://powerfromthesun.net.

PROBLEMS

7.1. For 2014, how many megawatts of power tower were installed in the world?

7.2. For 2014, how many megawatts of linear CSP were installed in the world?

7.3. For the year with the latest data available, how many megawatts of dish, point focus CSP were installed in the world?

7.4. What is the largest power tower project in the world, MW rating? Year started operation?

7.5. What are the advantages and disadvantages of using CSP to produce electricity?

7.6. For the parabolic trough, SEGS, California, what are the total megawatts installed? What are the estimated kilowatt hours/year produced?

7.7. What is the capacity factor for SEGS, California?

7.8. What is the estimated capacity factor for PS20 in Spain?

7.9. Reference 1 lists CSP projects by country. List the first five countries and megawatts installed.

7.10. What are average efficiencies, sun energy to electricity, for a power tower? For a linear parabolic trough?

7.11. What is the average efficiency, sun energy to electricity, for a dish/engine?

7.12. Why are Fresnel lenses used for CSP?

7.13. Why was Solar Two shut down?

7.14. Why would CSP be used with PV?

7.15. Contrast efficiency and cost between flat-plate PV and CSP PV.

7.16. What is the advantage of the dish and Stirling engine for producing electricity rather than the dish and thermal generation of electricity?

7.17. In a good solar regime, estimate annual kilowatt hours for a one-unit dish/engine for Stirling Energy Systems. Be sure to include the collector area of the unit.

7.18. In a good solar regime, estimate the annual kilowatt hours for one-unit point focus at Amonix. Be sure to include the collector area of the unit.

7.19. Dishes were used for process heat. Are there any such systems on the commercial market today?

7.20. Are there any operating solar ponds in the world today? Go to the Internet and briefly describe one project. If you cannot find a current project, describe a past project.

7.21. Calculate the maximum theoretical efficiency for a solar pond.

7.22. Calculate the maximum theoretical efficiency for a power tower.

7.23. Calculate the maximum theoretical efficiency for a parabolic trough system.

7.24. What is the status of the Toeele Army Depot dish/engine plant? What are the megawatts, number of units, rated power per unit, and operational status?

7.25. What is the land area/megawatt needed for a power tower? linear system? dish/engine? For a power tower and linear system, you need to include the steam turbine and storage area.

8 Solar Systems

8.1 INTRODUCTION

The design of solar systems is based on engineering (use of math), architecture, and research by scientists. The design then is reduced to guidelines for builders, contractors, homeowners, and so on. Of course, if you have an architect, the architect will do an integrated design with detailed plans for you. Today, many of the applications are in terms of computer simulations and spreadsheets. Generally, the structure combines one or more systems.

Finally, if you want to design, build, implement, or do it yourself, there is a lot of information available. A number of homes with one or more renewable energy systems have been constructed in all regions of the United States. The Southwest probably has more passive solar homes (Figure 8.1) and buildings than other regions of the country. Most people are receptive to sharing and even showing their solar or renewable energy systems. Be sure to ask about performance and any difficulties or problems they have with their renewable energy systems. In any case, passive solar homes or renewable energy systems need to be designed and sized for the local climate. This chapter is primarily for description of homes, including underground homes since most have a south-facing side that has windows for direct gain. Finally, there are homes that are completely self-sufficient, off the grid. You can obtain information on solar home tours by state from the American Solar Energy Association (http://www.ases.org/solar-tour/) or from state solar organizations (http://www.ases.org/local/). For example, the Northern California Solar Energy Association has information on education and systems (http://www.norcalsolar.org).

8.2 PASSIVE SYSTEMS

If you want to design and construct your own passive system or do most of the work, the following are excellent sources: Passive Solar Home Design Checklist (http://nccleantech.ncsu.edu/wp-content/uploads/PassiveDesignChecklist.pdf) and Edward Mazria's *The Passive Solar Energy Book* [1]. The 27 patterns in *The Passive Solar Energy Book* provide more detailed information on design and sizing of components.

The following information is for those locations in the world with sufficient sunlight for heating during the winter and different techniques for cooling during the summer. Directions for north and south should be interchanged for the Southern Hemisphere and shading of windows in the equatorial zone has to be on both the north and south. There are three main components for a passive solar energy home: reduction or small heat loss in the winter and reduction or small heat gain in the summer (well insulated home), sunlight to enter space for heating in the winter and for daylighting all year, and thermal mass to reduce heat fluctuations on daily basis and 3–5 days with no sunlight. The thermal mass can also be used for keeping the house cool in the summer.

FIGURE 8.1 Passive solar home with direct gain; Trombe walls and large thermal mass provide most heating needs of an adobe home in Santa Fe, New Mexico. (From EERE, http:// www.eere.energy.gov.)

Indoor temperature in a passive solar home may fluctuate from 6°C to 15°C, and the amount of fluctuation is a function of the amount of building insulation (rate of heat loss); location, size, and type of windows; the amount of thermal mass; and the color of the interior surfaces. The solar heating cannot be turned on or off (unless you have moveable shutters and shades), and there is little control of the flow of heat inside the home. To prevent overheating, shading is used to reduce solar gain in summer, or excess heat is vented through open windows or by an exhaust fan. Most passive solar homes will have an auxiliary heat source to help reduce temperature fluctuations and to provide heat for long periods of overcast days. However, the size of an auxiliary heating system can be smaller due to the passive components than the size needed for the same floor area of conventional home.

Different systems, for example, direct gain, indirect gain, clerestory, skylights, Trombe wall, and sunroom, have different design limitations, so one or a combination of systems can be used. Choose a system or systems that satisfy most of the heating and/or cooling requirements of each space. Examples and designs of passive solar homes are available in the literature and the Internet.

8.2.1 Location, Orientation, and Shape

Place building on the site or lot that receives sunlight from 09:00 in the morning to 03:00 to 04:00 in the afternoon during the heating season. Remember there is a possibility of shading from future growth of trees or even buildings on adjacent lots. The simplest shape for the building is a rectangle, with the long side on an east–west axis. This long axis can be up to 15°–20° off of the east–west axis. If you need more heating earlier in the day, you would orient the building toward the northeast.

Reduce heat loss by having the north side slope toward the ground, by berms, or by building into the side of a south-facing slope. Reduce heat gain through the west side of the building from the late afternoon sunlight during the summer. This can be accomplished through small windows, external shading by deciduous trees, or even some type of shutters.

8.2.2 INDOOR SPACE

Those spaces that are occupied or used the most during the day, for example, the living room, should be placed along the south face of the building to receive sunlight, which provides heat and daylighting. Other rooms can be placed to the southeast and southwest, for example, bedrooms. Spaces with minimal heating and lighting requirements, for example, closets, garages, and laundry rooms, should be placed along the north face of the building. However, laundry rooms need heat in the winter to protect water pipes from freezing.

To reduce heat loss and/or heat gain, the main entrance could be a vestibule or foyer to provide a double entry or air space. Entrance should be located away from prevailing winter winds or shaded to reduce wind velocity. Entrance area can also be a storage space for coats.

8.2.3 WINDOWS

Major windows to let in sunlight will be located on the south face, with possible additional major windows on the southeast and southwest according to the internal requirements of each space. On the west and the north side of the building, window area should be small. In most cases, windows should be double pane or double pane, low-e. Some designs used recessed windows on the north to reduce heat loss in the winter due to wind velocity. Once again, remember large windows on the west let in too much heat during the late afternoon in the summer. A number of houses with the front facing west have aluminum foil placed on the inside of those windows during the summer. Just because the road runs north–south, the long axis of a passive solar home should not be oriented along a north–south axis.

In cold climates (average winter temperatures of −7°C to 0°C), the south window area should be between 0.2 and 0.4 m^2 for each 1 m^2 of floor area. In temperate climates (average winter temperatures of 2°C–7°C), the south window area should be between 0.12 and 0.26 m^2 for each 1 m^2 of floor area. With sufficient thermal mass, this will let in enough solar radiation during most of the winter to keep the interior of the building at an average temperature of 18°C–24°C. In most cases, use the larger number for window area to reduce daily fluctuations of temperature and to have enough stored energy for 3–5 days of overcast skies.

8.2.4 DIRECT GAIN

The most common system, especially for the south walls, is direct gain, and it will provide 30%–75% of the heat in the winter. That means in the spring and fall it could provide all the heat needed. All the south wall could be windows; however, in general that would not be a good design as it would provide too much heat (some sunrooms overheat) and would have a higher heat loss at night, and then there is the cost differential of windows versus insulated wall. Design must include interior surfaces with colors to eliminate or reduce glare.

To retrofit an existing building with a direct gain system is difficult, unless the building has a clear southern exposure and has some thermal mass, or it is fairly easy to increase the thermal mass (see Example 8.2). One way to increase thermal mass is to use water storage; however, water storage has some potential problems, primarily freezing and leaks.

8.2.5 THERMAL MASS

The thermal mass for the storage of solar heat can be in the interior walls and floors constructed of masonry materials (minimum of 10 cm thickness; however, 20 cm is preferred for less temperature fluctuations and for 3–5 days storage during overcast days). For a direct gain system, to reduce glare and store the heat, the direct sunlight can be diffused over the surface area by using a translucent glazing material, by a number of small windows that admit sunlight in patches, or by reflecting direct sunlight off a light-colored interior surface onto other surfaces. For the thermal mass, use a medium-dark color for masonry floors, any color for masonry walls, light color for construction of little thermal mass, and do not cover the thermal mass floors with insulation (carpets). However, some small rugs may be all right as long as most of the sunlight is being absorbed. If interior water storage is used, then lightweight construction can be used for walls and floors.

Mazria [1] gives three case studies for thermal mass with direct gain system which are listed as follows:

Case 1: A dark-colored concrete mass is placed against the rear wall or in the floor in direct sunlight. The surface area of the concrete exposed to direct sunlight over the day is 1.5 times the area of the glazing. Space temperature fluctuation during the day was about 40°F (p. 137).

Case 2: A dark-colored concrete mass is placed against the rear wall or in the floor in direct sunlight. An increase in masonry thickness beyond 20 cm (8 in.) results in little change in system performance. The surface area of the concrete exposed to direct sunlight over the day is 3 times the area of the glazing. The temperature fluctuation during the day is 25°F (p. 138).

Case 3: The entire space, walls and floor, becomes the thermal mass. The surface area of the concrete exposed to direct sunlight is 9 times the area of the glazing. Sunlight strikes a white surface first and then diffuses over the entire space. An increase in masonry thickness beyond 10 cm (4 in.) results in little change in system performance. The temperature fluctuation during the day is 13°F (p. 139).

An interior water wall for heat storage for a direct gain system should be located so that it receives direct sunlight from south facing windows. The surface of the container exposed to direct sunlight should be a dark color, at least 60% solar absorption. Use around 300 liters of water for each 1 m² of window area. The Alternative Energy

Institute had some water storage tubes (see Figure 8.5), however top covering was not good, food dye stratified, and tubes collected dust.

Thermal mass for an indirect gain system (glazing, storage wall, and then living space) could be a Trombe wall or a water wall (see Section 5.4.2). The system would be south facing, double-glazed, and thermal storage (masonry or water). In cold climates (average winter temperatures of −7°C to 0°C), the south glazing area should be between 0.4 and 1.0 m² for 1 m² of floor area for masonry wall and 0.3 and 0.8 m² for 1 m² of floor area for water wall. In temperate climates (average winter temperatures of 2°C–7°C), the south glazing area should be between 0.22 and 0.6 m² for 1 m² of floor area for masonry wall and between 0.16 and 0.43 m² for 1 m² of floor area for water wall. The outside face of the wall should be a dark color. In cold climates, vents (see Figure 5.8) are placed at the top and bottom of a masonry wall to increase the system performance. The ratio of the area of each row of vents to the wall area should be approximately 1 to 100. Dampers are used on the upper vents to prevent reverse airflow at night.

If a masonry wall for thermal storage is exposed to the exterior, insulation needs to be placed on the exposed side. At the perimeter of foundation walls, use 5 cm of rigid waterproof insulation from top to 40 to 60 cm below grade.

8.2.6 CLERESTORIES AND SKYLIGHTS

Clerestories admit direct gain into spaces that are not on the south side of the building. The ceiling of the clerestory and reflecting walls should be in a light color. In general, skylights, light pipes, and light chimneys are used for daylighting. One problem with skylights is heat gain in the summer as it is more difficult to shade skylights.

8.2.7 SUNROOM (SOLAR ROOM) AND ATTACHED GREENHOUSE

Sunrooms and attached greenhouses are commercially available, and there are many kits and design plans for those who want to do the construction. The solar gain from the sunroom or attached greenhouse can be enough to heat that structure and the home, especially during fall and spring. However, the sunroom or attached greenhouse has major problems in controlling the flow of heat to the rest of the home and overheating in the summer. Without sufficient thermal mass, they will even overheat during spring and fall. Another problem with a sunroom or an attached greenhouse is that people will want to extend the growing season of plants and so they have auxiliary heating.

8.2.8 PASSIVE COOLING

Make the roof a light color or reflective material. High temperatures in attics are a major problem, so passive ventilators (which can be covered in winter) are very useful or PV powered ventilators. Now there are reflective paints available for the inside of roofs.

In climates with hot-dry summers, open the building up at night (operable windows or vents) to ventilate and cool interior thermal mass. Arrange large openings of roughly equal size so that inlets face the prevailing nighttime summer breezes and outlets are located on the side of the building directly opposite the inlets or in the low-pressure areas on the roof and sides of the building. Close the building up when you get up in the morning to keep the heat out during the daytime.

I (Nelson) used passive cooling for my home in Canyon, Texas, where the climate is temperate, semi-arid. In general, the summer nights are cool because the elevation is around 1,100 m. The average low and high temperatures, °C, in the summer months are June 16.4, 30.8; July 18.5, 32.8; and August 17.7, 31.5; and there are an average of 5.2 days/year with temperatures above 37.8°C (100°F). There is quite a bit of thermal mass since I added a passive room to my home (see Example 8.2). By opening the windows at night and closing them the first thing in the morning, the air conditioner was only used two to three times during the summer, which meant my electrical bill was low compared to my neighbors who keep their homes closed and had the air conditioner on all summer. Note that many women do not like this type of operation because of the dust that enters the home, especially in our area, which has lots of wind and is semi-arid.

In climates with hot-humid summers, open the building up to the prevailing summer breezes during the day and evening. Arrange inlets and outlets as outlined above; only make the area of the outlets slightly larger than the inlets.

8.2.9 OTHER

To improve the performance of a passive system may require some manual control, for example, opening and closing windows on a daily basis and opening and closing vents on a seasonal basis. One way to improve performance during the winter is to use movable insulation on the inside of the building and/or shutters on the outside of the building for all glazed openings to reduce heat loss when the Sun is not shining. Single glazing gives a net energy gain in temperate climates and insulation would improve that performance. In cold climates, movable insulation should always be used on single glazing. The insulation must make a tight and well-sealed cover of the glazed opening.

A horizontal reflector roughly equal in width and 1–2 times the height of a vertical glazed opening can improve performance. As noted, Steve Baer had a moveable (manual) horizontal reflector with insulation to improve performance.

Vertical glazing on the south should always have overhangs for shading of the summer sun. This still allows daylighting and reduces the direct gain. Software programs are available for calculating the amount of overhang by location and time of year.

Before designing or constructing a solar system, obtain climate information and do your background collection of renewable energy systems you might want to employ. Finally, find out types of systems installed in your area and talk to the owners or operators about the performance. The examples show a few of the homes and systems in the area of Canyon, Texas.

Example 8.1

This example is of a passive solar home (Figure 8.2) located southeast of Canyon, Texas, on Cemetery Road. The passive systems are direct gain through a sunroom and two clerestories, direct gain on bedrooms situated to the southeast and

FIGURE 8.2   Passive solar home; notice overhang for sunroom, clerestories, and bedrooms.

southwest, an entryway into the house shielded from northern breezes, and only one small window on the north. Thermal mass is from 30 cm (1 ft) wide adobe walls, with the blocks made on site. Note that the clerestories have passive shutters on the inside of the house, with skylids from Zomeworks. The heat from the Sun drives liquid from one container to another, changing the balance. Previously, the southeast and southwest bedrooms had additional indirect gain with water barrels (direct gain), which were underneath the direct gain into the room. The water barrels were removed due to corrosion, and the glazing was removed and the openings were closed as the rooms received enough heat from the direct gain. Southwest bedroom direct gain is now completely shaded by a tree during summer.

Example 8.2

This example is of a remodel of a home in Canyon, Texas. The garage at my residence was converted into living space (Figure 8.3a), which has passive heating (direct gain) due to double-pane, low-e windows. The original overhang gives shading in the summer. A slate floor absorbs sunlight, and thermal mass for storage is provided by a slate floor, concrete slab (8 in.), and a brick wall (4 ft high, 24 ft long; the brick has magnesium for higher heat capacity) on the north side of the room (Figure 8.3b). There is a wide opening (6 ft) to the rest of the home, and auxiliary heating and cooling is from an existing HVAC (heating, ventilation, air conditioning) unit. On most sunny days in the winter, the room is solar heated and even provides some heat for the rest of the home. During early fall and late spring, the room provides a substantial portion of heat for the home. During the summer, windows are opened at night for cooling and then closed by 07:00 to 08:00 in the morning, so the air conditioner is needed only a few times during the summer. Because the elevation is 1,100 m and there is low humidity, in general the nights are cool. The original home had windows on the south in the dining room and bedroom, which in the past provided passive solar heating; however, a partial shading screen over the front and a large evergreen tree in the yard have reduced this heating. For better solar performance, both should be removed.

FIGURE 8.3 (a) Outside view at solar noon, May 19; notice shade from overhang is at bottom of wall. (b) Sunlight on slate floor during winter, January 16. There is a little extra radiation due to reflection off snow. (c) Inside view, solar noon, May 19; no direct sunlight, but more than ample daylight. Notice brick wall on north for additional thermal mass.

8.3 HYBRID SYSTEMS

Hybrid systems combine one or more design concepts of solar or renewable systems to provide heat, cooling, daylighting, or electricity. In general, passive systems, either direct or indirect gain, are the most cost effective. Again, seek local examples and information on the Internet. A self-sufficient house will be a hybrid system consisting of two or more systems: photovoltaic (PV), wind, solar (hot water and heat), and maybe earth berm or even geothermal. Only high-efficiency appliances, smaller number of electronics such as televisions, substitution of manual equipment for small appliances such as electric can openers, and even a smaller number of lightbulbs are used because of the cost of off-grid electricity and cost for storage (batteries). Again, there will be many local examples, so talk with the owners to find out the best and most cost-effective systems.

FIGURE 8.4   Renewable energy demonstration project; PV yard light at right.

Detailed information is presented on the renewable energy demonstration building, Alternative Energy Institute (AEI), West Texas A&M University. The renewable energy demonstration building (Figure 8.4) was located at the AEI Wind Test Center, on the north side of the West Texas A&M University campus. In 2010, the building and Wind Test Center equipment and towers were moved to the Nance Ranch as the university needed the previous location. Information on AEI and the Wind Test Center is available at http://www.windenergy.org.

The three main objectives of the renewable energy demonstration project were as follows:

Work space for research projects at the Wind Test Center

Demonstration of renewable energy systems for builders, contractors, and architects

Education for the general public about renewable energy systems

The performance of the building was monitored, and the information was used in research projects. The metal building is divided into a 30 × 50 ft workroom on the east and a 30 × 25 ft tech room on the west. *No auxiliary sources of heating or cooling were installed*; however, the building is connected to the utility grid. The renewable energy systems are as follows:

Passive solar heating, direct gain, in workroom and tech room

Two active flat-plate (vertical), hot air solar collectors (one with glass and the other with fiberglass reinforced plastic [FRP] glazing), with a common rock storage under 80% of the workroom floor

Passive daylighting for workroom

Passive cooling in summer with manual control of windows
Solar hot water
Stand-alone PV power for a sign and three PV power yard lights
Ceiling fans for circulating air in the workroom
Generation of electricity
  A 10-kW wind turbine connected to the grid through an inverter
  A 1.9-kW PV array connected to the grid through an inverter

The passive aspects of the building were designed by L. M. Holder III, an architect. The passive solar heating and cooling and daylighting systems were selected and optimized using computer simulations. The anticipated maximum temperature was lowered 5°F, and the minimum temperature was raised 10°F using the simulation. Using the computer model, it was ascertained that R19 insulation in the walls and R30 insulation in the roof would be optimum. Due to budgetary constraints, the insulation in the walls of the workroom was limited to R13 batts.

The active hot air solar collectors were designed by Jimmy Walker, extension agent, Oldham County, who also assisted AEI in their construction. AEI personnel installed the rest of the systems—solar hot water, wind turbine, PV array, stand-alone PV for lighting and sign—along with providing the design and development of data acquisition and storage and real-time display of the performance of the building and electric generation.

8.3.1 BUILDING

The east–west building is oriented 15° from south, so the south-facing windows and hot air collectors capture more early morning gain during the winter. An air infiltration barrier (Tyvek) was placed around the perimeter of the building; however, the construction contractor did not use proper methods for installation. A ventilated ridge and soffit vent along with a painted radiation barrier on the underside of the roof reduces the impact of the Sun during the summer and decreases black sky radiation on clear winter nights. On the north side of the building, there is a minimum number of windows, which are placed at a higher level, primarily for nighttime flushing of hot air in the summer. All windows are double pane, low emissivity with wood frames. The foundation is insulated down to 2 ft with 2-in thick Styrofoam board. Because of improper maintenance, the frames of some of the wood windows rotted, and the windows had to be replaced.

8.3.2 PASSIVE HEATING AND COOLING

The tech room has only passive heating with direct gain. The room has a dropped ceiling and 6 in. of insulation with R19 batt, and the walls are covered by 1/4-in. wood paneling. The thermal mass consists of water tubes next to the south windows (Figure 8.5), the concrete slab floor (6 in.), and in addition a concrete block wall filled with vermiculite on the north (Figure 8.6). Because of more insulation and the dropped ceiling, this room is around 5°F warmer on cold days in the winter than the workroom.

FIGURE 8.5 Water tubes for more thermal mass and added food color became stratified. Note water level due to evaporation because of poor cover. Lower sash on windows has manual opening.

FIGURE 8.6 Concrete block wall filled with vermiculite, 6 ft high, on north wall. High windows on north and low windows on south open for nighttime flushing during summer.

In the tech room, two windows on the west and the large window on the north were for viewing of wind turbines at the Wind Test Center. The area of these windows was above the window area recommended from the design simulation, which would have reduced west and north window area to a minimum.

The workroom has direct gain windows, daylighting, and operable windows on the south and north for cooling. The thermal mass for the workroom is a thick concrete slab for the floor and the rock storage. The lower section of the direct gain windows and high windows on the north are operated manually for passive cooling during late afternoon and through the night. All windows should be closed early in the morning. Manual control is a problem when personnel are not in the building at the end of the working day or the next morning. Manual control by students or personnel who do not have an interest in the project is even more of a problem.

The large east-facing windows in the workroom are for daylighting and passive gain during the winter. They were shaded with angled slats for the summer to keep out the early morning Sun; however, they still provide daylighting. Ceiling fans provide increased air movement inside the workroom during the summer.

8.3.3 Solar Hot Air System

Two active solar air collectors (234 ft² area) are part of the south face of the workroom (Figure 8.7). One collector has tempered glass glazing (west) and the other FRP glazing (east), as can be seen in Figure 8.4. The FRP has become less translucent over the 20 years of exposure to the Sun. Circulating fans, activated by a differential temperature controller, connect to plastic pipe in the rock bed, which consists of 1- to 1.5-in. washed river rock 1 ft deep and with an area of 1,250 ft². The 6-in. diameter inlet and outlet pipes (polyvinyl chloride, PVC) are 6 ft apart, with 1-in. diameter holes on 12-in. centers on each side. Even though there is one set of pipes for each collector in the rock storage, the rock storage is shared thermal mass. Additional thermal mass consists of an 8-in. concrete slab over the rock bed.

The collector (Figure 8.8) consists, from front to back, of glazing, 2-in. dead air space, selective surface copper absorber, 2-in. air channel, duct board insulation with aluminum foil, and 3/4-in. plywood, which is the interior wall.

8.3.4 Solar Hot Water

A 32-ft² flat-plate collector is mounted on the roof (see Figure 8.4). The storage tank has a heat exchanger for the circulating fluid for the solar collector, and there is no auxiliary heater for the hot water tank since the demand for hot water is low. The solar tank and the two inverters are housed in the tech room and are visible through patio doors.

8.3.5 Daylighting

There is a row of daylighting windows along the top of the workroom. These give direct gain in the winter and are shaded by an overhang in the summer. With the direct gain windows, the daylighting windows, the white insulation, and the metal

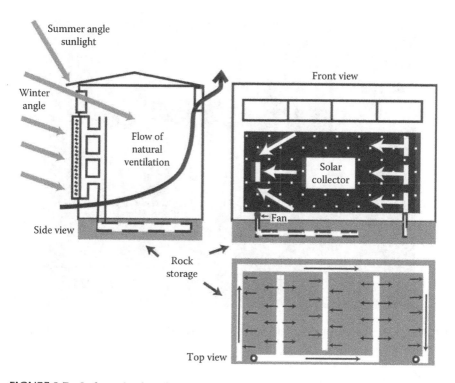

FIGURE 8.7 Left, angle of sunlight for summer and winter plus diagram of airflow at night during summer. Right, diagram of one of the hot air collectors and flow of air during winter collection of solar heat.

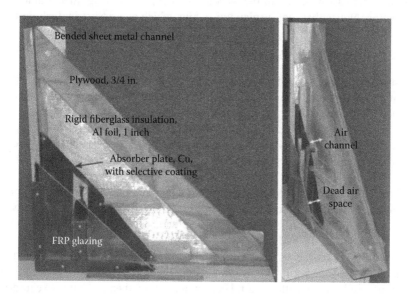

FIGURE 8.8 Cross section of hot air collector.

FIGURE 8.9 Interior of workroom showing that electrical lights are not needed due to daylighting.

beams painted white, lights are only necessary at night, even on overcast days (Figure 8.9). At 15:30 on June 24, the south wall of the building is completely in the shade (no direct sunlight striking the surface) because of the orientation of the building. The east windows in the workroom also provide a significant amount of daylighting. Notice that roof insulation is white, which also assists in daylighting.

8.3.6 ELECTRICAL GENERATION

A wind turbine (10 kW) and a PV array (1.9 kW) are connected in parallel with the building load through inverters (Figure 8.10). Two utility meters (ratchet) measure the energy purchased from the utility company and the energy fed back to the grid. Since installation in 1991 through 2010, the project was a net electrical energy producer, at a ratio of around 4 to 1. This does not count the displaced energy used on site, which was much larger when an electric van was part of the load at the building (ratio of 3/1).

The wind turbine is a 10-kW permanent magnet alternator with variable voltage and frequency. It is connected to the utility grid though an inverter (12 kW), which changes the output to constant frequency. The wind turbine is mounted on an 80-ft lattice tower next to the building.

A PV 1.9-kW$_p$ array consists of two panels (each panel has sixteen 60-W modules) mounted on the roof with 4-in. clearance underneath for cooling. The two panels are connected in series with a neutral common between them. The positive terminal on one module and the negative terminal on the other module are connected to the

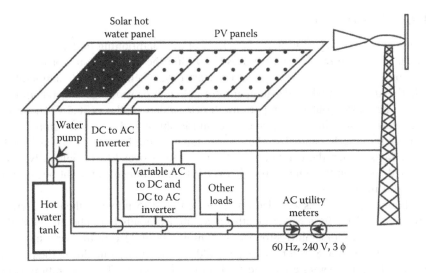

FIGURE 8.10 Diagram of electrical generation.

inverter, which converts the DC (direct current) power to constant frequency for the grid.

There are three stand-alone PV systems for yard lights, which use high-efficiency, low-wattage fluorescent bulbs. One stand-alone PV system provides fluorescent lighting for the demonstration project sign (Figure 8.11), which is lit for 4 h per night. These stand-alone systems have a PV panel, controller, and battery. The yard lights have been a problem, with the battery and lights needing to be replaced within 6 years.

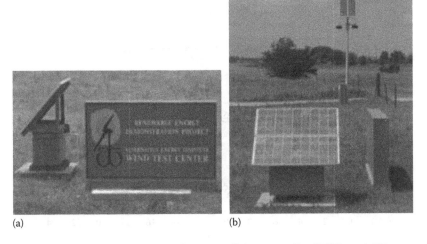

(a)   (b)

FIGURE 8.11 (a) Sign illuminated by fluorescent light powered by (b) PV panel. PV-powered yard light is in the background.

8.3.7 Performance

There is no auxiliary heating in the winter or air conditioning in the summer for the building. The inside temperatures are within the comfort zone for the functions of the two areas. In the winter, there are days when the outside temperature is 0°F and periods when the temperature is below or near freezing for up to a week. The average outside temperature for December 1998 was 36.3°F. However, in general during the winter, when the Sun is shining, the outside temperature will rise into the 40s and 50s. The average temperature for July 1998 was 77.8°F, and in the summer, there are days with temperatures over 100°F. Generally, even on hot days when the temperature is over 100°F, it cools off at night because of the 1,100-m elevation.

The temperature was sampled and averaged over 15 minutes, then an average day was calculated for a winter month and a summer month. For February 1998, the average outside temperature dropped to a minimum of 32°F in the early morning (Figure 8.12). As the Sun comes up, the building starts to heat up due to the passive solar, and the temperature reached 60°F in the afternoon. The temperature again drops in the evening hours. There is not much difference between the tech and workroom temperatures, but the workroom temperature is always lower because of the higher ceilings and less insulation in the walls. However, there is more thermal mass in the workroom due to the rock storage.

For August 1998, the temperature of both the workroom and tech room reached 87°F in midafternoon, which is only 3°F cooler than the average outside temperature (Figure 8.13). However, it must be remembered that, on the days when the afternoon temperature is even higher, the inside temperature is still around 87°F. Manual nighttime flushing was not done during this period, as can be seen from the cooler outside temperatures from 22:00 at night to 08:00 the next morning. If this had been done, then the inside temperatures during the day would have been lower.

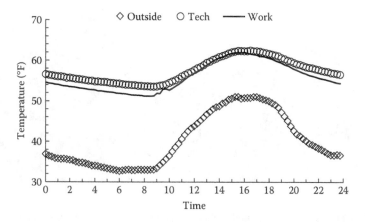

FIGURE 8.12 Average temperature by time of day for February 1988.

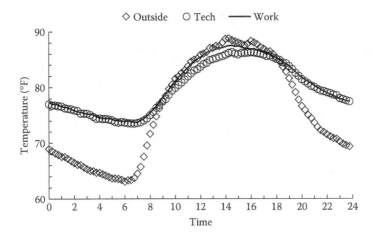

FIGURE 8.13   Average temperature by time of day for August 1988.

Example 8.3

An off-grid home near Amarillo, Texas (Figure 8.14), is discussed here. The home is 150 m² (1,600 ft²); the exterior walls were built from tires (rammed earth fill), with passive space heating and cooling (Figure 8.15). The following renewable energy systems were incorporated into the home:

Electric: 0.5-kW PV, 0.9-kW wind turbine, electricity stored in eight deep-cycle golf cart batteries, 4-kW inverter for AC (alternating current) power.
Space heating: Direct gain: thermal mass, concrete slab floor, wood stove backup for cold, cloudy days (around 1 cord/year).
Space cooling: Overhang shades windows in summer; no air conditioning. Generally, nights are cool, so the house can be opened during the night. Because of the thermal mass, the highest summer temperature was 28°C (83°F).

FIGURE 8.14   Off-grid home near Amarillo, Texas. South windows for passive, direct gain; PV; wind turbine; and tank for rainwater collection. (Courtesy of David Stebbins.)

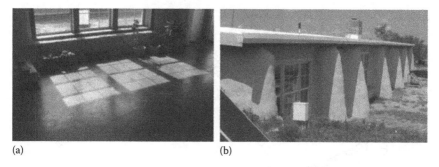

(a)   (b)

FIGURE 8.15 (a) Winter sun inside and (b) summer sun completely shaded by overhang. (Courtesy of David Stebbins.)

> *Solar hot water*: Indirect system, solar power pump, 0.2-m³ (50-gal) tank, 6-gal propane RV (recreational vehicle) heater backup for extended cloudy periods.
> *Rainwater collection*: Metal roof drains into two 6-m³ (1,600-gal) poly tanks; water is pumped through two filtration systems and an ultraviolet purifier. Water from carport drains into another 6-m³ tank for unfiltered water for vegetable garden. Annual average rainfall is 48 cm (19 in.).

Energy use is around 2 kWh/day. As stated, energy conservation is important for off-grid homes, so energy conservation measures include compact fluorescent bulbs, ultraefficient refrigerator, home design that eliminates air conditioning and electric or gas heating, and a front-loading washing machine. Phantom loads are eliminated, such as digital clocks, DVD players, televisions, microwaves, and wireless telephones. The owner has all those items and uses them but plugs them into a power strip that is turned off when the appliance is not in use.

8.4 ACTIVE SYSTEMS

On-demand hot water for residential and commercial users consumes over 3 EJ of energy per year in the United States. Solar hot water systems are the most common active systems, with over 1.5 million installed in homes in the United States, with around 6,000 added annually. Since solar hot water only supplies 2% of total energy now used for heating water, the opportunity for growth is large. System paybacks are 5 to 10 years when displacing electric hot water heaters. Hawaii now requires solar hot water on all new single-family dwellings.

Other countries have higher rates of installation; for example, 9 out of 10 homes in Cyprus have solar hot water, and Israel has around 40% of homes with solar hot water, and laws require all new residences to install solar hot water heaters. Tropical climates are a natural for passive thermosiphon systems. The International Energy Agency estimates that by 2050 solar heating could supply around 16% (8.9 EJ) of the world's total energy use for heating, equivalent to a capacity of around 3,500 GW$_t$. Solar collectors for low-temperature process heat in industry (<120°C) could reach a capacity of 3,200 GW$_t$, with an additional capacity of 200 GW$_t$ for swimming pools.

8.5   UNDERGROUND HOMES

A number of underground homes have been built, so generally you can find one or more homes (Figures 8.16 and 8.17) within a local region. The advantages of underground homes or earth-sheltered homes are as follows:

Conservation of energy
Small range of temperature variation (both heating and cooling)
Low maintenance
Noise insulation
Hazard protection (tornadoes and security, but a fire exit is needed)
Insurance ought to be less
High-density development possible, efficient use of land, can use land unsuitable for general residential housing
Less dust and longer life

(a)

(b)

FIGURE 8.16   (a) Boatwright home, Canyon, Texas, and (b) the northern side of home.

(a)

(b)

FIGURE 8.17   (a) Markham home on Cemetery Road, south of Canyon, Texas, and (b) northern side of home.

The disadvantages are as follows:

Cost of construction is 10%–20% more. Stronger construction is needed for roof to support weight of earth.

Proper water barrier is needed.

Leaks are hard to find from the inside as actual location of leak can be some distance away.

Quality installation of butyl sheet or neoprene membranes (bituthene) on outside of concrete is needed.

Marketability and financing are a problem.

There is visual isolation; the back rooms only have artificial lights.

Obtaining a contractor may be difficult.

The type of soil must be known. Clays will expand and shrink with water absorption to give large forces.

Notice that on the Markham underground home (Figure 8.17) that the large overhang on the south eliminates passive solar for heating. Also, original construction of concrete had too much heat loss on the south, so concrete was covered with 10 cm Styrofoam and then with stucco. One underground home in the area had so many problems with water leaks that they built a roof over the home. Water leaks were due to poor construction techniques, and entrance of leaks on the outside is difficult to find from the inside.

FIGURE 8.18 (a) Carlson home on Hunsley Hills Road, Canyon, Texas. This is an earth-bermed (west and north side) home built on an abandoned quarry (cheap land). Note that front of house has passive solar, and entrance is a vestibule. (b) View from southwest, front of house is to the right. (c) View to the northwest.

An earth berm can also save energy as it provides a heat sink for part of the structure (Figure 8.18). Heat transfer through the earth for 20 to 30 cm dampens the daily air temperatures, while 2 m of earth essentially gives a constant temperature. Soil temperatures 2 m below ground can be obtained from the local office of the Soil and Water Conservation Service.

There are a number of considerations in designing and building an underground home. On the roof and near-exposed areas near the surface, 5 to 10 cm of Styrofoam insulation is needed. Roof spans need to be less than 7.5 m and can be steel, wood, concrete, or precast concrete. The internal walls need to be load bearing for the extra weight. There is a need to get light into the structure by windows (it is good to have south-facing glazing for solar heating), skylights, or an atrium. An open courtyard in the middle would provide daylighting for all areas of the home.

8.6 COMPUTER SOFTWARE

A number of computer software tools are available. Many tools for whole-building analysis, including renewable energy, are available from the Energy Efficiency and Renewable Energy (EERE) Network, Department of Energy. Architects and builders will also have software for home designs that incorporate solar systems.

RETScreen International provides free software for almost all aspects of renewable energy [2].

The RETScreen Clean Energy Project Analysis Software is a unique decision support tool developed with the contribution of numerous experts from government, industry, and academia. The software, provided free of charge, can be used worldwide to evaluate the energy production and savings, costs, emission reductions, financial viability and risk for various types of Renewable-energy and Energy-efficient Technologies (RETs). The software (available in multiple languages) also includes product, project, hydrology and climate databases, a detailed user manual, and a case study based college/university-level training course, including an engineering e-textbook.

The software has been expanded from renewable energy, cogeneration, and district energy to include a full array of clean power, heating and cooling technologies, and energy efficiency measures. Climate data have been expanded to cover the entire surface of the planet, including central-grid, isolated-grid, and off-grid areas, and software has been translated into 35 languages that cover roughly two-thirds of the population of the world.

8.7   OTHERS

8.7.1   STRAW BALE HOUSE

In this house, the straw bales, which can be load- or non-load-bearing walls, are plastered on the inside and outside. The advantages are as follows:

Superinsulation, with R values as high as R50
Good indoor air quality and noise reduction
Speedy construction process (walls can be erected in a single weekend)
Construction costs as low as \$10 per ft^2 (depending on owner involvement)
Natural and abundant renewable resource used that can be grown sustainably
   in one season

One problem that has to be avoided is moisture collecting in the straw, which will result in decomposition (rot). Another problem is that bales must be sealed against rodents and even insects.

8.7.2   ADOBE AND RAMMED EARTH

Adobe is used throughout the world, even in wet climates, and is popular in the southwestern United States. Adobe, in combination with passive solar design, makes for effective heating in cold winter areas. The use of high-mass walls, insulation, and passive solar can reduce energy use in January by 60% or more. High-mass earth walls also cut cooling costs in hot desert locales.

Rammed earth is a process by which walls of a house are constructed by compacting an earth-cement mixture into forms. The forms are then removed, leaving solid earth walls 45–60 cm thick, which can be load- or non-load bearing (http://www.rammedearthworks.com/).

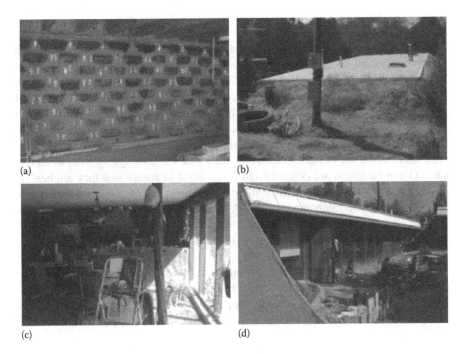

(a)   (b)

(c)   (d)

FIGURE 8.19 Zavatsky home, next to Camp Harrington on Ranch Road 1541 near Canyon, Texas. Passive solar home (17 by 6 m) and adjoining workshop (9 by 6 m). Both have tire walls and are bermed on three sides; roofs have R55 insulation and are designed for catching rainwater. (a) Unfinished tire wall in workshop. Notice cans for space filler. (b) Roof of home showing the berm; cistern catchment on left. (c) Interior of the home. (d) Exterior of home; notice overhang for shading during the summer. Photos taken March 27, 2004.

8.7.3 TIRE HOUSES, EARTHSHIPS

Tire houses are constructed from used tires filled with earth (Figure 8.19); however, a lot of labor is needed to pack the tires with earth. Outside and inside can then be plastered with a mixture of adobe and cement. Sometimes, bottles and cans are used for space filler for non-load-bearing walls.

8.7.4 DOUBLE-ENVELOPE HOUSE

Few double-envelope houses have been built because of higher construction costs and problems with performance. The design employs a double envelope with a continuous airspace of at least 15–30 cm on the north wall, south wall, roof, and floor. The south wall is heated by the Sun, and during the day, the circulating air warms the inner envelope of the house and gives up heat to the earth or rock storage in the crawl space. At night, the thermal mass releases the heat. The east and west walls are single, conventional walls.

8.7.5   GREEN BUILDING

A significant change is that now green building has become marketable, with builders, associations, and architects promoting green building. The World Green Building Council is a network of national councils from over 100 countries [3], with over $1.2 * 10^9 \, m^2$ of building space registered (2014). Of course, a major aspect of green building is to use solar energy within the local climate. Green buildings are resource efficient in terms of energy and construction materials, including recycled, renewable, and reused resources to the maximum extent practical [4]. Energy efficiency means the use of passive solar and other renewable energy systems when feasible. Green buildings are designed and constructed to ensure that they are healthy for their occupants and are typically more comfortable and easier to live with due to lower operating and maintenance costs. A number of programs are available; again, there is a lot of information on the Internet. The Green Home Building site (http://www.greenhomebuilding.com/index.htm) has information on all aspects of different types of homes: earth, straw, and so on.

The Leadership in Energy and Environmental Design (LEED) is a green building certification program to transform how buildings and communities are designed, constructed, operated, and maintained (http://www.usgbc.org/leed). In general, solar home tours will have LEED homes on the tour from certified to platinum level.

REFERENCES

1. E. Mazria. 1979. *The Passive Solar Energy Book*. Rodale Press, Emmaus, PA.
2. RETScreen International. http://www.retscreen.net/ang/home.php.
3. World Green Building Council. http://www.worldgbc.org/index.php?cID=1.
4. U.S. Green Building Council. *An Introduction to LEED and Green Building*. http://go.usgbc.org/Intro-to-LEED.html.

RECOMMENDED RESOURCES

LINKS

Passive Solar Design

Passive Solar Design. http://www.nmsea.org/Passive_Solar/Passive_Solar_Design.htm.
Passive Solar Design. http://apps1.eere.energy.gov/buildings/publications/pdfs/building_america/29236.pdf.
Passive Solar Design Guidelines for Northern New Mexico. http://www.nmsea.org/Education/Homeowners/Detailed_Passive_Solar_Guidelines.php.
Sun Plans. http://www.sunplans.com.

Examples of Solar Homes

Connective Loop for Transporting Heat. http://www.ecohome.com/hernikl/.
Maine Solar House. http://www.solarhouse.com/main.htm.
North Carolina Solar Center. http://www.ncgreenbuilding.org/site/ncg//index.cfm?
Passive House Case Studies. http://www.duluthenergydesign.com/Content/Documents/GeneralInfo/PresentationMaterials/2014/Day2/coulca-passhou.pdf.

Residential Case Studies of Passive Strategies. http://www.slideshare.net/aiahouston/williams-residential-case-studies.

Superinsulation. http://www.motherearthnews.com/green-homes/green-building-techniques-zmaz86sozgoe.aspx#axzz3Ew3BOLTG.

Components

Natural Home Building Sources. http://www.thenaturalhome.com/passivestuff.html.

Sierra Solar Systems. http://www.sierrasolar.com/ssstore/store.htm.

Solar Components Corporation. http://www.solar-components.com/solprod.htm.

Solar Hot Water

Florida Solar Energy Center. http://www.fsec.ucf.edu/en/consumer/solar_hot_water/index.htm.

IEA. http://www.iea.org/publications/freepublications/publication/technology-roadmap-solar-heating-and-cooling.html.

Solar hot water. http://energy.gov/energysaver/articles/solar-water-heaters.

Underground Homes

Home Sweet Earth Home. http//www.undergroundhomes.com/home.html.

Performance Building Systems. http://www.earthshelter.com.

R. Roy. Earth-sheltered houses. *Mother Earth News.* http://www.motherearthnews.com/green-homes/earth-sheltered-homes-zmaz06onzraw.aspx#axzz3Ew3BOLTG.

A sampling of homes by Davis Caves. http://www.daviscaves.com/homes.htm.

Underground Buildings: Architecture and Environment. http://www.subsurfacebuildings.com/default.asp.

Software and Tools

Building Energy Software Tools Directory. http://apps1.eere.energy.gov/buildings/tools_directory/.

Design Tools from Sustainable Design. http://www.susdesign.com.

DOE sponsored tools. http://apps1.eere.energy.gov/buildings/tools_directory/doe_sponsored.cfm.

R. Hendron. December 2008. *Building America Research Benchmark Definition*, technical report NREL/TP-1550–44816. http://apps1.eere.energy.gov/buildings/publications/pdfs/building_america/44816.pdf.

OTHER RESOURCES

S. Banks and R. Heinichen. 2006. *Rainwater Collection for the Mechanically Challenged.* Tank Town, Dripping Springs, TX.

Design and Construction of a Straw Bale House, Lowell, VT. http://www.cavedogs.com/building2.html.

D. Easton. 1996. *The Rammed Earth House.* Chelsea Green, White River Junction, VT.

R. A. Ewing and D. Pratt. 2009. *Got Sun? Go Solar.* Pixiejack Press, Masonville, CO.

Harvesting, Storing and Treating Rainwater for Domestic Indoor Use. Go to Alliance for Water Efficiency, http://www.allianceforwaterefficiency.org/default.aspx, search documents.

Home Power Magazine. http://www.homepower.com.

W. H. Kemp. 2009. *The Renewable Energy Handbook.* Aztext Press, Tamworth, ON, Canada.

S. O. MacDonald and M. Myhrman. 1998. *Build it with Bales: A Step by Step Guide to Straw Bale Construction.* 2nd ed. Treasure Chest Books, Tucson, AZ.

E. Paschick and P. Hendricks. 1995. *The Tire House Book.* Sunstone Press, Santa Fe, NM.

J. Schaeffer. 2007. *Solar Living Sourcebook.* Galam Real Goods, Louisville, CO.

Straw bale. http://www.strawbale.com.

A. Swentzell Steen, B. Steen, and D. Bainbridge, with D. Eisenberg. 1994. *The Straw Bale House.* Cheslea Green, White River Junction, VT.

PROBLEMS

There are two 9-ft (2.75-m) by 5-ft (1.5-m) windows for passive solar heating at my home. The wall is 8 ft (2.4 m) tall; the bottoms of the windows are 2 ft (0.6 m) above the ground. The windowpane is 53 in. (1.24 m) tall as the frame takes up some space.

8.1. Estimate the width of overhang for the window to be completely shaded at solar noon on August 31. Do for a latitude of 35° N, 102° W, in Canyon, Texas. For design information, http://www.susdesign.com, go to overhang design tools.

8.2. Compare advantages and disadvantages of tire houses and straw bale houses.

8.3. What should be the orientation and relative shape for a passive solar home?

8.4. If you were building a solar home, what systems would you use? Why?

8.5. For an off-grid home, what are the primary considerations for use of electric energy in the home?

8.6. For an off-grid home, what are the primary considerations for space heating and cooling in the home?

8.7. How many underground or earth-sheltered homes are in your area? Pick one and describe it.

8.8. The new term is a *zero-energy home or net zero home.* What is it? Describe it.

8.9. Are there any green builders in your area? Get information on one and briefly describe the builder's product or service.

8.10. What was the coldest outside temperature during the working day for the AEI demonstration building?

8.11. What was the hottest outside temperature during the working day for the AEI demonstration building?

8.12. You want to design and build a solar hot water system. Describe the system and components and note the source of the design information.

8.13. Find an example of a Trombe system, give its location, describe the system size, and estimate the performance or obtain information on actual performance. Do not use examples given in the text.

8.14. What are the two major problems in constructing an underground home?

8.15. For a home for 4 people, what size solar hot water system would you need? Estimate cost and payback time. Hint: http://energy.gov/energysaver/articles/estimating-cost-and-energy-efficiency-solar-water-heater.

8.16. For new home construction in your country, what renewable systems should be promoted? Give reasons for your answer as if you were head of the national energy department.

9 Wind Energy

9.1 INTRODUCTION

Before the Industrial Revolution, wind was a major source of power for pumping water, grinding grain, and long-distance transportation (sailing ships). Even though the peak use of farm windmills in the United States was in the 1930s and 1940s, when there were over 6 million, these windmills are still being manufactured and used in the United States and around the world [1].

The advantages and disadvantages of wind energy are similar to most other renewable energy resources. The advantages are it is renewable (nondepleting) and ubiquitous (located in many regions of the world) and does not require water for the generation of electricity. The disadvantages are that it is variable and a low-density source, which then translates into high initial costs. In general, windy areas are distant from load centers, which means that transmission is a problem for large-scale installation of wind farms. Other perceived disadvantages are visual, noise, and mortality of birds and bats.

The main applications are the generation of electricity and water pumping (Table 9.1). Except for the installed capacity for wind farms, the other numbers are best estimates, as data are difficult to acquire. Applications for the generation of electricity are divided into the following categories: utility-scale wind farms; small wind turbines (≤100 kW) distributed-community; wind-diesel; village power (generally hybrid systems); and telecommunications (high-reliability hybrid systems) and street lighting. Of course, there will be overlap between the categories. For example, another aspect of utility scale is cooperatives or individual farmers installing wind turbines and selling to the grid. Large businesses have also installed utility-scale projects, some are independent power producers and others are distributed wind. There are wind hybrid systems and some wind power systems for village power. In some cases, village power has diesel/gas for the backup. Many village power systems use photovoltaic panels with battery storage, 1–3 days. Stand-alone systems generally have batteries for storage.

Wind-assist systems are where two power sources work in parallel to produce power on demand, and stand-alone systems. All wind turbines connected to the utility grid are wind-assist systems. In terms of size, wind turbines range from the utility-scale megawatt turbines for wind farms to small systems (stand alone and connected to the grid) to the 20–300 W remote units for sailboats and households, primarily in the developing world. Some people refer to these as micro wind turbines. Be skeptical of vendors claiming that micro and small wind turbines could produce electricity for utility scaled cheaper than large wind turbines, as all you need to do is to connect a large number of them together.

The rapid growth of wind power has been due to wind farms with 369 GW installed by the end of 2014; in addition, there are around 1.5 GW from other applications. There will be overlap between large and small wind turbines in the diverse applications of

TABLE 9.1

Wind Energy Installed in the World, Estimated Numbers and Capacity, End of 2014

Application	No.	Capacity (MW)
Wind turbines (primarily in wind farms)	210,000	369,000
Around 10,000 MW in offshore		
Distributed[a]-community	2,000	1,500
Wind diesel	300	100
Village power	2,100	50
Small wind turbines	900,000	~1,000
Telecommunication	3,500?	2–5
Farm windmill[b]	310,000	Equivalent, 155
Production ~ 4,000/year		

[a] The overlap between distributed wind turbines and wind farm installations is difficult to distinguish. For example, in Denmark, the large number of distributed units is counted as part of the national capacity.

[b] Farm windmills are being replaced by electric and PV pumps. Production primarily replaces 30- to 40-year-old windmills.

distributed and community wind, wind-diesel, and village power (primarily hybrid systems). Wind turbines for stand-alone and grid-connected systems for households and small businesses, telecommunications, and water pumping are primarily small wind turbines. The numbers installed and capacity are estimates, with better data for wind farms and rougher estimates for the other applications.

As of 2014, over 100 countries had installed wind power as most countries are seeking renewable energy sources and have wind power as part of their national planning. Therefore, countries have wind resource maps, and others are in the process of determining their wind power potential, which also includes offshore areas.

During the 1930s, small wind systems (100 W to 1 kW) with batteries were installed in rural areas; however, these units were displaced with power from the electric grid through rural electric cooperatives. After the first oil crisis in 1973, there was a resurgence of interest in systems of this size, with the sale of refurbished units and manufacture of new units. Also as a response to the oil crisis, governments and utilities were interested in the development of large wind turbines as power plants for the grid. Then, starting in 1980s the market was driven by distributed wind in Denmark and the wind farm market in California, which led to the significant wind industry today.

9.2 WIND RESOURCE

The primary difference between wind and solar power is that power in the wind increases as the cube of the wind speed:

$$\frac{P}{A} = 0.5 * \rho * v^3, \ \text{W/m}^2 \tag{9.1}$$

where:
  ρ is the air density
  v is the wind speed

The power/area is also referred to as wind power density. The air density depends on the temperature and barometric pressure, so wind power will decrease with elevation, around 10% per 1,000 m. The average wind speed is only an indication of wind power potential, and the use of the average wind speed will underestimate the actual wind power. A wind speed histogram or frequency distribution is needed to estimate the wind power/area. For citing of wind farms, data are needed at heights of 50 m and maybe even up to hub heights of 100 m. Since wind speeds vary by hour, day, season, and even years, 2–3 years of data are needed to have a decent estimate of the wind power potential at a specific site. Wind speed data for wind resource assessment are generally sampled at 1 Hz and averaged over 10 min (sometimes 1 h). From these wind speed histograms (bin width of 1 m), the wind power/area is determined.

9.2.1 WIND SHEAR

Wind shear is the change in wind speed with height, and the wind speed at higher heights can be estimated from a known wind speed. Different formulas are available [2, Section 3.4], but most use a power law.

$$\frac{v}{v_0} = \left(\frac{H}{H_0} \right)^{\alpha} \tag{9.2}$$

where:
  v is the estimated wind speed at height H
  v_0 is the known wind speed at height H_0
  α is the wind shear exponent

The wind shear exponent is determined from measurements; in the past, a value of 1/7 (0.14) was used for stable atmospheric conditions. Also, this value meant that the power/area doubled from 10 to 50 m, a convenient value since the world meteorological standard for measurement of wind speed was a height of 10 m.

  In many continental areas, the wind shear exponent is larger than 0.14, and the wind shear also depends on the time of day (Figure 9.1), with a change in the pattern from day to night at a height around 40 m. This means that wind farms will be producing more power at night when the load of the utility is lower, a problem for the value of the energy sold by the wind farm. There is more power in the wind at 40 m and higher heights than determined by data taken at a height of 10 m (Figure 9.2). The pattern of the data at 50 m for Washburn (not shown on graph) was similar to the 50-m data at White Deer; however, there was some difference between the two

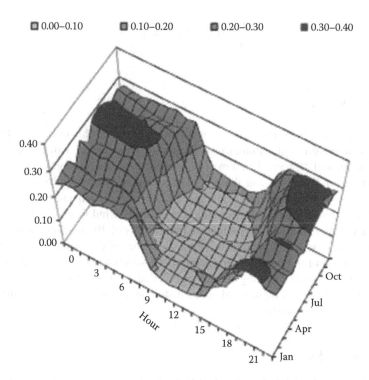

FIGURE 9.1 Annual average wind shear, 10–50 m, by month and time of day for Dalhart, Texas.

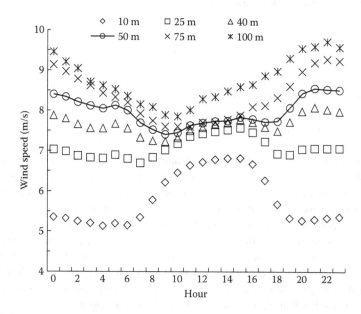

FIGURE 9.2 Average annual wind speed by time of day for White Deer, Texas (10, 25, 40, 50 m), and Washburn, Texas (75, 100 m).

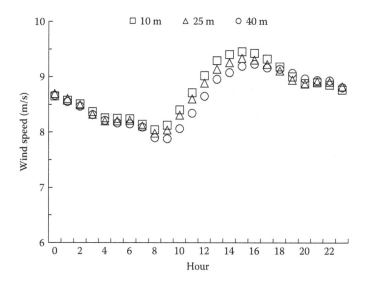

FIGURE 9.3 Average annual wind speed by time of day for Guadalupe Pass, Texas.

sites. Both sites were in the plains, around 40 km apart. This shows that wind power is fairly site specific. Wind data for wind farms have to be taken at heights of at least 40 m to 50 m as at these heights and above the wind pattern will be same, and the wind speeds at higher heights at the same site can be estimated using Equation 9.2. There are some locations, such as mountain passes, where there is little wind shear (Figure 9.3), so taller towers for wind turbines would not be needed.

9.2.2 WIND MAPS

Wind power maps (W/m²) are available [3] for many countries, regions, and states or provinces within countries. Early maps were for a height of 10 m with an estimate for 50 m using the power law for wind shear (Equation 9.2) and a wind shear exponent of 0.14. The wind power map for the United States (Figure 9.4) shows large areas with wind class 3 and above. More detailed state maps are available. In addition, wind power potential is estimated using geographic information systems (GISs) with land excluded due to urban areas, highways, lakes and rivers, wildlife refuges, and land at a distance from high-voltage transmission lines. The wind power potential is large, so it is not a question of the wind energy resource but a question of locations of good-to-excellent wind resource, national and state policies, economics, and amount of penetration of wind power into the grid. For example, the capturable wind power potential of Texas is estimated at 223,000 MW [4], which is much larger than the 110,000 MW generating capacity of the state. Note that wind power capacity in Texas at the end of 2014 was around 14,000 MW, 13% of the total.

Computer tools for modeling the wind resource have been developed by a number of groups: the National Wind Technology Center, National Renewable Energy Laboratory (NREL) in the United States; National Laboratory for Sustainable

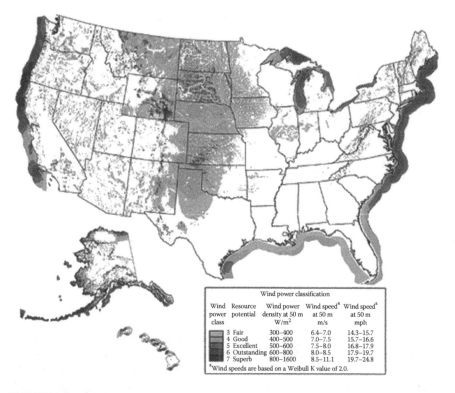

Wind power classification				
Wind power class	Resource potential	Wind power density at 50 m W/m²	Wind speed[a] at 50 m m/s	Wind speed[a] at 50 m mph
3	Fair	300–400	6.4–7.0	14.3–15.7
4	Good	400–500	7.0–7.5	15.7–16.6
5	Excellent	500–600	7.5–8.0	16.8–17.9
6	Outstanding	600–800	8.0–8.5	17.9–19.7
7	Superb	800–1600	8.5–11.1	19.7–24.8

[a]Wind speeds are based on a Weibull K value of 2.0.

FIGURE 9.4 Wind power map at 50 m height for the United States. Notice wind classes. (Data from National Wind Technology Center, NREL, http://www.nrel.gov/wind/resource_assessment.html.)

Energy (RISO) in Denmark; other government laboratories; and private industry. The Wind Atlas Analysis and Application Program (WAsP) has been employed in over 100 countries and territories around the world. Now, revised wind power maps (for heights of 50 and 80 m) are available that used terrain, weather balloon data, and computer models [5]; the maps were also verified with available data at heights of 50 m. These maps showed regions of higher-class winds in areas where none was thought to exist. Also, because of the larger wind shear than expected, more areas have suitable winds for wind farm development. Remember that 2%–7% accuracy in wind speeds means a 6%–21% error in estimating wind power, so data on site are still needed for locations of wind farms in most areas. These maps are a good screening tool for wind farm locations. Interactive wind speed maps by 3Tier [6] and AWS-Truewind [7] are available online for many locations in the world. Wind Atlases of the World contains links for over 50 countries [8].

Complete coverage of the oceans is now available using reflected microwaves from satellites [9,10]. Ocean wind speed and direction at 10 m are calculated from surface roughness measurements from the daily orbital observations mapped to a 0.25° grid; these are then averaged over 3 days, a week, and a month. Images of the data can be viewed on websites for the world, by region or selected area. Ocean

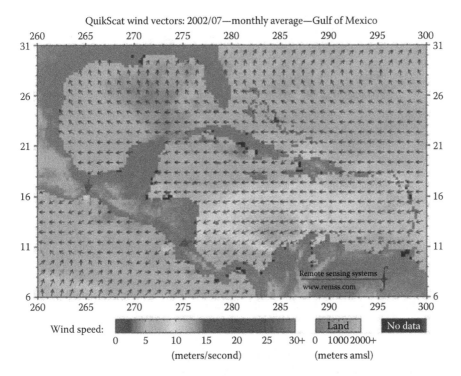

QuikScat wind vectors: 2002/07—monthly average—Gulf of Mexico

FIGURE 9.5   Ocean winds for July 2002. Two arrows on land indicate excellent onshore wind regions.

winds are not available within 25 km of the shore as the radar reflections of the bottom of the ocean skew the data.

Ocean winds will indicate onshore winds for islands, coasts, and some inland regions of higher winds (Figure 9.5). There are now wind farms in the Isthmus of Teohuantepec, Mexico, and the Arenal region of Costa Rica, where the northeast trade winds (average wind speeds of 10 m/s) are funneled by the land topography.

9.3   WIND TURBINES

Wind turbines are classified according to the interaction of the blades with the wind, orientation of the rotor axis with respect to the ground and to the tower (upwind, downwind), and innovative or unusual types of machines. The interaction of the blades with the wind is by drag, lift, or a combination of the two.

For a drag device, the wind pushes against the blade or sail, forcing the rotor to turn on its axis, and drag devices are inherently limited in efficiency since the speed of the device or blades cannot be greater than the wind speed. The maximum theoretical efficiency is 15%. Another major problem is that drag devices have a lot of material in the blades. Although a number of different drag devices (Figure 9.6) have been built, there are essentially no commercial (economically viable) drag devices in production for the generation of electricity.

FIGURE 9.6 Drag device with cup blades, similar to anemometer. (Courtesy of Charlie Dou.)

Most lift devices use airfoils for blades (Figure 9.7), similar to propellers or airplane wings; however, other concepts are Magnus (rotating cylinders) and Savonius wind turbines (Figure 9.8). A Savonius rotor is not strictly a drag device, but it has the same characteristic of large blade area to intercept area. This means more material and problems with the force of the wind on the rotor at high wind speeds, even if the rotor is not turning. An advantage of the Savonius wind turbine is the ease of construction.

Using lift, the blades can move faster than the wind and are more efficient in terms of aerodynamics and use of material, a ratio of around 100 to 1 compared to a drag device. The tip speed ratio is the speed of the tip of the blade divided by the wind speed, and lift devices typically have tip speed ratios around seven. There have even been one-bladed wind turbines, which save on material; however, most modern wind turbines have two or three blades. The power coefficient is the power out or power produced by the wind turbine divided by the power in the wind. A power curve shows the power produced as a function of wind speed (Figure 9.9). Because there is a large scatter in the measured power versus wind speed, the method of bins (usually 1 m/s bin width suffices) is used.

Wind turbines are further classified by the orientation of the rotor axis with respect to the ground: horizontal axis wind turbine (HAWT) and vertical axis wind turbine (VAWT). The rotors on HAWTs need to be kept perpendicular to the wind, and yaw is the rotation of the unit about the tower axis. For upwind units, yaw is by a tail for small wind turbines and a motor on large wind turbines; for downwind units, yaw may be by coning (passive yaw) or a motor.

VAWTs have the advantage of accepting the wind from any direction. Two examples of VAWTs are the Darrieus and Giromill. The Darrieus shape is similar to the

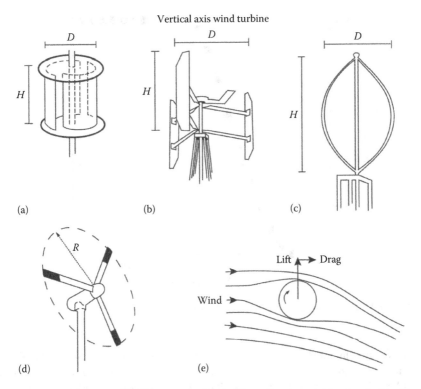

FIGURE 9.7 Diagram of different rotors for horizontal and vertical axis wind turbines: (a) Savonius, (b) Giromill, (c) Darrieus, (d) horizontal axis wind turbine, and (e) Magnus effect.

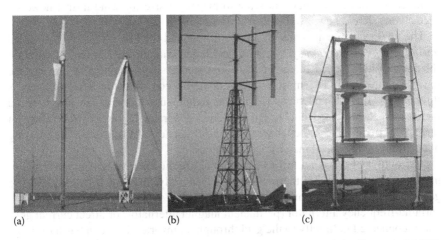

FIGURE 9.8 Examples of different wind turbines. (a) HAWT, 10 m diameter, 25 kW; Darrieus, 17 m diameter, 24 m tall rotor, 100 kW. (b) Giromill, 18 m rotor diameter, 12.8 m high, 40 kW. (c) Savonius, 10 kW. (Courtesy of Gary Johnson.)

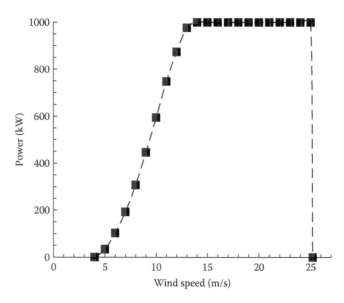

FIGURE 9.9   Power curve for a 1-MW wind turbine.

curve of a moving jump rope; however, the Darrieus is not self-starting as the blades have to be moving faster than the wind to generate power. The Giromill can have articulated blades that change angle, so it can be self-starting. Another advantage of VAWTs is that the speed increaser and generator can be at ground level. A disadvantage is that taller towers are a problem for VAWTs, especially for units of wind farm size. Today, there are no commercial, large-scale VAWTs for wind farms, although there are a number of development projects and new companies for small VAWTs. Some companies claim they can scale to megawatt size for wind farms; however, such claims should be viewed with skepticism.

The total system consists of the wind turbine and the load, which is also called *a wind energy conversion system*. A typical large wind turbine consists of the rotor (blades and hub), speed increaser (gearbox), conversion system, controls, and the tower (Figure 9.10). The most common configuration for large wind turbines is three blades, full-span pitch control (motors in hub), upwind with yaw motor, speed increaser (gearbox), and doubly fed induction generator (allows a wider range of revolutions per minute for better aerodynamic efficiency). The nacelle is the covering or enclosure of the speed increaser and generator.

The output of the wind turbine, rotational kinetic energy, can be converted to mechanical, electrical, or thermal energy. Generally, it is electrical energy. The generators can be synchronous or induction connected directly to the grid or a variable-frequency alternator (permanent magnet alternator) or direct current generator connected indirectly to the grid through an inverter. Enercon has built large wind turbines with huge generators and no speed increaser, which have higher aerodynamic efficiency due to variable revolution-per-minute operation of the

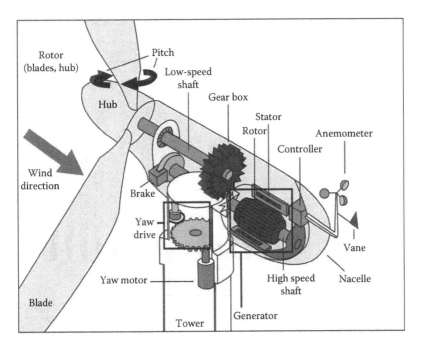

FIGURE 9.10 Diagram of main components of large wind turbine.

rotor. However, there are some energy losses in the conversion of variable frequency to the constant frequency (50 or 60 Hz) needed for the utility grid. There are a number of megawatt wind turbines with direct drive, permanent magnet generators.

9.4 WIND FARMS

The development of wind farms began in the early 1980s in California with the installation of wind turbines ranging from 20 to 100 kW as those were the only sizes available in the commercial market. This development of wind farms in California was due to U.S. federal laws and incentives (1980–1985) and due to the avoided costs for energy set by the California Public Utility Commission for electricity generated by those wind farms. As the wind farm market in the world continued, there was a steady progression toward larger-size wind turbines due to economies of scale; today, there are commercial multimegawatt units, with the larger ones developed for offshore.

Since then, other countries have supported wind energy, and by the end of 2014, there were 369 GW installed (Figure 9.11) from around 210,000 wind turbines. This capacity of wind turbines will generate around 800 TWh/year using a capacity factor of 25% (around 4% of total electricity generated). The primary regions in the world with installed wind are Europe, China, and North America (Table 9.2).

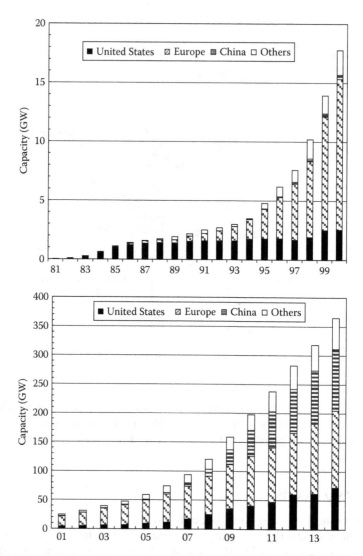

FIGURE 9.11 Wind power installed in the world, primarily wind farms.

Installation of wind turbines in Europe was led by Denmark in the early days, and its manufacturers captured a major share of the world market in the 1980s. Then, other Europe countries installed large numbers of wind turbines, and Germany became the world leader. In addition, there was consolidation of manufacturers, with both Germany and Spain becoming major players. In 2007–2008, the major wind farm installations shifted back to the United States, with a large number also installed in China (Figure 9.12). China is now ranked second in the world behind Europe and expects to continue installing large numbers of wind turbines in wind farms in the coming years due to the large increase in demand for electricity; wind

TABLE 9.2
World Installed Wind Capacity (GW) by Region for 2007, 2011, and 2014

	2007	2011	2014
Africa and Middle East	1	1	2.5
Asia	10	98	142
Europe	57	109	134
Latin America and Caribbean	1	4	8.5
North America	19	66	78
Pacific	1	3	4.4
World total	94	283	369

Source: Global Wind Energy Council.

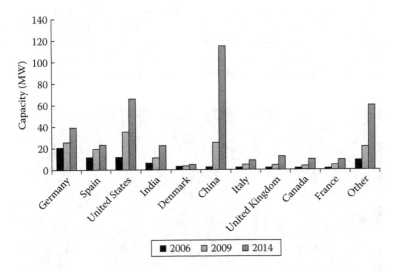

FIGURE 9.12   Cumulative wind power installed by country at the end of the year, primarily wind farms.

farms would also reduce the number of new coal plants due to the requirement of more nonfossil energy in the premier energy consumption for climate change. The United States led in number of wind turbines installed and electricity generated by wind in 2014 ($7.3 * 10^4$ TWh); however, wind energy accounted for only 1.3% of the total electricity generated. Note that wind power accounted for 31% of new electric power-generating capacity in the United States from 2009 to 2014. Other countries obtain a larger share of their electric demand from wind, and Denmark is the leader as over 30% of its electricity comes from wind power.

By the end of 2014, there were 74 offshore wind farms in Europe, 8,045 MW from 2,488 wind turbines, average size = 3.2 MW (some offshore installations are using ≥5 MW wind turbines). There are another 12 projects under construction with a capacity of 2,900 MW. Installations offshore are due to the high cost of land and the excellent wind resource. Information on European key trends and statistics is available from the European Wind Energy Association and the Global Wind Energy Council. Europe could reach 40,000 MW of offshore wind power by 2020 with new technology for deeper water and adequate regulations and incentives. In China, the first 100 MW offshore wind farm was completed in 2010, and around 1,000 MW were installed by end of 2014. Thirty GW of offshore wind farms are planned for by the year 2020. Offshore wind farms are being planned for other parts of the world (e.g., in the United States off the East Coast, off the Texas Gulf Coast, and in the Great Lakes).

There are economies of scale for installation of wind turbines for wind farms, and in general, most projects need 30–50 MW to reach this level. The spacing for wind turbines is three to four rotor diameters within a row and eight to ten rotor diameters from row to row. On ridgelines and mesas, there would be one to two rows with a two-rotor diameter spacing within a row. In general for plains and rolling terrains, the installed capacity could be 5–10 MW/km^2 and for ridgelines 8–12 MW/ linear km. Spacing for offshore may be larger, up to 10 rotor diameters between turbines; however, the units will be larger so capacity would be around 5 MW/km^2. Satellite images show the layout of wind farms (Figure 9.13); however, the maps

FIGURE 9.13   Satellite view of layout of part of the Sweetwater wind farm (south of Sweetwater, Texas). Notice distance between rows is larger than distance between wind turbines within a row. Rows are perpendicular to predominant wind direction during the summer (lower wind months).

may not show the latest installations. In Texas in 2009, there were five wind farms over 500 MW, and the largest was Roscoe at 782 MW. Large global wind farms in 2014 are Gansu (7,965 MW) in China, Alta Wind Energy Center (1,548 MW) in California, and Jaisalmer Wind Park (1,064 MW) in India, and the London Array (630 MW) in the United Kingdom is the largest offshore wind farm.

Wind power in the world grew at an average rate of 29% from 1995 through 2009; however, the rate declined to 12% in 2013 and rebounded to 16% in 2014. These numbers demonstrate the problem of exponential growth. Growth rates decline due to the base becoming larger, and at some point, a linear increase will become the norm, and finally, the production will be for repowering and replacement, essentially little growth. Texas surpassed California in installed capacity in 2008, and with over 14 GW installed by the end of 2014, Texas continues to lead the United States (Figure 9.14) in installed capacity [11]. Notice that the small number for installed capacity in 2013 is due to the late extension of the production tax credit.

There have been a number of estimates for the future installed capacity in the world, generally which have been on the low side. For example, one estimate in 2008 predicted a 100% increase to 240 GW by 2020 (Note that was surpassed in 2014). That projection was low compared to other projections, and it would be equivalent to a growth rate of around 7% per year, which is smaller than growth rates in the past 5 years for wind power. Of course, the growth rates of 30% will not continue and even a linear increase is possible.

Our estimate is that the world wind capacity will reach 600 GW by 2020, an increase of 230 GW over 2014. This estimate is due to changes in national policies promoting wind power, primarily in the United States, China, and Europe. Also, the estimate was based on continuing incentives for renewable energy and the

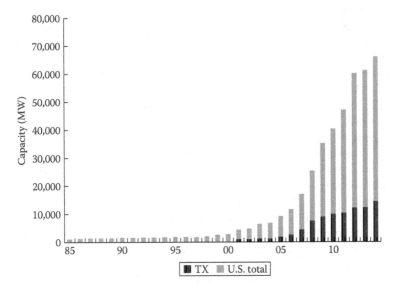

FIGURE 9.14 Wind power installed in the United States, primarily wind farms.

construction of high-voltage transmission lines from windy areas to load centers. Wind energy could produce 20% of U.S. electricity by 2030 [12], and if solar, bioenergy, and geothermal energy were included, then renewable energy would provide an even larger percentage of U.S. demand for electricity. The new mandate for China is 150 GW of wind power by 2020, and Europe plans an additional 100 GW by 2020. It is assumed that the rest of the world will install at least 100 GW by the end of 2020. The prospects for the wind industry are excellent, and this does not count the increased numbers of distributed, community, and small wind turbines.

Market forecasts for wind power were seen as overly optimistic at the time of the prediction and then were exceeded every time by the actual amount of installations. World wind capacity grew by over 170% over the 5 years from 2005 through 2009. In 2009, the Global Wind Energy Council forecast 409 GW by the end of 2014 [13] which can be compared to the actual value of 369 GW.

9.5 WIND INDUSTRY

Wind turbines for wind farms increased from the 100-kW to megawatt size due to economies of scale. There were two different tracks for the development of wind turbines for wind farms. The first was research and development (R&D) plus demonstration projects of large wind turbines for utility power in the 1970s and 1980s, primarily funded by governments. Only prototypes were built and tested [2, see Table 10.11]. The second track was wind turbines in the 50- to 100-kW size built by private manufacturers [14] to meet the distributed market in Europe and for wind farms in California. The manufacturers of the second track were successful in developing the modern wind turbine industry, while the units developed primarily by aerospace companies did not make it to the commercial stage.

There were a number of different designs built and sold in the wind farm market in California, including Darrieus wind turbines. In the United States, the most common designs for two blades were fixed pitch, rotor downwind, teetered hub, and induction generator, and for three blades were variable pitch, rotor downwind, and induction generator, of which U.S. windpower built over 4,000 units for the California wind farm market. In Europe, the three-blade, fixed-pitch, upwind rotor model was the predominant design. Now, the three-blades, upwind rotor, full-span, variable-pitch design with a wider range of revolutions per minute is the major type for wind farms. Enercon has a wind turbine with a large generator and no gearbox.

Today, wind turbines are available from one to eight megawatts with rotor diameters of 60 to over 100 m installed on 60- to over 100-m towers. Manufacturers designed and built wind turbines in the 5- to 10-MW size, primarily for offshore installations. The top 15 manufacturers had over 80% of the installed capacity of 35.5 GW in 2013 (Table 9.3). European manufacturers still maintain a major share of cumulative installed capacity; however, of the top 15 manufacturers, 7 were from China, 6 from Europe, and 1 each from the United States and India. Vestas is still the leader in terms of production and global installed capacity (over 62 GW, 52,000 turbines); however, there are a number of manufacturers with production of 3–4 GW

TABLE 9.3

**Estimation of Global Cumulative Capacities and Numbers Plus
2013 Market Share for Manufacturers of Large Wind Turbines[a]**

	Cum GW 2010	Share 2013 %	Cum GW 2014[a]	Cum Share %	Cum Turbines 1,000[a]
Vestas (DK)	45.5	13.1	62	17	52
Goldwind (PRC)	9.1	10.3	23	6	15
Enercon (GE)	22.6	10.1	33	9	22
Siemens (DK)	13.5	8.0	21	6	13
Suzlon Group (IND)	17.3	6.3	24	7	15
GE Wind (US)	26.9	4.9	28	8	18
Gamesa (S)	21.8	4.6	30	8	15
United Power (PRC)	2.4	3.9	12	3	8
Mingyang (PRC)	1.8	3.7	8	2	5
Nordex (GE)	7.0	3.4	11	3	6
XEMC (PRC)	1.1	3.2	6	2	3
Envision (PRC)	0.4	3.1	5	1	3
Sinovel (PRC)	10.0	2.3	17	5	11
Dongfang (PRC)	6.4	2.3	10	3	6
Sewind (PRC)	1.1	2.2	5	1	3
Others	18.8	18.6	70	20	25
Total	205		369		222

[a] Data obtained from manufacturers' websites. China data provided by Shi Pengfei, honorary chairman China Wind Energy Association.

per year. The domestic wind turbine industry in China accounted for 93% of the 2013 China market. The United States has only one manufacture (GE Wind) in the top fifteen, and Suzlon (India) was the other non-European and non-China manufacture, although in previous years, Mitsubishi (Japan) was one of the top 15 manufacturers. Notice with the large megawatt units that the number of units installed is around the half the number of megawatts installed. One problem with the data presented is the absence of good information on repowering of wind farms and decommissioning of old wind turbines. Other major international companies are buying existing manufacturers of wind turbines or starting to manufacture wind turbines for the wind farm market.

As an example of a large wind turbine installation, a Vestas V90, rated at 3 MW, 90-m diameter on an 80-m tower is located north of Gruver, Texas (Figure 9.15). Twenty trucks were needed to haul an 800-metric ton crane to the site, and another 10 trucks were needed for the turbine and tower. The weight of the components were as follows: nacelle = 70 metric tons, rotor = 41 metric tons, and tower = 160 metric tons. The foundation required 460 m^3 of concrete and over 40 metric tons of rebar.

FIGURE 9.15   Vestas V90 3-MW wind turbine. Notice minivan next to the tower.

9.6 SMALL WIND TURBINES

There are a number of different configurations and variations in design for small wind turbines (watts to 100 kW). Many of the small wind turbines have a tail for both orientation and control in high winds. There is an overlap of small wind turbines with village power systems, distributed-community systems, and wind diesel. Most of the wind turbines for village power are less than 100 kW, and it is also the same for some of the distributed and wind-diesel systems. So the numbers reported for the production and installation of small wind turbines will include all areas. In the United States and Europe, much of the small wind turbine capacity is grid connected, while in China and other developing countries, most are stand-alone systems for households, 50–300 W. A very rough estimate for global number of small wind turbines (100 kW or less) that has been produced is over 1,100,000 units with a capacity around 1,100 MW, very impressive numbers that show the impact of small wind. For numbers currently installed, reduce that by 200,000, and reduce capacity by 100 MW because of replacement, upgrade to larger turbine, lifetime, and operational failures (see Table 9.1). The rough estimates are due to unknown accuracy of the production in China and due to the difficulty of obtaining data from manufacturers in all parts of the world with small production. In all the reported data, the average size of units has been increasing.

TABLE 9.4
Estimation of Small Wind Turbine Production as of 2013 by Region[a]

	Major Companies	Cumulative Number	Cumulative MW	2014 Avg Size KW
China	34	845,000	790	0.9
United States	4	181,000	248	3.0
Europe	? 8	80,000	115	2.6
Other	? 4	15,000	15	
Total		1,121,000	1,168	

[a] Includes units that are exported.

Another unknown in China is how much of the present production is for replacement of old wind turbines and how many are upgrades from the 50–100 W size units to 200–500 W. So the estimate for China used a reduction of 30% on reported numbers and then a reduction of 100,000 due to lifetime of old turbines and replacement. Even in 2011, the largest number produced was for 300 W wind turbines and the average size was 0.76 KW.

There are approximately 330 manufacturers [15]; however, the number of major manufacturers is around 50 with over 20 in China (Table 9.4). The largest production was in China, with 116,000 units in 2011, of which 57,000 were exported [16], primarily to other Asian countries. In the past, most of the Chinese production was 50–100 W wind turbines for remote households (Figure 9.16). The small wind turbines

FIGURE 9.16 Small wind turbine, 50 W, isolated household, Inner Mongolia, China. Notice rope on tail for manual control; however, this control mechanism is not recommended by manufacturer.

provide enough electricity for a couple of lights, a radio, and a small black-and-white TV. Marlec and Ampair in the United Kingdom and Southwest Windpower (in 2013, not in business) in the United States have produced large numbers of micro and small wind turbines, and units for sailboats are a big market.

The trend for the installation of small wind turbines for the three major countries (China, the United States, and the United Kingdom) was the same from 2005 through 2013, increased numbers and increased size (Figure 9.17). The UK Market Report [17] includes wind turbines from 100 to 500 kW. Over the period from 2004 through 2011, 2,549 units were mounted on buildings 10% of the number installed. Kestrel, South Africa, has sold over 1,000 wind turbines from 2010 to 2014, and two-thirds of them are exported to other countries.

The National Wind Technology Center (NWTC), National Renewable Energy Laboratory (NREL), has a development and testing program for small wind [18]. The American Wind Energy Association has a distributed section (formerly small wind), which includes reports by year from 2007 through 2012 [19]. The U.S. Roadmap (2002), now outdated for distributed wind, estimated that small wind could provide 3% of U.S. electrical demand by 2020.

The American Wind Energy Association promulgated a program for the independent certification of small wind turbines; a working group was established in 2006, and then in 2008, the Small Wind Certification Council (SWCC) was established [20]. SWCC was established because of the problems for adoption of small wind turbines, which included the following: non-standardized performance specifications and optimistic and inconsistent claims by suppliers; lack of consumer-friendly tools to compare small wind turbines and to accurately estimate energy performance; need for greater assurance of safety, functionality, and durability for consumers and agencies providing financial incentives; and field testing conducted for less than half of the small wind turbine models on the market.

The SWCC is an independent certification body that certifies small wind turbines that meet or exceed the performance and durability requirements of the AWEA standard. The Small Wind Turbine Performance and Safety Standard of AWEA is a common North American framework. SWCC issues labels for rated annual energy output, rated power, and rated sound level and also confirms that the turbine meets

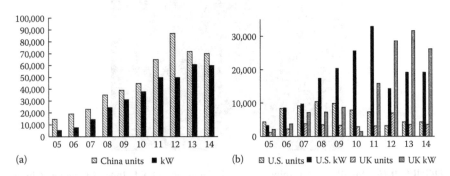

FIGURE 9.17   (a) Production of small wind turbines in China. (b) Production of small wind turbines in the United States and United Kingdom.

(a)   (b)

FIGURE 9.18 PV-wind powers streetlight and flashing red lights at stop signs, McCormick Road on I27 between Canyon and Amarillo, Texas.

durability and safety requirements. SWCC publishes power curves, annual energy performance curves, and measured sound pressure levels. The rated power is at 11 m/s, and the rated annual energy is calculated from an annual average wind speed of 5 m/s (11.2 mph), a Rayleigh distribution [2, Section 3.11], and the wind turbine power curve.

A fairly new market for small wind turbines is for street lighting (Figure 9.18). Even though there may be a transmission line nearby, the cost of the transformer, meter and billing, and electricity is more than the cost of electricity from a PV-wind system. Sizes of the PV-wind systems are around 50 W for the PV and 100 W for the wind turbine.

Another new market is for telecommunications (Figure 9.19). It is estimated there are over 3,000 telecommunication stations having small wind turbines as part of the power supply. In China alone, there were around 2,500 wind turbines installed for telecommunication sites from 2009 to 2012. The number of telecomm sites will be less as sites could have more than one wind turbine, for example, one site had 16, 1 kW units. This is a growing market due to the increased use of cellular phones, especially in the more remote areas of the world. Check the websites of different manufacturers for examples of these two applications.

9.7 DISTRIBUTED-COMMUNITY WIND

Distributed systems are the installation of wind turbines on the retail side of the electric meter for farms, ranches, agribusiness, and small industries. In the United States at the end of 2013, there were approximately 72,000 units with a capacity around 850 MW [19]. Note that these numbers include small wind turbines connected to the grid, which are also counted in small wind. The installations will

FIGURE 9.19 Two 1.5-kW wind turbines at telecommunication system, China. (Courtesy of Charlie Dou.)

range from one or more small units to mid-size (100–1,000 kW) to megawatts for large industries. As an example, two 10 kW units were installed at a liquor store near Amarillo, Texas, and in Lubbock, Texas, the American Wind Power Center and Museum installed a 660 kW unit and a cottonseed oil plant installed ten 1 MW units. Distributed wind turbines for farmers, ranchers, and agribusinesses will be somewhat similar to the farm implement business, for example, a large tractor costs over $200,000. Average costs for installations ranged from $2,500/kW for utility-scale turbines, $2,800/kW for mid-sized turbines, $7,000/kW for small wind turbines, and $4,100/kW for refurbished small wind turbines.

The definition of community wind varies and again there is overlap in reported number of turbines and capacity for utility scale, small wind, and distributed wind. The American Wind Energy Association (AWEA) defines community wind as projects that incorporate local financial participation and control. Another definition is that community wind are projects using turbines over 100 kW in size and completely owned by villages, towns, cities, commercial customers, and farmers, but excluding publicly owned or municipal utilities. In this definition, the three 100 kW wind turbines at North Texas University and the eight 60 kW units at three school districts near Lubbock, Texas, would be classified as distributed wind, not community wind.

Community wind projects leverage local distribution grids for the economic and environmental benefit of the local community, while they provide investment opportunities for community members to become financial partners. As with larger wind farms, community wind brings rural economic development with local participation.

In the United States, the market for farm/industrial/business and community was estimated at 3,900 MW by 2020. Distributed wind systems will have an impact especially on smaller utilities and electric cooperatives. The international market is difficult to measure, as much of that market would be in village power and remote systems; however, for farm/industrial/business, the market was estimated at 600 MW by 2020.

AWEA has a Community Wind U.S. Projects Database (2011–2012) and in this database, installed numbers and capacity for the large projects and large (>100 kW) wind turbines will also be in the AWEA quarterly market reports. For example, the Wind Ranch (78.2 MW; 34, 2.3 MW wind turbines) of the Golden Spread Electric Cooperative in the Panhandle of Texas is listed in the community wind database. Others such as Kotzubue and Kodiak, Alaska, are examples of wind-diesel projects.

Windustry has a section on community wind [21] that includes overview, projects, and toolbox. The toolbox offers practical information for farmers and rural landowners looking to develop commercial-scale projects. The core content came from the *Community Wind Development Handbook* [22], which was developed on behalf of the Rural Minnesota Energy Board. *Community Wind 101* [23] is a primer for policymakers and clean energy advocates that looks at economic benefits, examines obstacles facing community wind developers, and has some examples.

9.8   VILLAGE POWER

Village power is another large market for small wind turbines as approximately 1.3 billion people do not have electricity; extension of the grid is too expensive in rural and remote areas with difficult terrain. There are around 2,000 village power systems with an installed capacity of 55 MW. Village power systems are minigrids, which can range in size from small microgrids (<100 kWh/day, ~15 kW) to larger communities (tens of megawatt hours per day, hundreds of kilowatts). Today, there is an emphasis on systems that use renewable energy (wind, photovoltaic [PV], mini- and microhydro, bioenergy, and hybrid combinations). These systems need to supply a reliable, however limited, amount of energy, and in general, much of the cost has to be subsidized. The other components of the system are controllers, batteries, inverters, and possibly diesel or gas generators. In windy areas, wind turbines are the least-cost component of the renewable power supply, and one or multiple wind turbines may be installed in the 10- to 100-kW range. There are software programs for modeling hybrid systems. Notice the primary difference between village power and wind–diesel is in the size of the system, although there will be overlap.

China leads the world in installation of renewable village systems (around 1,250), of which 100 include wind [2, Section 10.6.1]. Their Township Electrification Program in 2002 installed 721 village power systems with a capacity over 15 MW (systems installed: 689 PV, 57 wind/PV, and 6 wind). An example is the village power system (54 kW) for Subashi, Xinjiang Province, China (Figure 9.20), which

FIGURE 9.20 China village power system (PV/wind/diesel), 54 kW. (Courtesy of Charlie Dou.)

consisted of 20-kW wind, 4-kW PV, 30-kVA diesel, 1,000-Ah battery bank, and a 38-kVA inverter. The installed cost was $178,000 for power and minigrid, which is reasonable for a remote site.

9.9 WIND DIESEL

For remote communities and rural industry, the standard for electric generation is diesel power. Remote electric power is estimated at 12 GW, with 150,000 diesel gensets ranging in size from 50 to 1,000 kW. In many locations, these systems are subsidized by regional and national governments.

Diesel generators have low installed costs; however, they are expensive to operate and maintain, especially in remote areas, and major maintenance is needed from every 2,000 to 20,000 h, depending on the size of the diesel genset. Even with diesel generators, for many small villages electricity is only available for a few hours in the evening. Costs for electricity were in the range from $0.20 to $0.50/kWh; however, as the cost of diesel increases, the cost per kilowatt hour increases.

Wind turbines can be installed at existing diesel power plants at a low (fuel saver as the diesel does not shut down), medium, or high penetration (wind power supplies more of the load, which results in better economics as diesel engines may be shut down). The wind turbines may be part of a retrofit, an integrated wind-diesel system, or wind/PV/diesel hybrid systems for village power. Rough estimates indicate that there are over 220 wind-diesel systems in the world, ranging in size from 100 kW to megawatts.

Data on wind-diesel turbines installed and commissioned from the Alaska Energy Authority [24] show that through 2012 there were 63.8 MW and 144 turbines installed at 27 locations, starting in July 1997. Manufacturer of the wind turbine, rated power, date commissioned, and costs are also provided, with a wide range of installed costs from $3,000 to over $20,000 per kW. Cost of diesel was around $1.20/L ($5/gal), which translates to $0.38/kWh for the fuel cost, assuming 3.4 kWh/L (13 kWh/gal) from the genset. There were 48.5 MW installed in 2012 primarily due, eleven 1.5 MW wind turbines at Fire Island and twelve 2 MW wind turbines at Eva Creek. Pillar Mountain was a hydro-wind-diesel system with a 1.5 MW wind turbine, which provides around 10% of the power. Online performance of the wind-diesel systems is available which includes power curves. Unalakleet expects to displace 90,000 gallons of diesel fuel per year from 600 kW of wind turbines.

Wales, Alaska, had a high-penetration system (Figure 9.21) with battery storage. The system at Wales has not worked properly since 2006 as problems have arisen with the high penetration and it essentially needs to be reworked. The University of Alaska, Fairbanks, has a wind-diesel applications center, and two important aspects are a wind-diesel best practices guide and a wind hybrid simulator. The simulator includes two 200 kW diesel gensets, 100 kW wind turbine, wind turbine

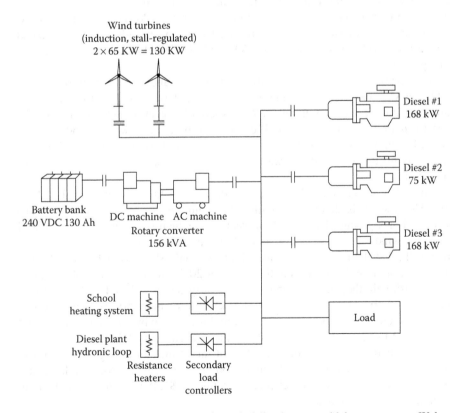

FIGURE 9.21 Diagram of high penetration, wind diesel system with battery storage, Wales, Alaska.

simulator, grid simulator, battery bank, secondary load control, synchronous condenser, inverter, controls, and of course data acquisition.

A case study of a wind/diesel system [25] is provided by the Kotzebue Electric Association (KEA), Alaska, whose grid has five diesel generators with a combined capacity of 11.04 MW. The annual average load is about 2.5 MW, with a peak load around 3.9 MW, and the minimum load is around 1.8 MW. Loads are greatest during the winter months for heating and lighting. KEA maintains a high reserve capability to prevent loss of power during the winter. Critical electrical loads include the regional hospital, airport, and water system, and critical heating loads include the town water supply. Typically, KEA runs one generator continuously during the winter, with the rest as backup. In 2012, KEA used around, 5.3 million liter (1.4 million gal) of diesel fuel, with an average efficiency of 3.8 kWh/L (14.6 kWh/gal). The fuel costs for the diesel generators were estimated at $0.23/kWh, as the delivered diesel cost was $0.80/L ($3.39/gal). Fuel costs are around 60% of the operational cost. KEA receives its annual fuel supply during the short summer season when the river is navigable by barge, and the diesel is stored in two 3.78 million liter (1 million gallon) steel tanks.

The current wind farm (2012) consists of 19 turbines due to the addition two EWT (900 kW) wind turbines in 2012, which increased the capacity to 2.94 MW. This is a high penetration system and storage and/or dump loads are needed to allow excess power to be absorbed and then released during peak loads or used for thermal applications, so KEA plans to install a 500 kW/3.7 MWh flow battery and electrical dump loads.

The first three turbines were installed in July 1997, and seven more turbines were added in May 1999. The 10 wind turbines (Atlantic Orient, 66 kW, 15 m diameter) are located on a relatively flat plain 7 km south of Kotzebue and 0.8 km from the coast (Figure 9.22). The site is well exposed to the easterly winter winds and the westerly summer winds, with an annual average wind speed of 6.1 m/s. The cost of energy for the wind turbines was estimated at $0.13/kWh for the first 2 years of operation.

The 10 wind turbines should reduce the annual fuel consumption by about 340,000 L, which is about 6% of normal fuel requirements. In the year 2000, the 10 wind turbines produced 1.1 MWh of electricity, which saved 265,000 L of diesel fuel. The wind turbines were shut down during part of the summer due to construction on the distribution system, so availability was only 85% during that period. KEA added two more AOC turbines in the spring of 2002. Because of the cold-weather, high-density air, they had to change the control system to reduced peak power output. A Northern Power wind turbine (100 kW), three more 50 kW units, and one remanufactured by Vestas, V15, 65 kW, were installed, so by 2007 there was a total of 17 wind turbines at the site. In 2007, they generated 667,580 kWh of energy, which resulted in a savings of 172,240 L of diesel fuel. Installing foundations in permafrost and operating in cold climates present problems not found at lower latitudes. With the price for diesel fuel escalating to $1.25/L in 2008, wind–diesel becomes more economical.

With the addition of the two EWT wind turbines (900 kW) in 2012, the wind farm will generate around 4 million kWh/year, so the displacement of diesel is around

FIGURE 9.22 Wind turbines at Kotzebue, Alaska. Starting at left foreground: Northern Power, 100 kW, Vestas 65 kW; some Atlantic Orient wind turbines (downwind), 50 kW; and EWT, 900 kW. (Courtesy of Rich Stromberg. Alaska Energy Authority.)

1.05 million liters/year and with diesel at $0.80/L, that is, a saving of $800,000 per year. Of course, when diesel fuel is higher there is more dollar savings. A zinc-bromine flow battery (rated at 2.8 MWh) was installed; however, due to materials issues, commissioning problems, and manufacturer reorganization, the battery was returned and the project is currently on hold.

Wind turbines (3,250 kW) were added to the diesel system (four 1,200 kW) on King Island, between Tasmania and Australia, and wind power provided 18% of the electrical demand. In 2003, another 1,700 kW of wind power and a 200 kW battery and inverter system were added [26] to produce around 50% of the electrical demand. The large-flow vanadium redox battery reduces the variability of the wind energy; however, in 2012 the battery was out of service, and Hydro Tasmania is evaluating rectification on replacement of the storage system.

Powercorp [27] has case studies of 10 wind-diesel installations with 8.5 MW of wind power. To take care of the instabilities of integrating wind power with diesel, a flywheel that can absorb or release 1 MW of power within 0.5 ms is part of the system. Power from two wind turbines and percent electricity is online for the Mawson research station [28] in Antarctica. Ross Island, Antarctica, has three 300 kW wind turbines installed in 2004 which provided 22% of the electricity and provided 11% fuel saving (463,000 liters/year). In the second phase, summer 2010 and 2011, 13 wind turbines (approximately 4 MW) were installed. From 1992 through 2010, Vergnet [29] installed 7.7 MW of wind power, with turbines ranging in size from 12 to 275 kW. Case studies are given for two locations: Devil's Point, Vanuatu, has eleven 275 kW wind turbines and Coral Bay, Australia, has three 275 kW wind turbines and this system has flywheel storage. Danvest Energy [30] lists eight projects (also on Facebook) with a total of 4.6 MW of wind power, with fuel savings of 50%–85% per year. Site has downloads available describing wind diesel, service and controlling, desalination, dump load controlling, and other topics. Ramea (population 700), which is located on Northwest Island off the coast of Newfoundland, Canada, has a wind-diesel project [31] of six remanufactured Windmatic turbines (total 390 kW).

Wind-diesel and wind hybrid systems are now available for village power, so the wind becomes an integral part of the original design. A number of wind turbine manufacturers have wind-diesel or wind hybrid options. These range from simple, no storage systems to complex, integrated systems with battery storage and dump loads.

Wind-hydrogen systems are similar to wind diesel if the hydrogen is used to power a genset, with the added advantage that fuel does not have to be transported to remote locations. Of course, the hydrogen can also be used as a fuel for heating and cooking. In 2009, wind-hydrogen [32] was added to the Ramea project, which consists of 300 kW wind and a 250 kW hydrogen genset (hydrogen produced by electrolysis of water). Seven projects (includes Ramea) using wind for hydrogen production are listed in presentation at workshop [33]. On the island of Utsira [34], Norway, 10 households in a pilot project are supplied power from two 600 kW wind turbines, a 5 kWh flywheel storage, an electrolyzer with a peak load of 48 kW, and a 55 kW engine.

9.10 OTHERS

Innovative wind turbines have to be evaluated in terms of performance, structural requirements, operation and maintenance, and energy production in relation to constraints and cost of manufacturing. Most innovative wind systems are at the design stage, with some even making it to the prototype or demonstration phase. If they become competitive in the market, they would probably be removed from the innovative category. Some examples are tornado type, tethered to reach the high winds of the jet stream; a tall tower to use rising hot air; torsion flutter; electrofluid; diffuser augmented; and multiple rotors on the same shaft. There have been numerous designs, and some prototypes have been built that have different combinations of blades or blade shapes.

There are also companies that are building wind turbines to mount on buildings in the urban environment [2, Section 9.1]. There is an Internet site for urban wind [35] with downloads available. The wind turbine guideline includes images of flow over buildings and example projects. An unusual design for a building is the incorporation of three wind turbines (225 kW each) on the causeways connecting two skyscrapers in Bahrain [36].

The farm windmill is a long-term application of the conversion of wind energy to mechanical power, and it is well designed for pumping small volumes of water at a relatively high lift. It is estimated that there are around 300,000 operating farm windmills in the world, and the annual production of farm windmills is around 4,000. The rotor has high solidity and a large amount of blade material per rotor-swept area, which is similar to drag devices. The tip speed ratio is around 0.8, and the annual average power coefficient is 5%–6% [1].

Different research groups and manufacturers have attempted to improve the performance of the farm windmill and to reduce the cost, especially for developing countries [2, Section 10.7]. A wind-electric system is more efficient and can pump enough water for villages or small irrigation. The wind-electric system is a direct connection of the permanent magnet alternator (variable voltage and variable frequency) of the wind turbine to a standard three-phase induction motor driving a centrifugal or submersible pump. Annual power coefficients are around 10%–12%.

9.11 PERFORMANCE

In the final analysis, performance of wind turbines is reduced to energy production and the value or cost of that energy in comparison to other sources of energy. The annual energy production (AEP) can be estimated by the following methods:

Generator size (rated power)
Rotor area and wind map values
Manufacturer's curve of energy versus annual average wind speed

The generator size method is a rough approximation as wind turbines with the same size rotors (same area) can have different size generators, but it is a fairly good first approximation.

$$AEP = CF * GS * 8,760, \text{ kWh/year or in MWh/year} \qquad (9.3)$$

where:
AEP is the annual energy production
CF is the capacity factor
GS is the rated power of the wind turbine
8,760 is the number of hours in a year

The CF is the average power divided by the rated power. The average power is generally calculated by knowing the energy production divided by the hours in that time period (usually a year or can be calculated for a month or a quarter). For example, if the AEP is 4,500 MWh for a wind turbine rated at 1.5 MW, then the average power = energy/hours = 4,500/8,760 = 0.5 MW, and the CF would be 0.5 MW/1.5 MW = 0.33 = 33%. So, the CF is like an average efficiency.

CFs depend on the rated power versus rotor area as wind turbine models can have different size generators for the same size rotor or the same size generators for different size rotors for better performance in different wind regimes. For wind farms, CFs range from 30% to 45% for class 3 wind regimes to class 5 wind regimes.

Example 9.1

Estimate the AEP for a 2-MW wind turbine in a class 4 wind regime. Since class 4 is a good wind regime, CF = 40% = 0.40. Use Equation 9.3.

$$AEP = 0.4 * 2\ MW * 8{,}760\ h = 7{,}000\ MWh/year$$

Availability is the time the wind turbine is available to operate, whether the wind is or is not blowing. The availability of wind turbines is now in the range of 95%–98%. Availability is also an indication of quality or reliability of the wind turbine. So, the AEP in the example is reduced to 6,650 MWh/year for an availability of 95%. If the wind turbine is located at higher elevations, then there is also a reduction for change in density (air pressure component) of around 10% for every 1,000 m of elevation.

Since the most important factors are the rotor area and the wind regime, the AEP can be estimated from

$$AEP = CF * A_r * W_M * 8.76,\ kWh/year \qquad (9.4)$$

where:
A_r is the area of the rotor (m^2)
W_M is the value of power/area for that location from the wind map (W/m^2)
8.76 h/year converts watts to kilowatts

Example 9.2

For a wind turbine with rotor diameter of 60 m in a region with 450 W/m^2. Assume the CF is 0.40.

$$Rotor\ area = \pi * r^2 = 3.14 * 30 * 30 = 2{,}826\ m^2$$

$$AEP = 0.40 * 2{,}826 * 450 * 8.76 = 4{,}450\ MWh/year$$

If the availability is 95%, then the AEP = 4,200 MWh/year.

The manufacturer may provide a curve of AEP versus annual average wind speed (Figure 9.23), where AEP is calculated from the power curve for that wind turbine and a wind speed histogram calculated from average wind speed using a Rayleigh distribution [2, Section 3.11].

The best estimate of AEP is the calculated value from measured wind speed data and the power curve of a wind turbine (from measured data). The calculated AEP is just the multiplication of the power curve value times the number of hours for each bin (Table 9.5). If the availability is 95% and there is a 10% decrease due to elevation, then the calculated energy production would be around 2,600 MWh/year.

The calculated AEP is the number from which the economic feasibility of a wind farm project is estimated and the number that is used to justify financing for the project. Wind speed histograms and power curves have to be corrected to the same height, and power curves have to be adjusted for air density at that site. In general, wind speed histograms need to be annual averages from 2 to 3 years of data; however, 1 year of data may suffice if it can be compared to a long-term database.

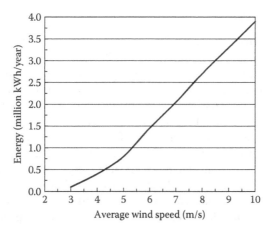

FIGURE 9.23 Manufacturer's curve for estimated annual energy production as function of average wind speed for a 1-MW wind turbine.

TABLE 9.5
Calculated Annual Energy Using Power Curve for 1-MW Wind Turbine and Wind Speed Histogram Data for White Deer, Texas

Wind Speed (m/s)	Power (kW)	Bin (h)	Energy (kWh)
1	0	119	0
2	0	378	0
3	0	594	0
4	0	760	171
5	34	868	29,538
6	103	914	94,060
7	193	904	174,281
8	308	847	260,760
9	446	756	337,167
10	595	647	384,658
11	748	531	396,855
12	874	419	366,502
13	976	319	311,379
14	1,000	234	233,943
15	1,000	166	165,690
16	1,000	113	113,369
17	1,000	75	74,983
18	1,000	48	47,964
19	1,000	30	29,684
20–24	1,000	20	39,540
>25	0	20	0
	Sum	8,760	**3,060,545**

Note: Bin width = 1 m/s; data adjusted to hub height of 60 m.

9.12 COMMENTS

Wind power from wind farms has become a major component for new power installations in many regions of the world. Annual CFs range from 0.30 to 0.45 at good-to-excellent wind locations. One limitation for wind power is that, in general, windy areas are distant from major load centers, and wind farms can be installed faster than construction of new transmission lines for the utility grid. As wind power increases beyond 20% penetration into the grid, stricter requirements similar to conventional power plants, such as wind power output forecasting, power quality, fault (low voltage) ride through, and so forth, will be required at power grid operators.

If storage becomes economical, then renewable energy and especially wind power will supply even more of the world demand for electric energy. In 2011, pumped storage is most economical and should be considered with long distance extra high voltage transmission lines during national electric power system planning to accommodate more wind and solar power penetration into large grids.

There is a growing market for small wind turbines for stand-alone and grid connection and midsize wind turbines for the distributed and community market. The market potential for village power is large; however, there are still problems, primarily institutional and economic costs for these communities.

Past estimates of future installation of wind power have been low, so now the planned installations are at least feasible, for example, the proposal of 20% by 2030 for the United States [12]. China has become a leader in manufacturing of large wind turbines and in installed capacity.

REFERENCES

1. V. Nelson and N. Clark. 2004. *Wind Water Pumping*. CD web format. Alternative Energy Institute, West Texas A&M University, Canyon, TX.
2. V. Nelson. 2013. *Wind Energy, Renewable Energy and the Environment*. 2nd Ed. CRC Press, New York.
3. National Wind Technology Center. http://www.nrel.gov/wind/resource_assessment.html.
4. V. Nelson. 2008. Wind energy. In: *Texas Renewable Energy Resource Assessment*, Chap. 4. http://www.seco.cpa.state.tx.us/publications/renewenergy.
5. U.S. northwest states, wind maps. http://www.windmaps.org/default.asp.
6. 3Tier. *FirstLook Prospecting*. http://www.3tiergroup.com/wind/overview.
7. AWS Truepower Navigator. https://www.windnavigator.com/cms/.
8. *Wind Atlases of the World*. http://www.windatlas.dk/World/About.html.
9. Remote Sensing Systems. http://www.remss.com.
10. Jet Propulsion Laboratory, Physical Oceanography DAAC. Ocean winds. http://podaac.jpl.nasa.gov/index.htm.
11. American Wind Energy Association. Market reports. http://www.awea.org.
12. *20% Wind Energy by 2030 Report*. http://www.20percentwind.org/20p.aspx?page=Report.
13. Global Wind Energy Council. http://www.gwec.net/global-figures/wind-energy-global-status/.
14. R. Lynette and P. Gipe. 1984. Commercial wind turbine systems and applications. In: *Wind Turbine Technology* (Spera, D. A., Ed.). ASME Press, New York, Chap. 4.
15. World Wind Energy Association. 2014. *Update Small Wind World Report*, http://small-wind.org/wp-content/uploads/2014/03/2014_SWWR_summary_web.pdf.
16. China Wind Energy Association.

17. RenewableUK. October 2013. *Small and Medium Wind UK Market Report.* http://www.renewableuk.com/en/publications/reports.cfm/page/2/.
18. National Wind Technology Center, NREL. www.nrel.gov/wind/smallwind/.
19. EERE. 2013. *Distributed Wind Market Report.* http://energy.gov/eere/wind/downloads/2013-distributed-wind-market-report.
20. Small Wind Certification Council. www.smallwindcertification.org.
21. Windustry, Community Wind Projects. http://windustry.org/community-wind-projects.
22. *Community Wind Development Handbook.* http://www.ontario-sea.org/Storage/28/2000_Community_Wind_Development_Handbook.pdf.
23. P. Mazza. 2008. *Community Wind 101: A Primer for Policymakers.* http://www.community-powernetwork.com/sites/default/files/CommWind_101_for_Policymakers%20copy.pdf.
24. R. Stromberg, Alaska Energy Authority. 2013. Also provided power points, Alaska wind energy barriers, Alaska Wind Energy Workshop; Unalakleet wind-diesel data analysis; Alaska Energy Authority Overview. http://alaskarenewableenergy.org/about/events/alaska-wind-working-group/.
25. Kotzebue Electric Association. http://kea.coop/news/renewable-energy.php.
26. *Hydro Tasmania's King Island Renewable Energy Integration Project.* www.kingislandrenewableenergy.com.au.
27. Powercorp, wind diesel. www.pcorp.com.au.
28. Australian Antarctic Division, Mawson Station Electrical Energy. www.antarctica.gov.au/living-and-working/stations/mawson/living/electrical-energy.
29. P. Larsonneur. 2010. Realities and promises of wind-diesel systems. *European Wind Energy Conference.* www.vergnet.com/pdf/vergnet_ewec.pdf.
30. *Danvest Hybrid Power, Wind Diesel.* www.danvest.com/home.pp.
31. C. Brothers. 2011. *Wind-Diesel and H_2 Activities in Canada.* www.uaf.edu/acep/alaska-wind-diesel-applic/international-wind-diesel/presentations/CarlBrothersAlaska_AdvancedWindDieselSystems_110310.pdf.
32. Nalcor Energy, Ramea Report. 2010. www.nalcorenergy.com/uploads/file/nalcorenergyrameareport_january2010.pdf.
33. Using hydrogen energy storage in remote communities. 2009. *International Wind-Diesel Workshop.* http://www.pembina.org/reports/wind-diesel-1-robert-mcgillivray.pdf.
34. *Ursira Wind Power and Hydrogen Plant.* www.iphe.net/docs/Renew_H2_Ustira.pdf.
35. Wind Energy Integration in the Urban Environment. http://www.urbanwind.org.
36. Bahrain World Trade Center. http://www.bahrainwtc.com/content/architecture-design.

RECOMMENDED RESOURCES

P. Gipe. 2009. *Wind Energy Basics Revised: A Guide to Home- and Community-Scale Wind Energy Systems.* Chelsea Green Publishing, White River Junction, Vermont.

R. Nolan Clark. 2014. *Small Wind, Planning and Building Successful Installations.* Academic Press, New York.

INTERACTIVE MAPS

Global Offshore Wind Farm Database. http://www.4coffshore.com/offshorewind/.

GWEC. *Wind Power Capacity in the World.* http://www.gwec.net/global-figures/interactive-map/.

International Renewable Energy Agency. *Global Atlas for Renewable Energy.* http://globalatlas.irena.org/default.aspx.

National Resources Defense Council. Renewable Energy for America. http://www.nrdc.org/energy/renewables/energymap.asp. Potential (wind, solar, cellulosic biomass, biogas, enhanced geothermal, non resource). Facilities; existing, planned (wind, solar, advance biofuel facilities, biodigesters, geothermal, low-impact hydroelectric).

Non utility purchasers of wind. http://www.aweablog.org/blog/post/googles-newest-purchase-highlights-the-wind-industrys-progress.

Renewable and Alternative Energy Projects in the UK. http://www.renewables-map.co.uk/?Status=1&mycompid=0.

The Telegraph. *Every Wind Farm Site in the UK.* http://www.telegraph.co.uk/news/earth/energy/9645593/Interactive-map-every-wind-farm-site-in-the-UK.html.

U.S. DOE. Wind farms through the years. http://energy.gov/maps/wind-farms-through-years#buttn. From 1975. Wind manufacturing facilities. http://energy.gov/maps/wind-manufacturing-facilities.

U.S. Geological Survey. *Wind Farm-Wind Turbine Interactive Web Map.* http://eerscmap.usgs.gov/windfarm/.

LINKS

Alternative Energy Institute. http://www.windenergy.org.

American Wind Energy Association. http://www.awea.org.

Danish Wind Industry Association. http://www.windpower.org/en/knowledge.html. This is a great site; check out Guided Tour and Wind with Miller.

European Wind Energy Association. http://www.ewea.org. Offshore statistics. http://www.ewea.org/fileadmin/files/library/publications/statistics/EWEA-European-Offshore-Statistics-2014.pdf.

Global Wind Energy Council. http://www.gwec.net.

Many countries have wind energy associations.

National Wind Technology Center, NREL. http://www.nrel.gov/wind.

PROBLEMS

9.1. What is the estimated installed capacity of wind farms in the world? Use the latest data available from http://www.gwec.net.

9.2. Besides being nondepletable, what are the other advantages of wind power?

9.3. From the wind power map for Massachusetts or the United States, what is the wind class offshore for Nantucket Sound, south of Cape Cod?

9.4. From the wind power map for the United States, what wind class for the great plains has the greatest area?

9.5. On the Internet, find a wind map for any country besides the United States. List country and approximate area with good-and-above wind resource.

9.6. Calculate power/area when density of air $= 1.1$ kg/m^3 for wind speeds of 5, 10, 15, 20, 30, and 50 m/s.

9.7. Calculate the wind speed at 100 m if the wind speed at 50 m is 8 m/s and the shear exponent is 0.21.

9.8. Why are wind turbines on tall towers?

9.9. For the latest data available, list the top five countries with wind turbines installed and their capacity.

9.10. What are two differences between drag and lift devices?

9.11. What would the wind capacity for the United States need to be to meet 20% of electricity needs from wind by 2030?

9.12. Are there any wind farms in your region? If yes, give location, name, megawatt capacity, and number of wind turbines.

9.13. When would wind be considered for village power?

9.14. What is the difference between low and high penetration for wind-diesel systems?

9.15. Estimate AEP for a 3-MW wind turbine using the generator size method.

9.16. Use rotor area and a wind map value of 400 W/m² to estimate AEP for a 110-m diameter wind turbine.

9.17. Use the manufacturer's curve (Figure 9.21) for a 1-MW wind turbine and estimate the AEP for an area with an average wind speed of 8.5 m/s.

9.18. A 120-MW wind farm produces around 347,000 MWh/yr. What is the CF for the wind farm?

9.19. Using the latest data available, what is the world installed capacity for offshore wind farms?

9.20. Go to Reference [12]: Approximately how many megawatts would have to be installed per year between 2010 and 2030 for the United States to meet the 30% goal?

10 Bioenergy

10.1 INTRODUCTION

The conversion of solar energy by the fundamental process of photosynthesis is the basis for almost all life. Although chemoautotrophic bacteria obtain their energy from chemical reactions, they are not important for the production and consumption of bioenergy. For example, the food chain in vents in the deep oceans is sustained by chemoautotrophic bacteria. Of course, humans are also dependent on biomass for food, fiber, and energy. In terms of the mass of the Earth, the thin layer of biomass is inconsequential, but it is significant in the regulation of the atmosphere and temperature of the Earth. There are three aspects for biomass: overall biomass (which is essentially steady state: growth, storage, and decay), food and fiber (Table 10.1), and bioenergy. In general, 13% of the total primary energy supply of the world is renewable energy, with solid biofuels (primarily fuel wood), accounting for around 75% of the renewables [IEA energy atlas]. In some developing countries, renewable energy can be 70%–90% of the energy supply, again with solid biofuel as the main source. Even in developed countries, the contribution from bioenergy can reach 25% due to a large forest industry, and in some of the developed countries, the contribution of bioenergy has been increasing. It is difficult to estimate the percentage of biomass for food, fiber, and bioenergy in the world, as in the developing world food, fiber, and sources of bioenergy are grown and traded locally.

Satellites are used to estimate global biomass production for land (54%) and oceans (46%), which means the land production (excluding areas with permanent ice cover) is around 430 g of carbon/(m^2/year), and for the oceans, it is around 140 g of carbon/(m^2/year). These numbers can be compared to average production per area for different sources of biomass: forests (tropical, temperature), cultivated crops, and microalgae.

The carbon cycle of the Earth is important, and the carbon production due to human activity is around $9 * 10^9$ tons/year, with combustion of fossil fuels at $7 * 10^9$ and deforestation at $2 * 10^9$ metric tons/year. The total carbon sink is around $5 * 10^9$ tons/year due to photosynthesis and soils (30%), the oceans (25%), and sediments and rocks (<1%), which leaves the difference of $4 * 10^9$ tons/year in the atmosphere. So, humans are affecting the carbon cycle, and since there is insufficient increase in the carbon sinks of biomass and oceans, then there is an increase of carbon dioxide, a greenhouse gas, in the atmosphere (see Figure 3.18). The question is at what level will carbon dioxide in the atmosphere result in serious climate impacts.

What was the climate of the Earth in past geological ages during high concentrations of carbon dioxide in the atmosphere? Therefore, geologists are able to indicate the coming general climate, temperatures, and sea levels due to increased

TABLE 10.1
Estimate of World Biomass: Amount and
Consumption/Production

Biomass	
Forests	$2 * 10^{12}$ tons
Land storage	3,000 EJ/year or 95 TW
Production	$104 * 10^{12}$ tons carbon/year
Food consumption	16 EJ/year
Bioenergy consumption	40–60 EJ/year

greenhouse gases in the atmosphere. For example, historical global concentrations of carbon dioxide of 400–650 ppm show sea levels around 22 m (72 ft) higher [1].

10.2   CLIMATE CHANGE

For meteorologists, climate refers to the average of 30 or more years. Of course the climate has changed and will change, for example, there was the snowball earth and the very warm earth when the sea level was over 30 m higher than today. Civilization developed in the Holocene, an interglacial period beginning around 10,000 years ago, and now we are entering the Anthropocene, the epoch where the impact of humans is noted on a global scale. The results of this uncontrolled experiment will be within physical laws, and global warming will affect the climate. The question is at what level will greenhouse gases in the atmosphere result in serious climate impacts and how will that affect the production of biomass.

The residual time of carbon dioxide in the atmosphere means that world will be warmer, even if we reduced carbon dioxide emissions to the 1990 levels. Therefore, the possible future paths are continued emissions as of today (worst case), abatement of emissions, mitigation, and adaptation (which will become more costly as we dither about abatement). The prediction is that the present increase of carbon dioxide in the atmosphere will increase the average temperature by around 3°C and sea level will rise by 0.6 m by 2100. If tipping points are reached due to more positive feedback and continued high rate of burning fossil fuels, then the temperatures and sea level rise will be higher. High scenarios would have a sea level rise of 10 m. Just think how many people live at an elevation within 10 m (33 ft) of sea level. Where will they go, what are the costs, and who will pay?

An increase in atmospheric CO_2 stimulates photosynthesis and assists plants in dry areas to use ground water more efficiently. This would increase the biomass and be a carbon sink. However, the amount of increase depends critically on water and nutrient availability. Warmer temperatures will shift growing seasons, regions for crop production, precipitation, and the possible detrimental effects of heat waves. Climate models show 30% or more reduction in precipitation in the subtropics, which means the major deserts of the world will increase in size. Even in wetter areas, extreme storms will increase which will mean more flooding, a negative effect. "Based on many studies covering a wide range of regions and crops, negative impacts of climate change on crop yields have been more common than positive impacts (*high confidence*)" [2, p. 4].

There has been a trend to use climate change because it does not have same negative impact (political) as global warming. Some in industry and many politicians deny that there is global warming and some accept that there is global warming but deny that it is due to humans. Industry and some politicians maintain that we cannot reduce the emissions of CO_2 because of economics and because the science for CO_2 and global warming is not completely certain. Remember that the same comments were said about the ozone problem and addiction to tobacco. The former U.S. policy under the G. W. Bush administration was in sharp disagreement with that of the other industrialized countries, and the reasons for the United States were that it would cost too much and not enough provisions were made to curtail future emissions from developing countries.

10.2.1 CLIMATE CHANGE-ANTHROPOCENE

The following facts are the background for information and comments on climate change-A:

1. The climate of past geologic ages has varied by a significant amount with temperatures ranging from −5°C to +5°C from today.
2. The amount of carbon dioxide in the atmosphere has increased since the industrial revolution, primarily due to humans.
3. Changing the concentration of the greenhouse gases in the atmosphere can and will change the energy balance, that is, the amount of heat retention and thereby the temperature of the atmosphere and the Earth (see Section 3.7).
4. Oceans absorb carbon dioxide. An increase of carbon dioxide in the ocean results in more carbonic acid.
5. Sea level has risen by 0.2 m (8 in) since 1900. Sea level is higher due to warmer water having more volume and the melting of ice sheets and glaciers.
6. Aerosols such as sulfur dioxide and other fine particles result in cooling due to reflection of incoming solar radiation. Major sources of aerosols are volcanoes and from the burning of coal.

Knowledge of the interactions among the components in the system is a difficult scientific endeavor. Climate models are based on equations about energy and matter to simulate the atmosphere, oceans, land and sea ice, and vegetation cover. Climate models can be tested by using past data to compare known climate with predicted climate. The models reproduce observed continental surface temperature patterns and trends including the larger increase since 1950 and the cooling due to large volcanic eruptions [3].

Today's models are a fairly good, which means that they have incorporated most of the important factors and their interactions; however, they are still models that are dependent on input.

10.2.2   INTERGOVERNMENTAL PANEL ON CLIMATE CHANGE

The Intergovernmental Panel on Climate Change (IPCC), a scientific body under the auspices of the United Nations, was established in 1988 to assess the scientific information on climate change [4]. The IPCC reviews and assesses the most recent scientific, technical, and socioeconomic information and provides assessment reports from working groups and a synthesis report for policymakers. The IPCC does not conduct any research nor does it monitor climate related data or parameters. Thousands of scientists from all over the world contribute to the process on a voluntary basis.

Working Group I assesses the physical scientific aspects of the climate system and climate change [3]. Topics include changes in greenhouse gases and aerosols in the atmosphere; observed changes in air, land and ocean temperatures, rainfall, glaciers and ice sheets, oceans and sea level; historical and paleoclimatic perspective; biogeochemistry, carbon cycle, gases and aerosols; satellite data and other data; climate models; climate projections, causes and attribution of climate change.

Working Group II assesses the vulnerability of socioeconomic and natural systems to climate change, negative and positive consequences of climate change, and options for adaptation. The information is considered by sectors (water resources, ecosystems, food and forests, coastal systems, industry, and human health) and regions.

Working Group III assesses options for mitigating climate change through limiting or preventing greenhouse gas emissions and enhancing activities that remove them from the atmosphere. The costs and benefits of the different approaches to mitigation are analyzed, along with the available instruments and policy measures.

10.3   BIOMASS PRODUCTION

So, photosynthesis primarily converts carbon dioxide and water to biomass, with efficiencies of 3%–6%, sunlight to stored chemical energy. The carbon content will vary for biomass, coal (note the big difference for lignite and peat), and even natural gas and petroleum (Table 10.2). Because of the lower carbon content per released energy, the combustion of natural gas is better than coal because there is less emission of carbon dioxide. Of course, moisture content will be higher for biomass, so drying or other methods are used to reduce moisture. Note that the moisture content of peat is fairly high.

In the past, biomass, primarily from wood, was the major source of energy in the world, and even today bioenergy provides around 12% (55 EJ) of energy consumption for the world. Around 2.6 billion people rely on fuel wood, charcoal, and dung for cooking and heating. Fuel wood consumption has increased 250% since 1960, faster than the growth in population in some countries. For example, in some countries in Africa and Asia, wood and charcoal provide 50% to over 90% of the energy (Table 10.3; see IEA Energy Atlas to obtain information for any country, http://energyatlas.iea.org/). The collection of fuel wood for direct use and for the increased consumption of charcoal leads to deforestation and degradation of the land and in some areas exacerbates the problems of drought and desertification. Also, collection of fuel wood

TABLE 10.2

Carbon Content of Fossil Fuels and Bioenergy Feedstocks, Terajoule (TJ)

	Metric Tons/TJ
Coal (average)	25.4
Oil (average)	19.9
Natural gas (methane)	14.4
Coal 1 ton = 746 kg carbon	
Gasoline, 1 U.S. gal = 2.42 kg carbon	
Diesel/kerosene, 1 U.S. gal = 2.77 kg carbon	
	%
Bioenergy feedstocks	50
Wood, wood waste	
Agriculture residue	45

is primarily the work of women and children. For example, in the Sahel region of Africa, women walk on average 20 km (12 mi) per day to collect wood, and in the towns, families spend a third of their income on wood or charcoal. However, it takes 10 kg of wood to make 1 kg of charcoal. Dung from cows, buffalos, yaks, and even camels is the other major source of energy for heating and cooking in rural areas. In some cases, the fresh manure may be mixed with straw and water, flattened into patties, and dried. Open fires in confined spaces present a major health problem, so efficient stoves save lives and energy; however, the problem for many poor people is the cost of the stoves.

When people think of renewable energy, most think of solar (photovoltaic, PV) and wind energy and do not realize that bioenergy is a major component of renewable energy, even in industrialized countries. Asia uses almost 50% of the global bioenergy followed by Africa at 26%, most of it as heat for cooking and space heating, 46 EJ (2011 data). In the European Union-28, renewable energy [5] is the largest single source of primary energy (31% of the total). In terms of renewable energy, the bioenergy (Figure 10.1) component is 57% (2013 data), providing 90% of the renewable heat and 19% of the electricity, 65% of which is produced in combined heat and power (CHP) plants.

In the United States, around 10% of the total energy consumption is from renewable energy (Figure 10.2) and of that total biomass is 50% with biofuels at 22%. There was an increase in renewable energy from 7% in 2008. Note that for 2014, the total energy consumption in the United States is slightly less and natural gas and renewables have increased while coal has decreased. In the United States, that 50% due to biomass is from wood, biofuels, and biowaste, both solid and liquid (Table 10.4). In the United States, the biomass resource is large (Figure 10.3); however, there are limitations in terms of converting present agricultural crops to bioenergy. For example,

TABLE 10.3
Wood as Percentage of Total Energy Use
in Some African and Asian Countries

Africa	%
Mali	97
Rwanda	96
Burkina Faso	95
Tanzania	95
Ethiopia	94
Central African Republic	93
Somalia	91
Burundi	90
Niger	86
Chad	85
Benin	85
Nigeria	85
Cameroon	83
Sudan	82
Madagascar	80
Sierra Leone	76
Angola	75
Ghana	75
Mozambique	75
Kenya	70
Cote d'Ivoire	46
Zambia	37
Asia	**%**
Nepal	98
Thailand	63
Sri Lanka	63
Pakistan	37
India	35
Malaysia	10

the Texas Panhandle is a major producer of crops, but because of the large confined animal feeding industry (feedyards), grain is imported for those feedyards and for ethanol plants located in the area. For the United States, biomass maps are available by source: crop, forest, primary mill and secondary mill residues; urban wood waste; and methane from landfills, manure, and wastewater. In some areas of the world, including the United States, fast-growing trees for bioenergy are now a crop. In Sweden, Latvia, and Finland, bioenergy provides around 25% of the total energy consumption.

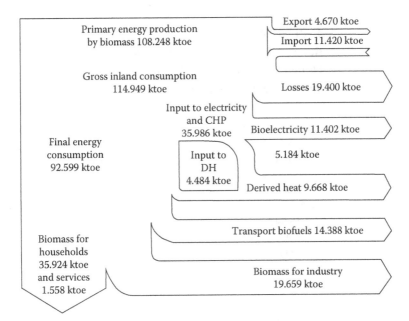

FIGURE 10.1 Bioenergy pathway (tons of oil equivalent) in the European Union-27 (2011). In general, data for 2014 will have larger numbers, but pathway will be similar. (Data from European Biomass Association.)

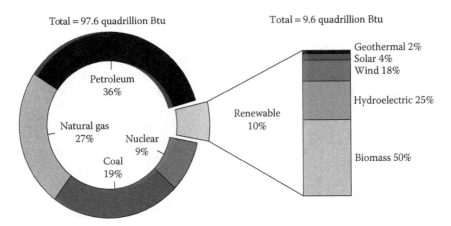

FIGURE 10.2 Energy and renewable energy consumption in the United States, 2014. (Data from Energy Information Administration, DOE.)

TABLE 10.4

Biomass Energy Production in the United States (2013) for Wood, Biofuels, and Waste (% of Biomass)

	%	%
Wood and derived fuels	43.6	
Biofuels	44.0	
Ethanol		24.4
Losses and coproducts		15.7
Biodiesel		3.7
Waste	10.1	
Landfill gas		4.6
Municipal solid waste		3.8
Other biomass		1.8

Total resources by county

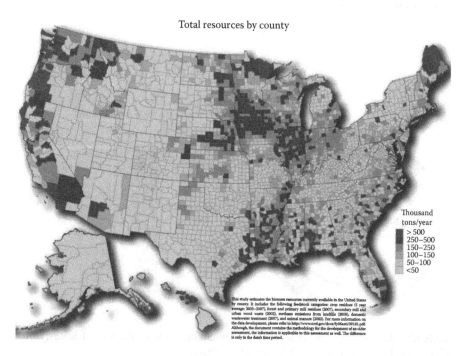

FIGURE 10.3   Biomass resource by county for the United States. (Data from NREL, http://www.nrel.gov/biomass.)

10.4 CONVERSION

Bioenergy is obtained from organic matter such as wood, crops, animal wastes, municipal solid and liquid wastes, and even algae and bacteria. The raw material (feedstock) is converted into a usable form of energy by combustion or biochemical or thermochemical processes (Figures 10.4 and 10.5). Besides combustion for heat, biomass can be converted to gas and liquid fuels, so a major area of concern is the production of liquid fuels, primarily for transportation.

The five major issues with bioenergy are as follows:

1. Power plants need to be located near the source of material to keep transportation costs from becoming too large.
2. Energetics has to be considered. What is the energy content of the product compared to the energy to make that product?
3. In the final analysis, the renewable energy source is the Sun, which means that bioenergy has the same attributes of low density and variability as solar and wind energy, so there is variability of yield and quality of crops per year. However, there is less variability for bioenergy than for solar and wind.
4. There may be a fairly high cost for the conversion plant and transportation. An advantage is the availability of stored energy in the biomass.
5. There is concern about land use related to food versus fuel, because dedicated energy crops will be produced in rotation with other crops, on marginal lands, or from forests. With good land resource management, this can be appropriately managed.

The general end products of bioenergy are heat for households and industry, biofuels (liquid or gas), and electricity. One concept is a biorefinery to produce biofuel, heat or power, and chemicals. A biorefinery could produce one or several low-volume, high-value, chemical products; a lower-value, high-volume liquid fuel; and heat for industrial use or for generation of electricity.

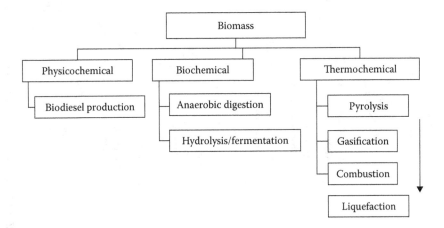

FIGURE 10.4   Diagram of bioenergy conversion. (Data from NREL, http://www.nrel.gov/biomass.)

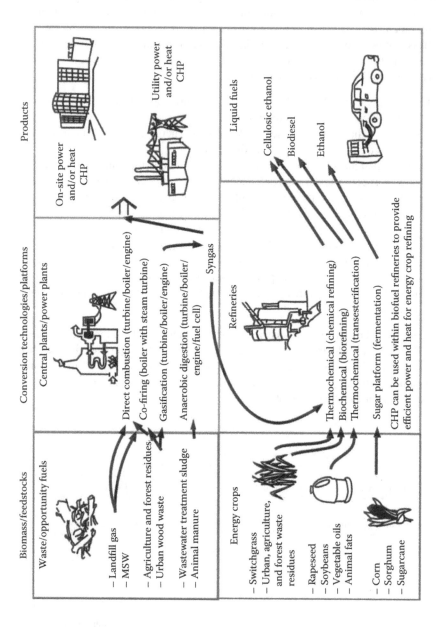

FIGURE 10.5 Conversion of biomass to bioenergy. (From Environmental Protection Agency.)

Bioenergy can be obtained from the following:

Crops: Corn, sugarcane, potatoes, beets, wheat, sorghum.

Oilseed crops: Largest source of fats and oils are cottonseed, soybean, rapeseed (canola), and palm oil; minor sources are sunflower, peanuts, flax, saf-
flower, sesame, jatropha, Chinese tallow, and castor.

Agriculture residues: Bagasse from sugarcane, corn stover, rice straw and
hulls, and nutshells.

Major research area: Production of ethanol from cellulose.

Wood: Sawdust, timber slash, mill scrap, paper trash, fast-growing trees like
poplars and willows.

Municipal solid waste (MSW)

Grasses: Fast-growing like switchgrass, elephant grass, and prairie bluestem.

Methane: Landfills, municipal wastewater treatment, manure, and lagoons
from confined animal feeding.

Biodiesel: Vegetable oils, animal fats, and recycled greases.

Petroleum precursors: Algae.

There may be two harvests per year for certain annual crops; however, there is still the
problem of year-round production for bioenergy from a cyclical source of raw material. In general, perennial oilseed crops (Table 10.5) have not yet been well adapted to
mechanical harvesting. However, once established, they have much annual higher oil
production potential per hectare than annual crops. Geneticists are trying to convert

TABLE 10.5
Oilseed Crops, Type, and Percentage Oil

Crop	Potential World	Season	Planting	Oil, %
Cotton	Major	Warm	Annual	17
Soybean	Major	Warm	Annual	18
Peanut	Minor	Warm	Annual	45
Canola	Major	Cool	Annual	40
Flax	Minor	Cool	Annual	35
Sunflower	Major	Warm	Annual	42
Safflower	Minor	Warm (and cool)	Annual	42
Sesame	Minor	Warm	Annual	50
Tung	Potential	Warm/subtropical	Perennial	35
Palm	Major	Warm/tropical	Perennial	35
Camelina	Potential	Cool	Annual	40
Brown mustard	Potential	Cool	Annual	40
Castor	Potential	Warm	Annual	50
Chinese tallow	Potential	Warm	Perennial	31
Jatropha	Potential	Warm/subtropical	Perennial	35

Source: Dr. David Baltensperger, Texas A&M University, Soil and Crop Sciences.

the major annual crops to perennial crops, which would reduce production costs. Also, if bioenergy crops can be grown on marginal land not used for food or fuel, that is an advantage if adequate yields can be obtained to economically support bioenergy production. A possible environmental problem related to oilseed production is that in the tropics more forests are being cleared for plantations of oil palms.

The production of electricity can be by direct combustion or combustible gas. Liquid fuels similar to diesel and petroleum can be obtained in a number of ways: ethanol, biodiesel, vegetable and nut oils, and by algae. Bacteria produce methane from liquid waste and landfills, and researchers are working on bacteria that produce hydrogen.

10.5 HEAT AND POWER

The industrial sector currently produces both steam or hot water and electricity from biomass in CHP facilities in the paper, chemical, wood products, and food-processing industries. CHP can improve energy efficiency by 35% over conventional power plants. The forest products industry, which consumes 85% of all wood waste used for energy in the United States, typically generates more than half of its energy from wood waste products and other renewable sources of fuel (wood chips and black liquor).

Global generation of electricity was around 370 TWh from biomass. In the United States (2013 data), there are 8,520 MW of wood and wood-derived products, which generated around 40 TWh/year of electricity, and 4,811 MW from other biomass, which generated around 30 TWh of electricity. In the European Union-28 (2013 data), biomass input was around 5 EJ for heat, power, transportation (1.3 EJ), and electricity (153 TWh).

Wood and biomass account for around half of renewable energy consumption in Europe with wood supplying 80% in Finland and Poland. Europe is the major producer in the world of wood pellets, $12.2 * 10^6$ t/year; however, EU consumes $18.3 * 10^6$ t of wood pellets (2013 data), the difference is primarily imports from the United States and Canada. In 2013, the United States exported 2.9 million tons of wood pellets to Europe.

Electric power plants can burn the wood or wood waste products directly (Figure 10.6), the biomass can be mixed with coal in small percentages in existing boilers, or the biomass is converted to gas, which is then burned. Commercial, cost-effective technologies for converting biomass feedstocks to electricity and heat currently available [4] are three types of direct-fired boiler systems (fixed bed, fluidized bed, and cofired) for converting woody biomass and then anaerobic digesters for animal waste or wastewater. In the United States, around 500 plants in the range of 10 to 50 MW use biomass as a fuel. Even though these are steam plants, their efficiency is generally in the 20% range, unless they are CHP plants. The biomass is usually low-cost feedstocks, like wood or agricultural waste, which also helps reduce the emissions typically associated with coal. The potential of using biomass for the production of electricity is large; for example, in Texas the burning of 1 million tons/year of cotton gin trash could produce $1.7 * 10^9$ kWh/year.

Gasification captures about 65%–70% of the energy in solid fuel by converting it into combustible gases. Fixed-bed gasifiers and fluidized bed gasifiers are becoming commercialized and are currently in limited use producing syngas for power and heat.

FIGURE 10.6 Power plant fueled by wood, Kettle Falls, Washington. (From NREL, http://www.nrel.gov/biomass.)

This gas is then burned for process heat or electricity or converted to synthetic fuels. The advantage of biogas over direct combustion is that the biogas can be cleaned and filtered to remove problematic chemicals. As an example, during World War II, some vehicles in Europe ran on gas from coal or charcoal. South Africa has had industrial coal to liquid fuel plants since 1955.

10.5.1 MUNICIPAL SOLID WASTE

The MSW industry has the following components: disposal (landfill) and then possible recycling, composting, or combustion to produce energy. MSW includes durable goods, nondurable goods, containers and packaging, food wastes, yard wastes, and miscellaneous inorganic wastes but does not include industrial waste, agricultural waste, and sewage sludge. Examples of MSW are appliances, newspapers, clothing, food scraps, boxes, disposable tableware, office and classroom paper, wood pellets, rubber tires, and cafeteria wastes.

The disposal of MSW in the United States is a huge problem [6] as 251 million tons of trash were generated in 2012, of which 53.8% was placed in landfills. Of the total, 34.5% was recovered through recycled material and compost (total of 87 million tons), and 11.7% was burned (bioenergy). Recycling and composting have increased from 10% in 1985 to 33% in 2008 and a slight increase to 34.5% in 2012. There are 86 facilities (capacity of 2.6 GW) for combustion of MSW; however, the ashes, around 10%, are then placed in landfills.

Other countries now have regulations that industrial products have to be built with recycling of the material at the end. In the United States, we have to reduce MSW as disposal space is becoming a problem. Where does New York City dispose of its garbage? What happens to old appliances and personal computers in the United States?

10.5.2 Landfill Gas

Landfill gas from the decomposition of MSW is about 50% methane, 50% carbon dioxide, and less than 1% other organic compounds. Around 12,200 m³/day of landfill gas is generated for every 1 million tons of MSW. Landfill gas is a major problem since methane is 20 times worse by weight than carbon dioxide as a greenhouse gas. Methane is at its highest level in the past 400,000 year and is 150% greater than in 1750. In the past, the landfill gas just seeped into the atmosphere, and for smaller landfills, it is still vented today. In the United States, landfills are the second-largest source of methane (23%) generated by humans. After a landfill is capped, release of landfill gas declines by 2% to 15%/year.

There is a large potential for landfill gas in the United States (Figure 10.7), and as of 2013, there were 621 operational projects, which generated $2.9 * 10^9$ m³ of landfill gas per year for direct use and production of electricity, $16 * 10^{12}$ kWh/year [7–9]. The production of electricity from landfill gas mostly uses internal combustion engines (70%) with rated output of 100 kW to 3 MW. Others use gas turbines (800 kW to 10.5 MW) and microturbines (30 to 250 kW). There is a 50-MW plant at Puente Hills, California, and a 50-MW plant at Inchon, South Korea. Of course, landfill gas can be used for any application that uses natural gas, such as injection into pipelines, burning in boilers, direct combustion, and vehicle fuel. The world's largest landfill at Altamont Pass in California is now converting landfill gas to produce 49,000 L/day of liquid natural gas for vehicle use.

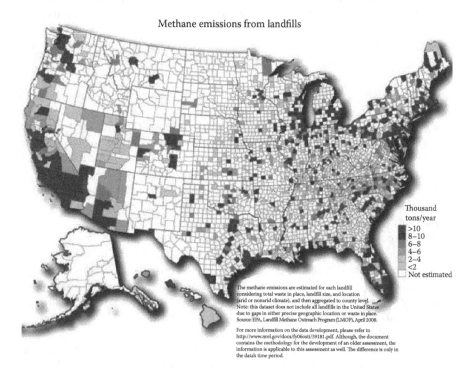

FIGURE 10.7 Resource of landfill gas by county in the United States. (Map from NREL, http://www.nrel.gov/biomass.)

In Latin America, there were 117 cities with a population of over 0.5 million that generate 75 million tons/year of MSW [10]. In 2006, only 1 city used landfill gas to generate electricity, and by 2013, there were 19 projects in Latin America. Seventy-one landfill gas projects around the world with funding from the World bank are available online [11].

10.5.3 BIOGAS

Anaerobic bacteria produce methane from liquid waste (municipal, industry, and manure), which can be used for heat and/or electricity. An interactive map of operational biogas systems (247 systems) at commercial livestock farms is available at the U.S. EPA (http://www.epa.gov/agstar/projects/). Farm project profiles for digester types, animals, and varying outputs are available for operating systems, dormant projects, and other project case studies.

Wastewater treatment facilities (WWTFs) that have anaerobic digesters generally use the biogas for operations and heat; however, a more efficient use would be CHP. In the United States, there are more than 16,000 municipal WWTFs, ranging in capacity from 10^4 m^3/day to more than10^6 m^3/day. Roughly 1,000 of these facilities operate with a total influent flow rate greater than $2 * 10^4$ m^3 per day; however, only 544 of these facilities employ anaerobic digestion to process the wastewater, and only 104 WWTFs utilize the biogas produced by their anaerobic digesters for combined heat and power (190 MW). If all 554 WWTFs that have effluent rates of $2 * 10^4$ m^3 per day were to install CHP plants, that would add 340 MW of electricity and would reduce CO_2 emissions by 3.2 million tons/year [12]. For $1.7 * 10^4$ m^3/day, enough biogas is generated to fuel around a 100-kW power plant. A state list of WWTFs and capacity (MW) with aerobic digesters is also given in Environmental Protection Agency [12]. General rules for considering CHP at wastewater facilities are as follows:

The facility processes $0.4 * 10^4$ m^3/day per person.

Approximately 0.3 m^3 of digester gas can be produced per person per day.

The heating value is around 600 Btu/ft^3. Notice that this is a lower heat content than natural gas.

In Europe, there are over 14,500 biogas plants (2013 data) with 7,857 MW of electricity, which produces around 40 TWh/year [13]. The total biogas production has increased from 3 million tons of oil equivalent (Mtoe) in 2004 to around 15 Mtoe in 2014, from landfill gas (over 50%), sewage sludge, and other sources. Germany leads with 9,035 biogas plants followed by Italy with 1,391 plants. Biomethane can be upgraded biogas or cleaned syngas from gasification of biomass and is pure enough to be injected into the natural gas grid. There are 282 biomethane plants in Europe that produce $1.2 * 10^9$ m^3/year (2013 data), and an interactive map of projects is available [14]. There is a list with overview and map of the 174 installations (heat/and or power, biomethane to grid) in the United Kingdom [15]. A plant in Britain uses 146,000 tons of slurry from 28 farms and waste from food processers to provide heat for 1.43 MW of electricity. Farmer cooperatives in Denmark and the Netherlands have central biogas CHP plants that use manure. Another example is the Lemvig Biogas, the largest biogas plant in Denmark which produces 21 GWh/year and surplus heat of 18 GWh/year is distributed to the Lemvig central heating plant. Input is slurry from around 75 farms and waste and residual products from industrial production.

Example 10.1

A palm oil facility (Figure 10.8) near San Pedro Sula, Honduras, was formed by 450 farmers. The facility produced 68,000 m^3 of effluent per year, which was discharged to open lagoons, which resulted in $3 * 10^6\, m^3$ of biogas emission per year. The anaerobic lagoons have been covered with plastic membranes (Figure 10.9), and the biogas now powers two 500-kW diesel generators, which produce around 5.5 GWh/year that is fed back into the utility grid. The carbon credits are sold to increase the economics of the electric generation.

For any bioenergy system, it takes energy to plant, harvest, and process the material to get the final product, so energy recovery should include all possible paths. In this case, it includes electricity from biogas and solid residue for burning and fertilizer.

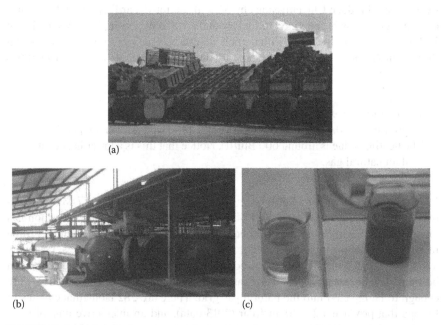

(a)

(b)   (c)

FIGURE 10.8 Palm oil plant near San Pedro Sula, Honduras. (a) Palm nuts from trucks are placed in hopper cars; (b) first step, steam heating of palm nuts in hopper cars; (c) final oil products.

FIGURE 10.9 Palm oil plant near San Pedro Sula, Honduras. Biogas lagoons have plastic covering (top left).

10.6 BIOFUELS

World production of biofuels has increased dramatically, and in 2014, the production was around 2.1 million barrels/day. For years, Brazil led the world in the production of biofuels from the production of ethanol from sugarcane (Figure 10.10); however, increased ethanol production since 2006 has made the United States number one in biofuels.

Projections indicate that by 2030 up to 30% of the annual petroleum demand in the United States could be supplied by approximately $1.1 * 10^9$ dry tons from forestry and agriculture biomass [16]. The U.S. Department of Energy anticipates that about $8 * 10^6$ million tons per year would be supplied from present crop residues and

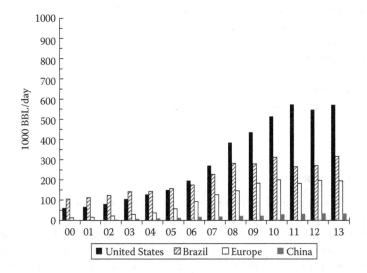

FIGURE 10.10 Leading countries for world production of biofuels. (Data from BP Statistical Review of World Energy June 2014.)

a new generation of dedicated bioenergy crops that are sustainable and integrated with existing food, feed, and fiber cropping systems. Also, over $3 * 10^6$ tons of forest resources will be needed. The main driver for the production of bioenergy is the farmgate or roadside price for the feedstock, with yield rate increases being second. The two account for the range of the projections from 1.1 to $1.5 * 10^6$ tons/year by 2030. Under the baseline assumptions, up to $9 * 10^6$ ha ($22 * 10^6$ acres) of cropland and $17 * 10^6$ ha ($41 * 10^6$ acres) of pastureland shift to energy crops at \$60 per dry ton. For crop residues, a minimum needs to be left behind for soil conservation (soil condition index), which will depend on region and yields. The sustainable retention coefficient (fraction of residue left on the field) depends on no till, around 0.5, and reduced till, around 0.8, while there is no removal of residue from fields with conventional till. In general, 20% of the crop residue is lost in collection, and a crop residue for bioenergy of less than 1 ton/ha is not economical.

Countries and individual states are mandating through policies and economic incentives a significant increase in the use of biofuels. For example, the renewable fuel standard in the United States will require $136 * 10^9$ L/year ($36 * 10^9$ gal/year) from biofuels by 2022. The European Union renewable energy directive (2009) is a policy for 20% renewables by 2020 through individual country targets. At least 10% of transport fuels must come from renewable sources that meet biofuels sustainability criteria. One of the criteria is that biofuels must achieve greenhouse gas savings of 50% by 2017 in comparison with fossil fuels. The new policy (2015) for Europe is 27% renewables by 2030, but there is no set amount for transport fuels. The China policy is 15 million tons of biofuel by 2020, primarily from ethanol. The International Energy Agency reports that biofuels could supply 27% of the world energy needed for transport sector by 2050 [17].

Biofuels generally require the following resources or feedstocks:

Sources of sugar and starches (nonstructural carbohydrates) through conventional fermentation
Lignocellulosic feedstocks use biochemical conversion of biomass to biofuels, which involves the following steps:
Pretreatment (size reduction, dilute acid, and steam)
Conditioning and enzymatic hydrolysis
Enzyme development
Sources of oils

10.6.1 Ethanol

The production of ethanol accounts for around 90% of the production of biofuels in the world, and production has increased significantly since 2000 due to increased ethanol production in other countries besides Brazil (Figure 10.11). World ethanol production increased from 0.9 million bbls/day in 2007 to 1.5 million bbls/day in 2014. Ethanol is produced by converting the starch content of biomass feedstocks into alcohol (Figure 10.12). The fermentation process of yeast and heat break down complex sugars into more simple sugars, creating ethanol. The liquid from the

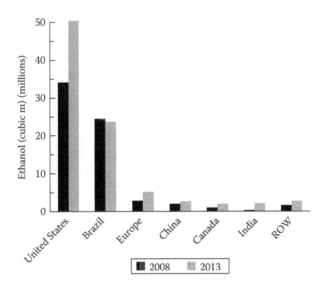

FIGURE 10.11 Ethanol production by major producing countries.

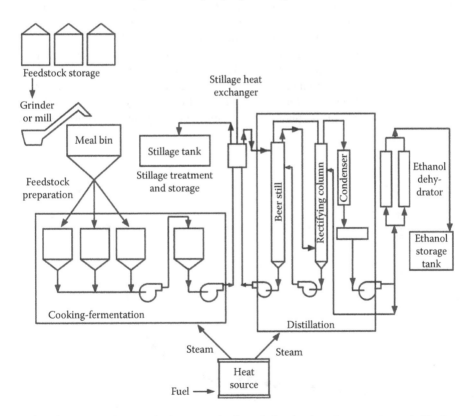

FIGURE 10.12 Diagram of ethanol production. Notice that heat for steam is needed in the process.

TABLE 10.6
Ethanol Yield from Crops

Crop	Liter/Ton	Liter/Ha/Year
Sugar beet		7,000
Sugarcane	70	400–12,000
Corn	360	250–2,000
Sorghum		3,500
Cassava (roots)	180	500–4,000
Sweet potatoes	120	1,000–4,500

fermentation is about 10% ethanol, which then needs to be distilled. The energy content of ethanol is 0.024 GJ/L.

Ethanol is primarily produced from maize (corn in the United States) and sugarcane (Table 10.6). Ethanol can be produced from the cellulose portion of biomass feedstocks like trees, grasses, and agricultural wastes. This process is relatively new; however, there is the potential for using a much wider variety of abundant, less-expensive, and nonfood feedstocks. It is estimated that the United States could produce around 20% of its demand for oil using cellulosic feedstock without affecting food supplies, although one source estimated it as high as 50% [18].

Ethanol provides a major part of the liquid fuel requirement in Brazil, and production has increased significantly in the United States (Figure 10.13). In the United States, the number of ethanol plants (Figure 10.14) increased from 56 with a capacity of $7.3 * 10^6$ m³/year in 2000 to 214 plants with a capacity of $53 * 10^6$ m³/year at end of 2014. Most of the plants are concentrated in the upper Midwest (http://

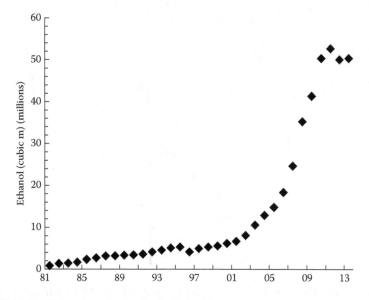

FIGURE 10.13 Ethanol production in the United States.

FIGURE 10.14   Ethanol plant near Herford, Texas.

www.ethanolrfa.org/bio-refinery-locations). Remember that a gallon of ethanol has about two-thirds of the energy content of a gallon of gasoline, and what is sold is gasohol, a 10%–20% mixture of ethanol and gasoline. Due to the economic crisis of 2008–2009, the construction of new ethanol plants decreased, and production was essentially flat from 2009 to 2014.

In the United States, the efficiency of production of ethanol from maize has increased such that the average energetics is now 2.3 (includes energy in distillers grains), while the energetics for ethanol from sugar cane in Brazil is 9–10. The energetics of ethanol produced from maize irrigated from wells of 100 m depth in semiarid areas is around 1, but it consumes a significant amount of water in growing and conversion. However, ethanol continues to have significant backing in Midwestern corn-producing states. In addition, you cannot take the entire crop residue off the land for lignocellulosic conversion to ethanol, because of soil quality impacts. However, emerging lignocellulosic biofuels production technology using dedicated energy crops such as energy cane has promise on improved energy efficiency and net benefits for carbon dioxide.

10.6.2   BIODIESEL

Biodiesel can be produced from beans and seeds, recycled vegetable oils, grease, and fat, and there is the possibility of dedicated fuel crops for biodiesel (Table 10.7), primarily palm oil (9.5 million tons in 2014) and jatropha. The use of soybeans (7.3 million tons in 2014) means competition for use as food and feed. World production of biodiesel was around 29 million tons, with biodiesel providing around 10% of diesel for transportation in Europe, with Germany the largest producer of biodiesel. In Europe, rapeseed was the largest feedstock. Biodiesel from nonfood feedstocks is now part of national policies; for example, China set aside a large land area for growing jatropha and other nonfood plants for biodiesel, and India has up to 60 million hectares of nonarable land available for jatropha. Other countries are also pursuing programs for the production of biodiesel from jatropha.

The United States produced around 40 million barrels of biodiesel in 2014, with around 3 more million barrels imported. The most common feedstock was soybeans; however, enough vegetable oil, soy oil, recycled restaurant grease, and other feedstocks are readily available in the United States to provide feedstock for about 40 million barrels/year of biodiesel.

TABLE 10.7

Dedicated Crops for Biodiesel and Possible Annual Production

Crop	Liter/Ha/Year
Palm	5,238
Jatropha	1,684
Castor	1,216
Canola	1,076
Sunflower	842
Soybean	561

10.6.3 BIOGAS DIGESTERS

Biogas digesters use animal manure, sewage sludge, and liquid waste and convert it into gas and liquid fertilizer (Figure 10.15). A biogas digester should produce 200–400 m³ of biogas with 50%–70% methane per dry ton input, around 8 GJ/ton input. This is less than the energy content of dry dung or sewage; however, the process produces clean fuel and disposes of fragrant (smelly) waste. The temperature needs to be at least 35°C, so heat is required for biodigesters in cooler climates.

Biogas is produced in a large number of small and medium biodigesters across the world, primarily in Asia and Africa, mostly for cooking and lighting; 30 million in China, 10 million in India, and over 50,000 biodigesters installed since 1992 in Nepal. In 2013, the Netherlands Development Organization renewable energy team celebrated the installation of over 500,000 biodigesters in 17 countries. In many developing countries, a biodigester is too costly for small farmers, and attempts to introduce community biogas digesters have faltered due to institutional problems. While solar power is excellent for lighting in rural areas, it is not well suited to

FIGURE 10.15 Floating roof biodigester in South Africa. (Courtesy of Solar Engineering, South Africa.)

providing energy for cooking. Traditionally, dung is burned directly as a fuel, or it is applied as a fertilizer to gardens. The processing of the cattle, or human, dung in a biogas digester provides both better quality gas and better fertilizer when compared with the raw manure product.

In the United States, there is a large amount of manure from confined animal feeding, from birds to pigs to cows. There was a plant to produce methane from manure near Guymon, Oklahoma; its name was Calorific Recovery by Anaerobic Processes, which gave them a great acronym. Presently in the United States, the primary commercial production of biogas is from landfills and large commercial agriculture operations.

10.6.4 Microalgae

Algae grow in aquatic environments and are classified as large, centimeter size, and micro algae. They are commonly seen in pods and the largest example are kelp in the sea. Microalgae are unicellular, micrometer in size, and grow in suspension in water. The advantages of using microalgae are rapid growth, high yield per land area, lack of sulfur in the biofuel produced, and nontoxicity and biodegradability. Some strains of microalgae have high levels, 25%–55% by weight of lipids, of precursors for oil, capable of realistically producing 4.5–7.5 m^3/ha/year (2,000–3,000 gal/acre/year) in open systems. Some algae can even produce hydrogen gas under specialized growing conditions. Production efficiency per land area must be improved and installed and O&M cost must be reduced before large-scale algal biofuel production can take place.

Algal biomass contains three main components: carbohydrates, proteins, and lipids/natural oils. Primary use of microalgae would be for biodiesel, with the possibility of using the residual mass for food, fertilizer, and even combustion for heat or electricity. Microalgae can double in size every 24 h and can regenerate in 48–72 h; cyanobacteria can regenerate in 5–20 h. These short generation times lead to the high potential for biodiesel production from algae.

Microalgae require three ingredients to grow: (1) high solar radiation, (2) carbon dioxide, and (3) brackish water or water high in salt content (up to 30,000 ppm). The logical location in the United States for growing algae under high levels of solar radiation would be the desert southwest.

Two possible systems for algae production [19] are raceway ponds and photo bioreactors. Raceway ponds provide for high production of algae and typically cost less per acre to construct; however, because they are open, they require control of contaminants and management of evaporation. Bioreactors are more costly to build per acre but can operate year round because they are enclosed, typically in glass or film tubes. After generation and production of lipids in the algae, the algae must be harvested, concentrated, and converted to fuel. Harvesting processes include processes such as pumping the algae to settling tanks and using rakes, skimmers, or centrifugal systems [20].

The lipid/algae water slurry must go through an oil separation and purification process. Chemical extraction and mechanical extraction are the primary methods for oil separation. Hexane is used successfully in separation applications but may be cost prohibitive. Centrifuge processes have also been successful but require high-energy inputs for large-scale production. Research is under way to develop high-capacity separation technologies. Algae production as a dedicated biodiesel feedstock is now an area of extensive research as academia, private industry, and governmental agencies are investigating microalgae in terms of processes to get to commercial operation. Major energy companies, including Exxon-Mobil, are now pursuing research and development, and there are a number of pilot projects. Because algae has the potential of producing the highest yield per hectare of any biofuel, it is not surprising that the U.S. Department of Defense is interested in the production of jet fuel for planes and bunker fuel for ships from algae.

U.S. DOE has an interactive map with selectable feedstock and project scale for integrated biorefineries [21]. By 2013, Sapphire's demonstration project produced 3.8 * 106 L/yr of biodiesel, and in 2014, Algenol's demonstration facility achieved a continuous production rate of 66 m3/ha (7,000 gal/ac) per year. A demonstration project using Algenol technology was installed at the world largest oil refinery in the state of Gujarat, India and by mid 2015 was being operated by Reliance engineers and biologists. There is a demonstration project of a photobioreactor In Mobile Bay, Daphne, Alabama that uses municipal wastewater and after the algae are harvested, clean water can be discharged into the bay without the risk of creating dead zones. A commercial facilities in Iowa has a capacity of 20,000 Mt/yr of oil from algae with potential expansion to 100,000 Mt/yr.

Example 10.2

This example concerns the Texas A&M AgriLife Research pilot project located 10 miles west of Pecos. The Permian basin (site of the pilot project) has superior characteristics for algae production: available nonarable land, high solar radiation, brackish water, and geologic carbon dioxide. There are two independent sets of ponds (Figure 10.16).

FIGURE 10.16   Algae production pilot project, Pecos, Texas. (Courtesy of Texas AgriLife Research, Texas A&M System, College Station, TX.)

FIGURE 10.17 Medium raceways. (Courtesy of Texas AgriLife Research, Texas A&M System, College Station, TX.)

FIGURE 10.18 Next size raceway, 3.8 m³.

Each set consists of four ponds that represent a step in the growing process. Each pond system includes the following:

Four medium raceways, bathtub size (Figure 10.17).
One 3.8-m³ raceway (Figure 10.18).
Two 7.6-m³ raceways.
One 22.7-m³ raceway.
Four settling tanks.
A third raceway system is a unique design that uses a pump rather than paddlewheels.

Algae are started in photobioreactors in the on-site laboratory and then moved from small to large raceways as density increases.

FIGURE 10.19 Photobioreactor for algae production. (Courtesy of Copyright Solix Biofuels, Fort Collins, CO.)

Example 10.3

Solix [22] has a pilot-scale photobioreactor (Figure 10.19) located near Durango, Colorado, on the Southern Ute Indian Reservation. Each panel in the basin is 36.6 m long and contains approximately 680 L of algal culture. There are 120 panels per basin with a capacity of approximately 82,000 L of algal culture, and there are three basins. Typically, growth rates at this location exceed 0.2 g/L per day during the growing season.

10.7 COMMENTS

The manipulation of bacteria and algae for the production of bioenergy and other products will have a significant impact on the future production of liquid fuels. The problem is still the same as most other renewable energy resources: low density, high capital and operating costs, and the logistics of transportation of biomass and end product. However, biofuels solve the problem of storage and logistics as distribution is already in place, and biofuels can also be used for power plants. Primary research areas are the use of lignocellulose rather than sugars for producing ethanol, bacteria for production of methane and hydrogen, and algae for production of biodiesel and jet fuel. Of course, commercial operations are seeking to improve the efficiency of their bioenergy operations.

The use of bioenergy for electricity and biofuels will increase in both developed and underdeveloped countries. In developed countries, electricity growth will be from landfill gas, MSW combustion, and wood and animal wastes. How much electricity would be from energy crops will primarily be dependent on tree farms. Even though there may be a shift from bioenergy in developing countries, the use of bioenergy will grow due to population growth. Therefore, bioenergy will remain the main domestic source for these countries. Although there is concern of shifting land from food and fiber to bioenergy production, some rotation with bioenergy crops is possible and, with the use of marginal land for farming, bioenergy could increase from today's 40–60 EJ/year to 100–200 EJ/year. The largest estimates are near 280 EJ/year by 2050 due to the large increase in biofuels.

If dedicated bioenergy crop production significantly increases, the switch of productive land from cultivation of food/feed crops to feedstock for bioenergy needs to be managed. However, bioenergy feedstocks would have to compete against food and feed crops, and farmers can receive significantly more money per unit for these crops than for bioenergy crops. How much land will be used per megawatt capacity for the following systems: trough collector, PV, wind, or bioenergy (corn)? Of course, use of marginal cropland for bioenergy and nonproductive land for wind and solar would be positive.

Besides food and fiber, the agricultural industry is now entering the energy industry. There have been economies of scale for wind farms and ethanol plants (Figure 10.14); however, there is the possibility to have modular bioenergy systems for farms and ranches, which not only would provide energy for the farm and ranch but also would provide excess energy to sell. A small-scale energy system could combine wind, PV, and biomass, and the farmer would be self-sufficient and could even sell excess energy not used at the farm.

Again, energetics and carbon have to be considered for bioenergy systems. The terms *energy balance, energy payback ratio, energy ratio, or energy return on energy invested* are also used. Energetics is the calculation of the energy content of the end product compared to the total energy input to produce that product. If the energetics is small, that energy system should be carefully scrutinized, especially if it receives economic subsidies. Again, it does not matter what the economics (dollars) are because in the final analysis physical laws triumph. The carbon economy may provide an important additional benefit for those bioenergy sources that can clearly demonstrate a net negative carbon balance, and in fact, carbon credits may produce additional revenue for bioenergy projects in states/countries who have adopted a low carbon fuel policy.

REFERENCES

1. G. L. Foster and E. J. Rohling. 2013. Relationship between sea level and climate forcing by CO_2 on geological timescales. *Proceedings of the National Academy of Sciences.* http://www.pnas.org/content/110/4/1209.full.
2. *Climate Change 2014. Impacts, Adaptation, and Vulnerability, Summary for Policymakers.* 2014. IPPC Working Group II. http://ipcc-wg2.gov/AR5/images/uploads/WG2AR5_SPM_FINAL.pdf.
3. *Climate Change 2013. The Physical Science Basis, Summary for Policymakers, Technical Summary and Frequently Asked Questions.* 2013. IPPC Working Group I. http://www.climatechange2013.org/images/report/WG1AR5_SummaryVolume_FINAL.pdf.
4. U.S. Environmental Protection Agency. 2007. *Biomass Combined Catalog of Heat and Power Technologies.* http://www.epa.gov/chp/documents/biomass_chp_catalog.pdf.

5. European Biomass Association. *European Bioenergy Outlook 2014.* http://www.aebiom.org.

6. *Municipal Solid Waste Generation, Recycling and Disposal in the United States: Facts and Figures for 2012.* http://www.epa.gov/epawaste/nonhaz/municipal/index.htm.

7. Environmental Protection Agency. 2013. *An Overview of Landfill Gas Energy in the United States.* http://www.epa.gov/lmop/documents/pdfs/overview.pdf..

8. Environmental Protection Agency. 2015. *Landfill Gas Energy Project Development Handbook.* http://www.epa.gov/lmop/publications-tools/handbook.html.

9. Environmental Protection Agency. Project profiles and spreadsheets of projects and suggested project sites are available. http://www.epa.gov/lmop/projects-candidates/index.html.

10. *World Bank-ESMAP Landfill Gas to Energy Initiative.* https://openknowledge.world-bank.org/bitstream/handle/10986/17972/esm3180PAPER001gastoenergy01PUBLIC1.pdf?sequence=1.

11. World Bank. *Projects and Operations.* Search landfill gas. http://www.worldbank.org/projects/search?langen&searchTerm=landfill%20gas.

12. Environmental Protection Agency. 2011. *Opportunities For and Benefits of Combined Heat and Power at Wastewater Treatment Facility.* http://www.epa.gov/chp/documents/wwtf_opportunities.pdf.

13. European Biogas Association. http://european-biogas.eu.

14. German Energy Agency. Biogaspartner. http://www.biogaspartner.de/en/europe.html.

15. Biogas. http://www.biogas-info.co.uk/maps/index2.htm.

16. U.S. DOE, EERE. U.S. 2011. *Billion-Ton Update: Biomass Supply for a Bioenergy and Bioproducts Industry.* http://www1.eere.energy.gov/bioenergy/pdfs/billion_ton_update.pdf.

17. International Energy Agency. 2011. *Technology Roadmap: Biofuels for Transport.* http://www.iea.org/publications/freepublications/publication/Biofuels_Roadmap_WEB.pdf. Foldout. http://www.iea.org/publications/freepublications/publication/Biofuels_foldout.pdf.

18. G. W. Huber and B. E. Dale. July 2009. Grassoline at the pump. *Sci Am,* Vol. 305, p. 52.

19. Y. Chisti. 2007. Biodiesel from microalgae, research review paper. *Biotechnol Adv,* 24, 294.

20. R. Avant. Biomass energy. In: *Texas Renewable Energy Resource Assessment, 2008,* Chap. 5. Texas State Energy Conservation Office, http://www.seco.cpa.state.tx.us/publications/renewenergy/.

21. U.S. DOE, EERE, integrated biorefineries. http://energy.gov/eere/bioenergy/integrated-biorefineries.

22. Solix BioSystems. http://solixbiosystems.com.

RECOMMENDED RESOURCES

L. R. Brown. 2011. *World on the Edge: How to Prevent Environmental and Economic Collapse.* W.W. Norton, New York.

L. R. Brown. 2012. *Full Planet, Empty Plates: The New Geopolitics of Food Security.* W.W. Norton, New York.

A. Hornung. 2014. *Transformation of Biomass: Theory to Practice.* Wiley, New York.

V. Smil. 2013. *Harvesting the Biosphere: What We Have Taken from Nature.* MIT Press, Cambridge, MA.

F. R. Spellman. 2012. *Forrest-Based Biomass Energy, Concepts and Applications.* CRC Press, New York.

L. Wang. Ed. 2014. *Sustainable Bioenergy Production,* CRC Press, New York.

INTERACTIVE MAPS

EurObserv'ER. GIS. http://observer.cartajour-online.com/Interface_Standard/cart@jour.
phtml?NOM_PROJET=barosig&NOM_USER=&Langue=Langue2&Login=OK&Pa
ss=OK.
European Biomass Industry Association. *European Map for Renewable Energy and Energy
Efficiency.* http://www.eubia.org/index.php/projects/ongoing-projects/10-projects/
ongoing-projects/66-repowermap.
Global Biofuel Information Tool. http://www.cifor.org/bioenergy/maps/.
International Energy Agency. Energy Atlas. http://www.iea.org/statistics/ieaenergyatlas/.
Natural Resources Defense Council. *Renewable Energy Map.* http://www.nrdc.org/energy/
renewables/map_geothermal.asp#map.
NREL. Bioenergy Atlas, BioFuels and BioPower. http://maps.nrel.gov/bioenergyatlas/.
U.S. DOE. *Bioenergy, Knowledge Discovery Framework.* https://bioenergykdf.net.
U.S. DOE, EERE. Integrated biorefineries. http://www.energy.gov/eere/bioenergy/integrated-
biorefineries.

LINKS

Anaerobic Digestion and Bioresources Association. http://adbioresources.org.
Biomass Combined Heat and Power Partnership. http://www.epa.gov/chp/technologies.html.
Biomass Energy Data Book. 4th Ed. http://cta.ornl.gov/bedb/index.shtml.
Biomass Power Association. http://www.usabiomass.org.
Biomethane Feed-in Project in Europe. http://www.biogaspartner.de/en/europe/list-of-
projects-in-europe.html.
Biomethane Injection Projects in Germany. http://www.biogaspartner.de/en/project-map/list-
of-projects-in-germany.html.
Department of Energy. *Energy Efficiency and Renewable Energy, Bioenergy Program.* http://
www.energy.gov/eere/bioenergy/bioenergy-technologies-office.
EurObserv'ER Report. 2014. *The State of Renewable Energies in Europe.* http://www.
energies-renouvelables.org/observ-er/stat_baro/barobilan/barobilan14_EN.pdf.
European Biofuels Technology Platform. http://www.biofuelstp.eu/index.html.
European Biomass Association. http://www.aebiom.org.
Global Renewable Fuels Alliance. http://globalrfa.org.
Google images. Many photos of dung for fuel are available.
International Energy Agency. http://www.iea.org/topics/renewables/subtopics/bioenergy/.
Technology roadmap, biofuels for transport.
List of commercial scale anaerobic digestion plants, world. http://www.iea-biogas.net/plant-
list.html.
National Algae Association. http://cta.ornl.gov/bedb/index.shtml.
National Biodiesel Board. http://www.biodiesel.org.
National Renewable Energy Laboratory. http://www.nrel.gov/biomass/.
Oilgae. http://www.oilgae.com.
Oilgae, algae energy. http://www.oilgae.com.
Renewable Fuels Association. http://www.ethanolrfa.org. Also has a map of ethanol plant
locations. Annual industry outlook, 2014 plus previous years. http://www.ethanolrfa.
org/pages/annual-industry-outlook.
U.N. Food and Agricultural Organization. http://www.fao.org/statistics/en/.
U.S. Energy Information Administration. *Renewable and Alternative Fuels.* http://www.eia.
gov/renewable/.
WBA Global Bioenergy Statistics 2014.
World Bioenergy Association. http://www.worldbioenergy.org.

PROBLEMS

10.1. OM (order of magnitude): Estimate the mass for humans on the Earth. Estimate amount of energy humans need in food per year.

10.2. OM: Estimate total annual production for the world in terms of mass and energy from each of the major food crops: corn, wheat, and rice.

10.3. What is average conversion efficiency of photosynthesis?

10.4. OM: For cultivated crops, use 1 ha of land. On an annual basis, what is the solar input, and what is the output? Be sure to state region and insolation (solar input) for your crop.

10.5. How is charcoal made? Why do people use charcoal for cooking and heating rather than wood?

10.6. Compare the energy content of very dry wood to coal; use units of energy/mass.

10.7. For your country, what percentage of the energy is derived from wood or wood products? Be sure to state the country and from where you obtained data (year).

10.8. What are the possible energy products from MSW?

10.9. Find the nearest landfill to your hometown. List the town and location. Is the methane vented or used? If used, if possible state how much is generated per year and what its uses are.

10.10. Why did ethanol production in the United States increase significantly?

10.11. Find the location of the nearest ethanol plant to your hometown. State the location and capacity.

10.12. OM: What is the energetics for producing ethanol from corn, which is irrigated from wells 80 m deep? Average rainfall is 50 cm/year.

10.13. How many CHP plants from WWTFs are there in the United States?

10.14. What is the typical size range (power) of CHP plants at WWTFs in the United States?

10.15. What two countries in the European Union use the most biodiesel?

10.16. If you live in the United States, pick any other country in the world that uses biogas. Give an estimate of amount and use.

10.17. In China and India, what is the primary use for biogas?

10.18. Why is a there a large research interest in microalgae?

10.19. List two advantages and two disadvantages for using microalgae to produce biodiesel.

10.20. What was the approximate world biofuel production in 2015? Make an estimate for 2020 and 2030 and give reasons for your answer.

10.21. OM: Suppose you wanted to supply 50% of the U.S. demand for gasoline with ethanol from corn. Approximately how many hectares would be needed? 1 ha = 2.5 acres.

10.22. OM: Suppose you want to supply 50% of the U.S. demand for diesel with biodiesel from algae. Approximately how many hectares would be needed? Be sure to state how much diesel per year is used in the United States.

10.23. OM: Suppose you want to supply 50% of the world demand for diesel with biodiesel from algae. Approximately how many hectares would be needed? Be sure to state how much diesel per year is used in the world.

10.24. OM: Suppose India wants to supply 80% of its demand for fuel for transportation from jatropha. Approximately how many hectares would be needed?

10.25. The goal of 25% energy from bioenergy by 2025 for the United States would require approximately what amount of resources by type?

10.26. Compare the two methods of algae production: open ponds and photoreactors.

10.27. OM: The U.S. military is interested in biodiesel from algae. Approximately how many hectares would be needed to supply that demand? Be sure to state how much diesel per year is used.

11 Geothermal Energy

11.1 INTRODUCTION

The temperature gradient in the Earth's crust is 17°C–30°C per kilometer of depth. For example, deep mines are hot, and most need cooling for the miners. Plumes of magma ascend by buoyancy and force themselves into the crust, generally along the edges of tectonic plates (Figure 11.1), which results in volcanoes. There are huge regions of subsurface hot rocks with cracks and faults that allow water to seep into the reservoir, which then results in hot springs, geysers, mud pots, and fumaroles. Two famous examples are Yellowstone Park and Iceland, which is an exposed section of the Mid-Atlantic Ridge.

Geothermal energy is not renewable in the same sense as solar, wind, and hydro energy, and the average heat flow of the Earth is a thousand times less than the low-density solar insolation. Another major difference is that solar and wind energy are variable on short time periods, and hydro is variable by season; however, geothermal energy only declines as heat is taken out, with lifetimes of 100 or more years. Even though the heat flow is small, there are many locations in the world with reservoirs of hot rock with water and steam that can be used for heating and for the generation of electricity. These regions have average heat flow around 300 mW/m^2 compared with a global average of 60 mW/m^2. In the generation of electricity, the heat flow of the Earth is much less than the removal of energy from the hot rock reservoirs, so it is similar to mining. But the geothermal reservoirs are large, and they will produce energy for years, although some fields have declined and are being recharged with water or new wells with fracking. Other geothermal fields need energy for pumping the fluids from the reservoir.

The heat content per unit mass is a function of pressure, volume, and temperature; reservoirs are classified according to temperature: high (water and steam at temperatures of 182°C and above); medium (100°C–182°C); and low (less than 100°C; essentially no steam). Around 87 TWh of electricity was generated from geothermal energy in 2013 from an installed capacity of around 12,000 MW. In addition, the amount of thermal energy for direct use is over double the electrical energy.

With reservoir temperatures of 120°C–370°C, hot water or steam can be used to generate electricity in a conventional power plant. Hot water that is trapped in underground reservoirs within 1–6 km of the surface can be tapped by drilling. Enhanced geothermal systems (EGSs) are hot rock reservoirs that have to be modified by hydraulic fracturing because they have low permeability and porosity. There is also renewed interest in the energy potential of geopressure-geothermal resources and the geothermal fluids found in oil and gas production fields as well as in some mining operations.

Shallow reservoirs of lower temperature, 20°C–150°C, are used for space heating, greenhouses, aquaculture, industry, and health spas. The macaques of northern Japan use hot springs for warmth in the winter (for some great photos, go to Google

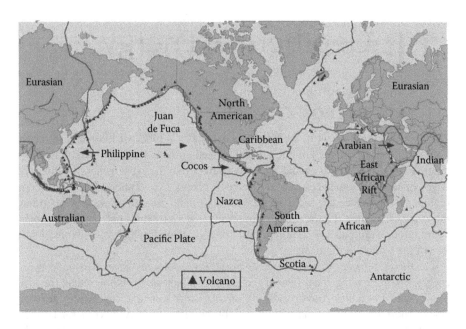

FIGURE 11.1   Tectonic plates of the Earth with volcanoes (historical).

images, search Macaques in Japan, http://www.google.com/imghp). Geothermal heat pumps (GHPs) use an electric heat pump to exchange heat with the ground or groundwater, instead of air, and can be used in almost all areas of the world. These systems for residences and larger buildings are now competing with conventional heating and cooling systems.

11.2   RESOURCE

The geothermal resource for direct use and the generation of electricity is located along the tectonic plate boundaries and magma plumes, such as in Hawaii and Yellowstone. The size of the resource (Table 11.1) could supply all the primary energy

TABLE 11.1
Geothermal Energy Resource Base for the World and the United States

Regime	World Continental (109 BOE)	United States (109 BOE)
Magma systems	2,400,000	160,000
Crustal heat	79,000,000	2,300,000
Thermal aquifers	130	9

Source: Geothermal energy, www.geothermal.org/GeoEnergy.pdf.
Note: BOE, barrel of oil equivalent.

for heat and electricity for the world. However, use is restricted due to location in relation to population, and of course, it is also restricted by economics. The total stored heat energy up to a depth of 5 km worldwide is estimated at $1.5 * 10^{26}$ J, and if 1% can be mined, the recoverable resource is on the order of $1.5 * 10^{24}$ J, which is way larger than the world use of energy in 2013, 530 EJ.

Regions and nations along the boundaries of the tectonic plates are using geothermal energy and have maps of the resource. In Europe (Figure 11.2), Italy, Iceland, and Turkey are the major areas. Iceland, with an area of 103,000 km² (Figure 11.3), has a geothermal resource estimated at 96,800 EJ within 3 km of the surface, and the technically capturable energy is 3,320 EJ, which at their current energy consumption would last for 40,000 years.

Information for regional and national geothermal resources is available in the Recommended Resources section of this chapter. An interactive map of the world with information on installations is available from Think Geoenergy, http://map.thinkgeoenergy.com. Two other sites for geothermal is the International Geothermal Association, http://www.geothermal-energy.org, and Geothermal Energy Association,

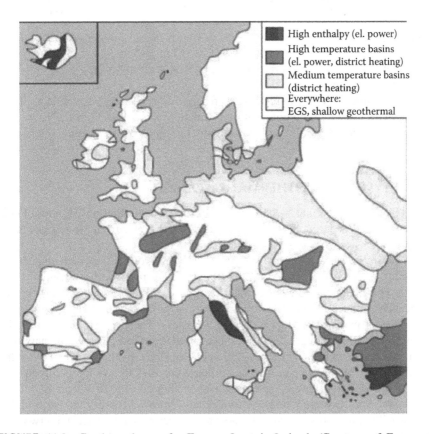

FIGURE 11.2 Geothermal map for Europe. Inset is Iceland. (Courtesy of European Geothermal Association.)

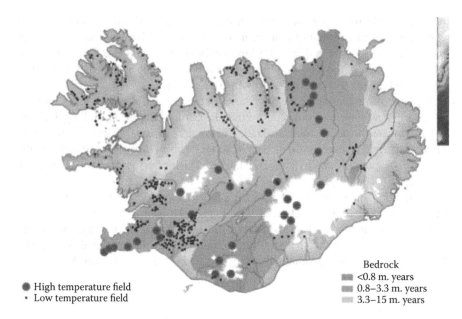

High temperature field
• Low temperature field

Bedrock
▨ <0.8 m. years
▨ 0.8–3.3 m. years
▨ 3.3–15 m. years

FIGURE 11.3 Geothermal map for Iceland showing high- and low-temperature fields.

the United States, which has resource maps for 14 states, http://geo-energy.org/gea_resource_maps.aspx.

In the United States, the resource is primarily in the west (Figure 11.4). There are 271 cities and communities with a population of 7.4 million in the 10 western states that could potentially utilize geothermal energy for district heating and other applications.

11.3 TYPES OF GEOTHERMAL RESOURCES

The types of geothermal resources are convective hydrothermal systems, EGSs, conductive sedimentary systems, coproduction with water from oil and gas fields, geopressure systems, and magma energy.

The convective hydrothermal system is a heat reservoir that has high enough permeability and porosity for the convection of water. The reservoir has a nonpermeable rock layer to retain the heat but enough fractures for recharge (Figure 11.5). Surface indications of such reservoirs are hot springs, geysers, and the like, and the reservoirs may or may not be associated with volcanoes. There is a limited distribution of these worldwide; however, they have been used throughout history and have been exploited for direct use and the generation of electricity. Estimated reserves for five western states plus Alaska in the United States are 10,000 MW for identified sites, 30,000 MW for undiscovered resources, and another 500,000 MW from EGS [1].

Regions for EGSs are found worldwide and feature hot conductive rock with low permeability and porosity. EGSs are now at the demonstration stage, with two projects in the United States (2.45 MW) and one project in Australia (1 MW). Boreholes to check the thermal gradient (Figure 11.6) have to be drilled, and then production

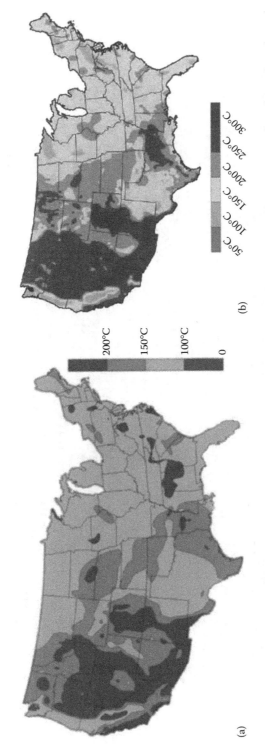

FIGURE 11.4 (a) Geothermal map for the United States, depth 6 km. (From EERE, DOE.); (b) geothermal resource map for enhanced geothermal systems for the United States, depth 10 km. (Courtesy of Geothermal Energy Association. March 2009.) *(Continued)*

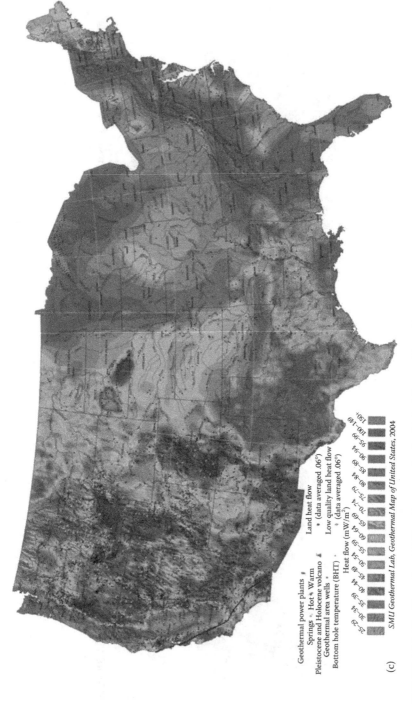

Geothermal power plants ⚡
Springs ◦ Hot ◦ Warm
Pleistocene and Holocene volcano ◢
Geothermal area wells •
Bottom hole temperature (BHT) •

Land heat flow
  • (data averaged .06°)
Low quality land heat flow
  • (data averaged .06°)
Heat flow (mW/m²)

25–29
30–34
35–39
40–44
45–49
50–54
55–59
60–64
65–69
70–74
75–79
80–84
85–89
90–94
95–99
100–149
150+

SMU Geothermal Lab, Geothermal Map of United States, 2004

(c)

FIGURE 11.4 (Continued)   (c) Heat flow map for the United States. (From Southern Methodist University.)

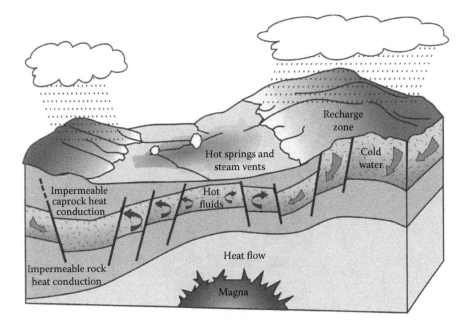

FIGURE 11.5   Hydrothermal hot reservoir.

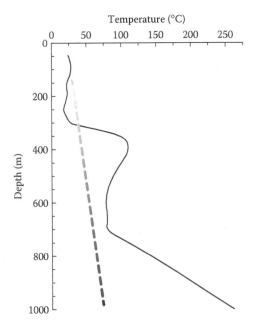

FIGURE 11.6   Thermal gradient in a borehole in geothermal area compared to typical linear gradient.

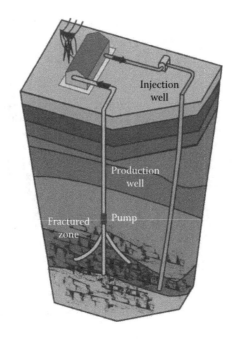

FIGURE 11.7 Enhanced geothermal system from formation that has been hydraulic fractured.

and injection wells have to be drilled and the formation fractured (Figure 11.7) with the same techniques used for oil and gas production. The primary challenges are creating a large fractured rock volume, providing enough flow for commercial production, minimizing cooling, and minimizing water loss. The main constraint is creating sufficient connectivity within the stimulated region of the EGS reservoir to allow for high production rates without reducing reservoir life by too rapid cooling.

Conductive sedimentary systems have basins with high heat flow, and in general, fracturing would be required. However, the resource is deep, so drilling costs will be high; at present, there are no commercial systems.

Coproduction of geothermal energy from oil and gas fields is a possibility. Some areas of oil and gas development have relatively high temperatures; for example, parts of east and south Texas and northwest Louisiana are characterized by temperatures of 150°C–200°C at depths of 4–6 km. In addition to temperature requirements, a geothermal project requires large-volume flows of water, about 4 m³/min per megawatt (depending on the temperature).

Many oil and gas wells in the United States are stripper wells, defined by production of oil less than 10 bbl/day and gas less than 2,100 m³/day. In some wells, the ratio of water to oil is higher than 10/1. Also, water flooding is a common technique for secondary recovery in old oil fields. In Texas and Oklahoma, the water production from oil fields is over 8,000 m³/min. There is the expense of disposal of this saltwater, primarily by reinjection. At an average temperature of 150°C, that would be an equivalent power of 8,000 MW. In general, these oil fields are not close to load centers, so the geothermal energy has never been used.

For depleted oil fields where subsurface temperatures are high enough, the wells could be converted to produce hot water instead of capping them, but the cost of the hot water would still be high due to pumping costs. There was one demonstration project in Wyoming, and one was planned for Florida. The Rocky Mountain Oil Test Center (http://www.rmotc.doe.gov/index.html) had a demonstration program near Casper, Wyoming. In August 2008, a 250-kW Ormat organic Rankine cycle (ORC) power unit was designed to use 40,000 barrels per day of 77°C water, and as of February 2009, the unit had produced 586 MWh of power from 3 million barrels of hot water [2]. System was shut down and then restarted and operated from September 2009 to end of February 2010 producing 478 MWh. The test center was closed as of October 2014 as the Naval Petroleum Reserve No. 3 (NPR-3) was sold. NPR-3 includes 9,481 acres, over 400 oil wells with current production of approximately 240 barrels/day.

For small producers, use of geothermal energy could reduce the cost of pumping the fluid. A demonstration project in Mississippi (June–October 2011) had a 30-kW unit on an oil well (2,900 m deep), which pumped 100 bbl oil and 4,100 bbl of water per day from a water flood field. The waste heat unit was estimated to pay about one-third of the pumping costs with a payback of 3–4 years as it would replace electric energy at $0.098/kWh for pumping.

Geopressure systems have reservoir pressures higher than hydrostatic pressure; however, they have limited distribution. In addition to thermal energy, there is kinetic energy and energy from methane. There is a large geopressure basin that runs along the Gulf Coast of the United States from the Mexican border to Mississippi with a geothermal resource of 46,000 EJ [3], and that does not count the energy in the natural gas (methane). To date, there are no commercial projects, although the geopressure characteristics have been studied extensively, wells were drilled, and there has been a feasibility project. From late 1989 until early 1990, a 1-MW$_e$ (megawatt electric) plant was operated on the Pleasant Bayou near Houston, Texas, from a well that produced hot water and natural gas. About half of the power was generated by a binary cycle plant and about half by a gas engine with a generator. The plant only operated for 6 months because of economics, primarily due to the low price of natural gas at that time. The well was flow tested for about 5 years with limited drawdown, so the reservoir is sufficiently large to sustain production for many years. There are several technical challenges for recovering geopressure energy. Magma energy is localized, and there are many technical challenges.

The U.S. geothermal resource has been estimated for the different types (Table 11.2). Also, the report [4] shows resource maps for different depths below the surface.

The hydraulic power (m³-m/time) is determined by the volume pumped and the dynamic head. The amount of power in the fluid is calculated from the mass flow, the enthalpy (energy/mass) at the input and output temperatures. Values for enthalpy are obtained from tables.

$$P = (m/t) * (Eh1 - Eh2) \tag{11.1}$$

where:

P is power
m/t is the mass flow (kg/s)

TABLE 11.2

Estimated Geothermal Resource by Type in the United States

Type	Resource to 10 km (EJ)
Convective	2.4–$9.6 * 10^{21}$
Enhanced geothermal	$1.4 * 10^{25}$
Conductive	$1 * 10^{23}$
Oil/gas field water	1.0–$4.5 * 10^{17}$
Geopressure	0.7–$1.7 * 10^{23}$
Magma energy	$7.4 * 10^{22}$

Eh1 is the enthalpy at the wellhead (heat/mass of saturated liquid, temperature 1, degrees kelvin)

Eh2 is the enthalpy at the output (heat/mass of saturated liquid, temperature 2, degrees kelvin)

Example 11.1

A well produces the following: $m/t = 32$ kg/s, Eh1 = 377 kJ/kg (90°C), Eh2 = 209 kJ/kg (50°C)

$P = 32$ Kg/s $* (377 - 209)$ KJ/kg $= 5,380$ KJ/s $= 5.4$ MW$_t$ (megawatts thermal)

Because of the confidence level for fluid production, use 0.75, then $P = 4.0$ MW$_t$ per well

If you have to pump the fluid to the surface, then that energy has to be subtracted from energy obtained in the fluid. The production for a geothermal field can be estimated [5] for the load data and use, especially peak demand. As noted, the design engineer should have a fairly good idea of the characteristics of the reservoir.

11.4 DIRECT USE

The direct use of geothermal energy in the world is estimated at 70 GW$_t$; however, the contribution of shallow reservoirs, especially for bathing, spas, local space heating, and other small installations, is more difficult to estimate. The main direct uses are for space heating, bathing, and swimming. For the different applications, capacity factors are estimated at 15%–72%. The leading countries for direct use are the United States (12.6 GW$_t$) and China (32.0 GW$_t$). Direct use in Europe is estimated at 20 GW$_t$, and the major countries making direct use are Sweden (4.5 GW$_t$), Norway (3.3 GW$_t$), Germany (3.2 GW$_t$), and Turkey (2.0 GW$_t$), primarily for district heating and swimming pools and spas [5, 2011 data]. In Iceland, direct use of geothermal energy is around 2.1 GW$_t$, and 90% of the homes are heated by geothermal hot water. The main use of geothermal energy in Hungary is for direct use with a production rate of 120 million m^3/year. However, due to the high production, the hydraulic head has decreased 50–70 m, so now reinjection is required.

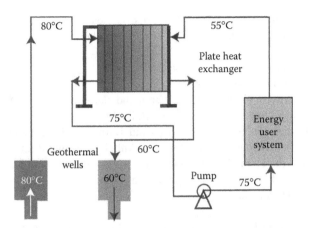

FIGURE 11.8 Diagram of direct use geothermal system with heat exchanger.

Bathing, spas, and local space heating are generally from hot springs, so no external energy may be needed to move the hot water. For other direct use systems, a well is drilled into the reservoir, and there are pumps to bring the hot water to the surface. Then, there are heat exchangers, pipes for distribution, controls, and a disposal system, either on the surface or reinjected (Figure 11.8). Surface disposal means that water quality standards must be met. Standard equipment is used in most direct use systems, with allowances for water quality and temperature as corrosion and scaling may lead to problems. Atmospheric oxygen must be prevented from entering energy district heating waters, which can be accomplished by heat exchangers. A peaking system may be necessary to meet maximum load, which can be a parallel source of energy or tank storage.

An interactive map of U.S. direct use projects is available (http://geoheat.oit.edu/dusys.htm). Twenty-one case studies for the United States are available from the Geo-Heat Center, Oregon Institute of Technology (http://geoheat.oit.edu/casestudies.htm).

11.4.1 SPRINGS, SPACE HEATING, AND OTHERS

The earliest use of natural hot springs was for bathing and cooking; today, resorts, swimming pools, and space heat are still major users of direct heat. Besides space heating for homes, the use of geothermal energy expanded into other areas: greenhouses, aquaculture, drying of agriculture products and lumber, desalinization, industrial processes that require heat, heated sidewalks, and even cooling.

There have been a few cases for which tubes are placed in the ground, and air is circulated through the tubes for heating and cooling. The problem is that a large mass of air needs to be circulated, and the heat exchange area with the ground also needs to be large.

11.4.2 DISTRICT HEATING

District heating is used in a number of places in the world, and geothermal energy is one of the sources. There are over 5,000 district heating systems in Europe, which includes over 240 geothermal systems (4.3 GW$_t$ capacity with a production of

12.9 TWh/year). Turkey uses 493 MW$_t$ for district heating [6], and there are a number of district heating utilities in Europe that use geothermal energy [7]:

France: 38 plants in the Paris region and others in Aquitaine
Italy: Po-plain (Ferrara), Tuscany
Germany: Northern Germany (e.g., Waren and Neustadt-Glewe), Munich area
    (e.g., Erding, Unterschleissheim, and Pullach)
Poland: Northern Poland (e.g., Pyrcyze) and Tatra foothills (Zakopane)
Austria: North and south of the Alps (e.g., Altheim and Bad Blumau)
Hungary: All the Pannonian basin (e.g., Hódmezövásárhely and Kistelek)
Others: Denmark (Thisted and Copenhagen), Sweden (Lund and Malmó), and
    Lithuania (Klaipeda)

The district heating utility in Reykjavik, capital of Iceland, is the largest in the world, as geothermal energy is used to heat the entire city and five neighboring communities. The geothermal power is about 780 MW$_t$, and 60 million m^3/year of hot water flow through the distribution system. Two of the low-temperature fields are located within the city limits, and the other two (high temperature) are 20 km northeast of Reykjavik. Initially, only the flow from springs and relatively shallow artesian wells was used, but in the 1960s and early 1970s, production wells were drilled in all the fields, and down-hole pumps were installed to increase the flow. Pumping in the low-temperature fields lowered the water level, and surface springs disappeared. By increasing water from the high-temperature fields, it was possible to reduce the pumping from the low-temperature fields, and the water level increased. About 70% of the energy used for district heating comes from the low-temperature fields; the rest is from the high-temperature fields. The water from the low-temperature fields is used directly for heating and tap water. Due to a high content of gases and minerals at the high-temperature fields, water and steam are used to heat freshwater. From 1998, electricity has been cogenerated from geothermal steam at another field.

The largest direct heating district in the United States is in Boise, Idaho [8,9]. There are four independent heating districts using geothermal energy in Boise. The Boise Public Works supplies heat to over 65 businesses in the downtown area, primarily space heating, but the supply also includes hot water for recreation, greenhouses, and aquaculture. Now, 100% of the water is injected back into the aquifer. The State operates the system that heats the State Capitol and several other buildings within the Capitol Mall Complex. The other two systems are the Veterans Administration and the Boise Warm Springs Water district for residential properties in the vicinity of Warm Springs Avenue.

The Oregon Institute of Technology campus in Klamath Falls has a district heating system. Since 1964, the campus has been heated by geothermal hot water from three wells. The combined capacity is 62 L/s of 89°C water, with an average heat utilization rate of 0.53 MW$_t$ and a peak rate of 5.6 MW$_t$. In addition to heating, a portion of the campus is cooled using an absorption chiller powered by geothermal hot water. The chiller requires a flow of 38 L/s and produces 541 kW of cooling capacity with 23 L/s of chilled water at 7°C. Plate heat exchangers have been installed in all buildings to isolate exposure to the geothermal fluids.

11.4.3 CASE STUDIES

1. Canby, California [10, Chapter 9]. Detailed analysis from inception to operation, including environmental aspects and costs. System design of 434 kW, however with computer control of flow rate to meet demand. Backup system is propane boiler, only used when down hole pump is off due to failure. Long-term production capacity estimated at 8.2 m^3/h; well depth 640 m, potential drawdown of 75 m, discharge to local river (filtration to remove mercury to 1 ng/l)
2. Riehen district heating, Basel, Switzerland. Production well, ~72 m^3/h from 1,547 m depth with water at 65°C is cooled to 25°C by heat pumps for distribution, then water is fed back into injection well, 1,247 m. http://www. geoener.es/pdf/ponencias2010/L-Rybach-Geothermal-energy-applications-in-buildings.pdf
3. In Klamath Falls, Oregon, there are over 550 geothermal wells serving a wide variety of uses. There is a city district heating system (8.5 MW$_t$ at 99°C) and snow melt systems (1.2 MW$_t$ at 99°C) for sidewalks; see "Geothermal in Oregon, Where It Is Being Used," http://www.oregon.gov/ ENERGY/RENEW/docs/tp124.pdf

11.5 GEOTHERMAL HEAT PUMPS

The installation of GHP, ground source heat pumps (GSHPs), or geoexchange systems has been growing rapidly across the developed world. GHPs can be used in almost all locations around the world. Since the ground temperature is fairly constant, around 10°C at 5 m depth, the ground can be used as a reservoir for heating and cooling with heat pumps, with heat taken out in the winter and injected in the summer. Remember for GHP that the heat pump is powered by electricity.

The types of GHPs are ground or ground source, groundwater source, and water source. Sometimes, the systems are referred to as geoexchange systems. The system can be a closed loop with boreholes, lateral bed, and ponds and lakes as the reservoir (Figure 11.9) and open loop (Figure 11.10) from an existing water source with

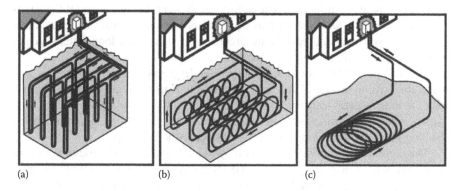

(a)   (b)   (c)

FIGURE 11.9   GHP, closed-loop systems: (a) borehole, (b) lateral, and (c) pond/lake.

FIGURE 11.10 GHP, open-loop system.

recharge or disposal. Heat pumps that use air as a reservoir have auxiliary resistance heating during freezing weather, so GHPs are more efficient.

The open-loop system uses well or surface body water as the heat exchange fluid that circulates directly through the GHP system. Once it is circulated through the system, the water is returned through the well, a recharge well, or surface discharge. This option is only practical if there is an adequate supply of relatively clean water, and all local codes and regulations regarding groundwater discharge are met.

Which system is used depends on climate, soil conditions, available land, and installed costs. The composition and properties of the soil and rock (which can affect heat transfer rates) affect the design as soil with good heat transfer properties requires fewer pipes. Also, the local conditions will determine whether lateral (Figure 11.11) or boreholes are used. Ground- or surface water availability (depth, volume, and water quality for open-loop systems) also plays a part in deciding what type of ground loop to use. Before an open-loop system is installed, be sure the hydrology of the site is known, so potential problems such as aquifer depletion and groundwater contamination can be avoided. Antifreeze fluids circulated through closed-loop systems generally pose little to no environmental hazard.

The amount and layout of the location of underground utilities or sprinkler systems also have to be considered in system design. Horizontal ground loops (generally the most economical) are typically used for newly constructed buildings with sufficient land. Vertical installations or slinky installations are often used for existing buildings because they minimize the disturbance to the landscape.

The biggest benefit of GHPs is that they use 25%–50% less electricity than conventional heating or cooling systems. GHPs can reduce energy consumption up to 44% compared to air-source heat pumps and up to 72% compared to electric resistance heating or conventional heating/cooling systems.

(a)   (b)

FIGURE 11.11 Examples of closed-loop system: (a) lateral with vertical slinky (Courtesy of Virginia Tech.) and (b) lateral with horizontal slinky. (Courtesy of Air Solutions.)

The heating efficiency of heat pumps is indicated by their coefficient of performance (COP), which is the ratio of heat provided per energy input. The cooling efficiency is indicated by the energy efficiency ratio (EER), which is the ratio of the heat removed to the energy input. Units should have a COP of 2.8 or greater and an EER of 13 or greater.

In the United States, around 1.8 million GHPs have been installed. In the past, the Energy Information Administration provides data on number of GHPs shipped (Figure 11.12) by type: water source, groundwater and ground source, direct geo-exchange, and others (Figure 11.13). Last data published by Energy Information Administration was 2009, after that annual report has been terminated. GHPs have a major impact on direct use in the United States as GHPs provide more energy than district heating. A ton is the unit used for heat pumps and air conditioners; 1 ton of cooling = 12,000 Btu/h = 3.516 kW.

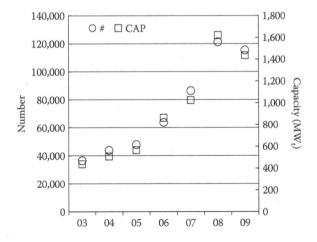

FIGURE 11.12 Number of GHPs in the United States, shipped and capacity.

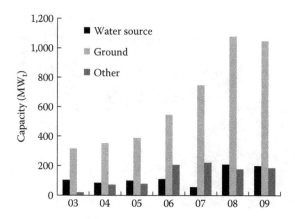

FIGURE 11.13 Capacity of GHPs in the United States, shipped by type.

For the United States, case studies (by type and number of cases) are available from the Geoexchange (http://www.geoexchange.org/library/case-studies): residential (7), commercial (5), office building (7), schools (15), affordable housing (5), and government facilities (11).

In Europe, the growth rate has been huge for GHPs [11], and by the end of 2014, over 1,000,000 units had been installed, equivalent to 12 GW_t of capacity. Installations are at the rate of 90,000/year, and Germany was the lead market; however, sales per year are lower after 2008 due to reduction in incentives. The use of GHPs increased dramatically in China since 2007, now around 32 GW_t. China reports building area heated, estimated at $5 * 10^7 \, m^2$ by end of 2015, rather than number of installations. The Olympic (2008) Village buildings used GHP, which used sewage water as the source.

11.6 ELECTRICITY

Electricity is generated in conventional power plants using the high- and medium-enthalpy heat reservoirs. The advantages of geothermal energy are as follows: It can provide peaking and base load power, and it is modular and in that more wells could be drilled if the reservoir is large enough. Some disadvantages are the high mineral content, which causes corrosion problems and environmental problems if downstream water is disposed on the surface; limited distribution of reservoirs (however, not a large problem if EGSs are considered); overproduction with need for pumps or reinjection of fluid; and a long period between starting of project and commercial operation.

World installed capacity of electric power from geothermal energy is around 12 GWe, and production is around 80 TWh/year (Figure 11.14), and around 2 GW are under construction in 14 countries [12]. The United States has the largest number of projects and largest capacity (Table 11.3) (3,442 MW_e in 2013); however, the geothermal component is only 0.4% of U.S. electric production. Other countries (Figure 11.15) obtain a larger percentage of their electricity from geothermal sources; for example, Iceland 29%, El Salvador 25%, Philippines 15%, New Zealand 14%, and Indonesia 11%. In the European Union-28, the installed capacity is 952 MW and

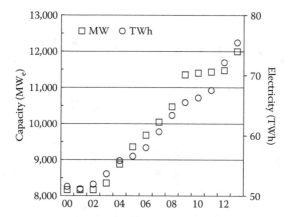

FIGURE 11.14 World installed capacity for electric power and energy from geothermal sources.

TABLE 11.3
Electric Capacity (MW) of
Geothermal Power Plants, 2013

United States	3,442
Philippines	1,904
Indonesia	1,333
Mexico	1,005
Italy	901
New Zealand	895
Iceland	664
Japan	537
Turkey	275
Kenya	237
Costa Rica	208
El Salvador	204
Nicaragua	104
Russia	97
Papua New Guinea	56
Guatemala	42
Portugal	29
China	28
France	17
Germany	29
Ethiopia	8
Australia	1
Austria	1
Thailand	0.3

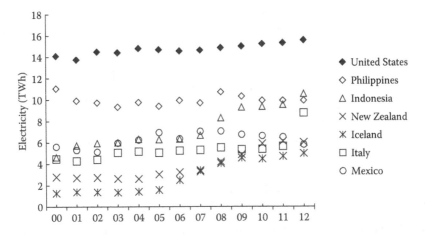

FIGURE 11.15 Leading countries, production of electric energy from geothermal sources.

the production was 6 TWh in 2013. The types of geothermal systems for producing electricity are dry rock (steam), flash, binary, and others with 25%, 58%, 14%, and 3%, respectively, of the world geothermal electric power plants (around 500). For more technical aspects see Glassley [10], Chapter 9, which also has a detailed case study of the Geysers plant in California. Average plant size was 44 MW for dry steam, 31 MW for flash, and 3.2 MW for binary. Theoretical thermodynamic efficiency for the generation of electricity from conventional steam plants is determined by temperatures (Equation 2.5). A more efficient system would be combined heat and power (CHP), with the lower-temperature heat used before the fluid is reinjected into the reservoir. An EGS 1.5-MW$_e$ pilot project is located in Soultz, Alsace, France [13], and the United States has five EGS demonstration projects with details on each project [14].

One aspect that was not given much consideration for EGSs was occurrence of earthquakes. A project in Basel, Switzerland, was suspended because more than 10,000 seismic events measuring up to 3.4 on the Richter Scale occurred over the first 6 days of water injection [15]. To stimulate the reservoir for a proposed hot, dry rock geothermal project, approximately 11,500 m^3 of water was injected at high pressures into a 5-km deep well from December 2 to December 8, 2006. A six-sensor borehole array, installed at depths between 300 and 2,700 m around the well, recorded more than 10,500 events during the injection phase.

11.6.1 DRY STEAM

Power plants using dry steam systems (Figure 11.16) were the first type of geothermal power generation plants built. They use steam from the geothermal reservoir as it comes from wells and route it directly through turbine/generator units to produce electricity. An example of dry steam generation can be found in the Geysers region (http://www.geysers.com) in northern California; this is the largest geothermal project in the world. The area was known for its hot springs in the mid-1800s, and the first well for

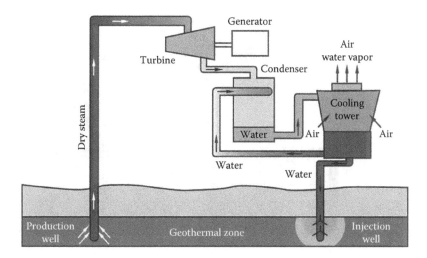

FIGURE 11.16 Diagram of dry steam geothermal system.

power production was drilled in 1924. Starting in the 1950s, deeper wells were drilled, and 26 power plants had been built by 1990, with a capacity of more than 2 GW.

Because of the rapid development in the 1980s and because of the operation of surface discharge, the steam resource started declining in 1988. Today, the operating capacity is 725 MW; however, the Geysers facilities still meet around 60% of the average electrical demand for northern California. The plants use an evaporative water-cooling process to create a vacuum that pulls the steam through the turbine, producing power more efficiently. However, this process loses 60%–80% of the steam to the air, with no reinjection. Although the steam pressure was declining, the reservoir is still hot. To rectify the situation, the Santa Rosa Geysers Recharge Project involved transporting 42,000 m³ per day of treated wastewater from neighboring communities through a 64-km pipeline and injecting it into the ground to provide more steam. The project came online in 2003, and further expansion is planned to increase the wastewater to nearly 76,000 m³ per day [16]. One concern with open systems like the Geysers is that they emit some air pollutants. Hydrogen sulfide—a toxic gas with a highly recognizable *rotten egg* odor—along with trace amounts of arsenic and minerals are released in the steam.

11.6.2 FLASH

Flash steam plants (Figure 11.17) are the most common type of geothermal power generation plants in operation today. They use water at temperatures greater than 182°C that is pumped under high pressure to the generation equipment at the surface. On reaching the generation equipment, the pressure is suddenly reduced, and some of the hot water is converted (flashed) into steam, which is used to power the turbine/generator units. The remaining hot water not flashed into steam and the water condensed from the steam are generally pumped back into the reservoir. An example of an area using the flash steam operation is the CalEnergy Navy I flash geothermal power plant at the Coso geothermal field.

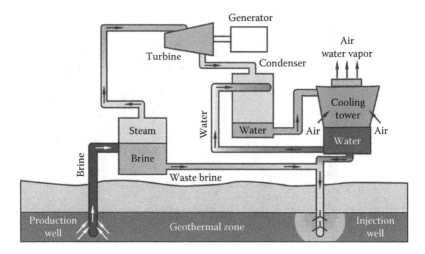

FIGURE 11.17 Diagram of a flash geothermal system.

FIGURE 11.18 Yangbajain, Tibet, geothermal field in 1990.

The most important field in China is the Yangbajain field in Tibet (Figure 11.18), which has eight double flash units for a capacity of 24 MW. There are 18 wells with an average depth of 200 m in the water-dominated shallow reservoir at 140°C–160°C. The field extension is only 4 km², although there are indications of a thermal anomaly of 15 km². The annual energy production is approximately 100 GWh, about 30% of the needs of the Tibetan capital, Lhasa. A deep reservoir has been discovered beneath the shallow Yangbajain field, characterized by high temperatures (250°C–330°C), with an estimated potential of 50–90 MW. A 2,500-m exploratory well was drilled in 2004, reaching the deep reservoir at

1,000–1,300 m. Other plants are installed in Langju, West Tibet (two double flash units, 1 MW each, 80°C–180°C), and a 1-MW binary power station (60°C–170°C) is operating in Nagqu. In other regions of China, two small 300-kW plants are operating in Guangdong and Hunan.

The Wairakei, New Zealand, power plant was first commissioned in 1958, and the present output is 140 MW_e, with annual production averaging 1,250 GWh at a capacity factor of 93%. At least 150 wells have been drilled in the field, which generally produce fluid at temperatures between 209°C and 261°C. About 5,300 tons of fluid, 1,500 tons of steam, and 3,000 tons of water (130°C) per hour are currently taken from the reservoir. Some of the steam is taken directly from shallow dry steam production wells (up to 500 m deep) and piped to the turbines. About half of the separated water is now reinjected, and half is discharged to the Waikato River. All steam condensate is discharged to the river. There had been some subsistence in some areas as the fluids were removed. In the Wairakei geothermal field, the Te Mihi power plant, 166 MW_e, will replace elements of the historic Wairakei station. Information on all geothermal fields in New Zealand is available from the New Zealand Geothermal Association (http://www.nzgeothermal.org.nz/index.html).

11.6.3 BINARY PLANTS

Binary, also known as ORC, geothermal power generation plants differ from dry steam and flash steam systems because the water or steam from the geothermal source never comes in contact with the turbine/generator units. In the binary system, the water from the geothermal reservoir is used to heat another working fluid, which is vaporized and used to turn the turbine/generator units (Figure 11.19). The advantage of the binary cycle plant is that it can operate with lower-temperature water, 107°C–182°C using working fluids that have a lower boiling point than water. They also do not produce air emissions. An example of a binary cycle power

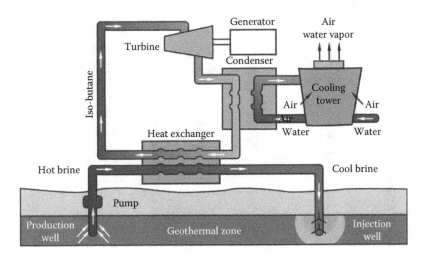

FIGURE 11.19 Diagram of a binary geothermal system.

generation system is the Mammoth Pacific binary geothermal power plants at the Casa Diablo, California, geothermal field.

11.6.4 COMBINED HEAT AND POWER

A CHP plant fueled by geothermal sources is similar to any other CHP in that the lower-temperature fluid after the production of electricity is used for other purposes (Figure 11.20). For example, separated water from the Wairakei field in New Zealand is used to provide hot fluids for the Netcor tourist facility and provides heat for a prawn farm adjacent to the power station. In Iceland, a system has 45 MW$_e$ and 150 MW$_t$ for district heating for an airport and nine cities. There are two operating CHP plants in Austria and three in Germany. Check with European Geothermal Energy Council for more details (http://www.egec.org).

Landau, Germany: Two wells, 3,300 m, 160°C, flow rate 250 m³/h, 3 MW$_e$
Unterhaching, Germany: Two wells, 3,200 m, 122°C, 3.6 MW$_e$
Neustadt-Glewe, Germany: One well, 2,250 m, 122°C, 230 kWe, heat for
    1,300 homes

11.6.5 OTHER SYSTEMS

Other systems are power plants of triple flash, back pressure, flash/binary hybrid, EGS, and other geothermal technology. EGS are also classified as infield (existing geothermal plant), nearfield (adjacent to existing geothermal plant), and greenfield (new EGS reservoir). There is a solar/geothermal hybrid power system at Stillwater geothermal project in Nevada, where 24 MW$_e$ of PV is connected to existing 47 MW$_e$ geothermal power plant. A demonstration project in China has a 400 kW generator

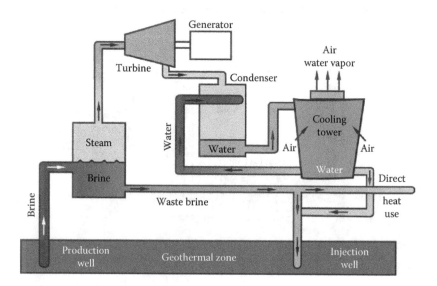

FIGURE 11.20 Diagram of geothermal system for combined heat and power.

using co-produced water at the Huabei oilfield. A modular waste heat to power system has been developed and the system is currently producing power from the brine stream used in the gold-leaching process at the Florida Canyon Mine in Nevada.

11.7 COMMENTS

Electricity is produced from geothermal sources in 24 countries, with 5 countries obtaining 11%–29% of their electric demand. There has been growth in the electric power capacity from geothermal sources since 2007, and by the year 2050 with existing technology, the projection is 70 GW; with EGSs, the projection is 140 GW. Some feel that the usable geothermal resource could be as high as 1,000 GW.

The growth rate of GHPs in the developed world is large, and there will be continued growth as China and other countries install GHPs. The projections for 2050 are 750 GW_t of GHPs and 50 GW_t for direct use (not GHPs). Remember that GHPs still require electricity, but this is more efficient than providing heating and cooling from conventional electric production from fossil fuels, and emissions would be reduced by 45%.

REFERENCES

1. *Assessment of Moderate- and High-Temperature Geothermal Resources of the United States.* 2008. http://pubs.usgs.gov/fs/2008/3082/pdf/fs2008-3082.pdf.
2. L.A. Johnson, E.D. Walker. 2010. *Ormat:Low-Temperature Geothermal Power Generation,* final report. 2010. http://www.geothermalcommunities.eu/assets/elearning/7.24.Ormat_report.pdf.
3. Geothermal Energy Association. March 2009. *U.S. Geothermal Production and Development Update.* http://www.geo-energy.org/reports/Industry_Update_March_Final.pdf.
4. *The Future of Geothermal Energy, Impact of Enhanced Geothermal Systems (EGS on the United States) in the 21st Century.* 2006. http://geothermal.inel.gov/publications/future_of_geothermal_energy.pdf.
5. *IEA Geothermal Implementing Agreement, Annual Report 2011.* http://iea-gia.org/wp-content/uploads/2013/10/2011-GIA-Annual-Report-Final-4Oct13.pdf.
6. O. Mettoglu. 2006. *Geothermal District Heating Experience in Turkey.* http://www.oit.edu/docs/default-source/geoheat-center-documents/quarterly-bulletin/vol-22/22-2/22-2-art3.pdf?sfvrsn=4.
7. Geothermal District Heating. http://geodh.eu/about-geothermal-district-heating/.
8. Geothermal heating district. http://publicworks.cityofboise.org/services/geothermal/.
9. *Direct Use Geothermal Heating District, Boise, Idaho.* http://operations.boisestate.edu/architecture-engineering/files/2012/02/FutureCitiesGeo11.pdf.
10. W. E. Glassley. 2014. *Geothermal Energy; Renewable Energy and the Environment,* 2nd Ed. CRC Press, New York.
11. *The State of Renewable Energies in Europe.* 2013. Observ'ER, http://www.energies-renouvelables.org/observ-er/stat_baro/barobilan/barobilan13-gb.pdf.
12. Geothermal Energy Association. 2014. *Annual U.S. and Global Geothermal Power Production Report.* http://geo-energy.org/events/2014%20Annual%20US%20&%20Global%20Geothermal%20Power%20Production%20Report%20Final.pdf.
13. J. Genter. 2008. *The EGS Pilot Plant of Soultz-sous-Forêts.* http://www.egec.org/target/strasbourg08/EGEC%20WS%20strasbourg%2007%20180608.pdf. Alsace, France, case study. http://www.soultz.net/version-en.htm.

14. US DOE. EERE, Geothermal Technologies Office. *Enhanced Geothermal Systems Demonstration Projects*. http://energy.gov/eere/geothermal/enhanced-geothermal-systems-demonstration-projects.

15. N. Deichmann, M. Mai, F. Bethmann, J. Ernst, K. Evans, D. Fäh, D. Giardini et al. 2007. *Seismicity Induced by Water Injection for Geothermal Reservoir Stimulation 5 km below the City of Basel*, Switzerland. http://adsabs.harvard.edu/abs/2007AGUFM. V53F.08D.

16. A. Braxton Little. July 2010. Clean energy from dirty water. *Sci Am*, 64.

RECOMMENDED RESOURCES

R. DiPippo. 2012. *Geothermal Power Plants, Principles, Applications, Case Studies and Environmental Impact*. 3rd Ed. Butterworth-Heinenmann, Oxford.

INTERACTIVE MAPS

Europe. *Very Shallow Geothermal Potential, Thermomap Viewer*. http://geoweb2.sbg.ac.at/thermomap/.

International Renewable Energy Agency. *Global Atlas for Renewable Energy*. http://irena.masdar.ac.ae.

Natural Resources Defense Council. *Renewable Energy Map*. http://www.nrdc.org/energy/renewables/map_geothermal.asp#map.

NREL. Geothermal prospector. http://maps.nrel.gov/gt_prospector.

Oregon Institute of Technology. *US Geothermal Projects and Resource Areas*. http://geoheat.oit.edu/dusys.htm.

Southern Methodist University. *National Geothermal Data System*. http://geothermal.smu.edu/gtda/.

Think GeoEnergy Map. http://map.thinkgeoenergy.com. Projects.

U.S. DOE. *Mapping Geothermal Heat Flow and Existing Plants*. http://www.energy.gov/articles/mapping-geothermal-heat-flow-and-existing-plants.

LINKS

Canadian Geothermal Association. http://www.cangea.ca/.

Energy Information Administration. 2010. *Geothermal Explained*. http://www.eia.gov/energy-explained/index.cfm?page=geothermal_home. International section has data on geothermal generation of electricity.

European Geothermal Energy Council. http://www.egec.org.

European Heat Pump Association. http://www.ehpa.org.

Geo Heat Center, Oregon Institute of Technology. http://geoheat.oit.edu/.

Geothermal District Heating, Europe. http://geodh.eu. Has a GeoDH Map (GIS), http://geodh.eu/library/.

Geothermal Education Office. http://geothermal.marin.org/.

Geothermal Energy Association. http://www.geo-energy.org. 2013. *Geothermal Power: International Market Overview*. http://geo-energy.org/events/2013%20International%20Report%20Final.pdf.

Geothermal Heat Pump Consortium. http://www.geoexchange.org.

Global Heat Flow Database of the International Heat Flow Commission. http://www.heatflow.und.edu/.

Ground Source Heat Pump: A Guidebook. 3rd Ed. 2007. http://www.erec.org/fileadmin/erec_docs/Projcet_Documents/RESTMAC/GSHP_brochure_v_2008.pdf.

Ground Source Heat Pump Information Center. http://www.virtualpet.com/portals/okenergy/gshp.htm.

IEA Geothermal. http://iea-gia.org; Trends in geothermal application. 2012. http://iea-gia.org/wp-content/uploads/2013/11/2012-Trend-Report-Weber-30Aug14.pdf.

International Geothermal Association. http://www.geothermal-energy.org.

International Ground Source Heat Pump Association. http://www.igshpa.okstate.edu/index.htm.

Market Trends Report. 2013 http://www1.eere.energy.gov/geothermal/pdfs/market-report 2013.pdf.

National Geothermal Data System. http://geothermaldata.org.

NREL, ground source heat pumps. http://www.nrel.gov/tech_deployment/climate_neutral/ground_source_heat_pumps.html.

Renewables 2014 Global Status Report. http://www.ren21.net/Portals/0/documents/Resources/GSR/2014/GSR2014_full%20report_low%20res.pdf.

SMU Geothermal Lab. http://www.smu.edu/Dedman/Academics/Programs/GeothermalLab. Site has a lot of material for teachers and students.

State of Renewable Energies in Europe. 2013. http://www.energies-renouvelables.org/observer/stat_baro/barobilan/barobilan13-gb.pdf.

Thermal springs of the United States. http://maps.ngdc.noaa.gov/viewers/hot_springs/.

U.S. DOE, EERE, Geothermal Technologies Office. http://energy.gov/eere/geothermal/geothermal-technologies-office.

Go to Internet web browsers to find local, regional, and national geothermal associations and other national and regional information on geothermal resource.

PROBLEMS

11.1. OM (order of magnitude): At the present rate of world consumption of energy, how long would geothermal energy last if 5% of world demand for electrical energy came from geothermal resources?

11.2. List two advantages and two disadvantages of geothermal energy for production of electricity.

11.3. List two advantages and two disadvantages of geothermal energy for direct use.

11.4. Go to the project map for the United States at http://geoheat.oit.edu/dusys. htm. Select Nevada. How many power plants are there? Find one with the largest rated power, then give specifications (location, etc.).

11.5. What is the average capacity factor for geothermal electric power plants for the world for year 2005? Year 2013?

11.6. How many stripper oil wells are there in the United States?

11.7. Briefly describe the process of hydraulic fracturing a rock formation.

11.8. Estimate the number of district heating projects in the United States. How much total energy do they provide per year? Use http://geoheat.oit. edu/dusys.htm to obtain data.

11.9. Find any district heating system outside the United States. State name, location, and energy provided per year.

11.10. For Amarillo, Texas, a home is 200 m². Estimate the size of GSHP using boreholes. How many boreholes and what depth would be needed?

11.11. In Indonesia, what percentage of their electricity is from geothermal resources?

11.12. What percentage of electricity in the United States is generated from geothermal resources?

11.13. For a binary plant producing electricity, what is the approximate thermodynamic efficiency?

11.14. If a GHP has a COP of 3.0, estimate how much electrical energy is needed for the heating season for a 200-m^2 house in Amarillo, Texas.

11.15. If a GHP has an EER of 11, estimate how much electrical energy is needed for the cooling season for a 200-m^2 house in Amarillo, Texas.

11.16. Estimate the geothermal power from the following well: 160°C, 250 m^3/h flow rate, 100°C out temperature.

11.17. Why is a binary system less efficient than a dry steam system? Give example numbers.

11.18. What is the main environmental problem associated with the Geysers geothermal plants in California?

11.19. Check with local installers and get information (type, size) on a GHP for a residence in your area. If possible, check with a homeowner who has a system to obtain their comments. If no systems are in your area, then get a brochure or information on a GHP system for a medium-size home.

11.20. What is the global installed capacity of geothermal power plants? Estimate proposed installed capacity for 2020.

12 Water

12.1 INTRODUCTION

Energy from water is one of the oldest sources of energy, as paddle wheels were used to rotate a millstone to grind grain. A large number of watermills, 200–500 W, for grinding grain are still in use in remote mountains and hilly regions in the developing world. There are an estimated 500,000 watermills in the Himalayas, with around 200,000 watermills in India [1]. Of the 25,000–30,000 watermills in Nepal, 8,349 water-mills were upgraded between 2003 and 2013 [2,3]. Paddle wheels and buckets powered by moving water were and are still used in some parts of the world for lifting water for irrigation. Water provided mechanical power for the textile and industrial mills of the 1800s as small dams were built, and mill buildings are found along the edges of rivers throughout the United States and Europe. Then, starting in the late 1800s, water stored behind dams was used for the generation of electricity. For example, in Switzerland in the 1920s there were nearly 7,000 small-scale hydropower plants.

The energy in water can be potential energy from a height difference, which is what most people think of in terms of hydro; the most common example is the generation of electricity (hydroelectric) from water stored in dams. However, there is also kinetic energy due to water flow in rivers and ocean currents. Finally, there is energy due to tides, which is due to gravitational attraction of the Moon and the Sun, and energy from waves, which is due to wind. In the final analysis, water energy is just another transformation from solar energy, except for tides.

The energy or work is force * distance, so potential energy due to gravitation is

$$W = F * d = m * g * H, \text{ J} \tag{12.1}$$

The force due to gravity is mass * acceleration, where the acceleration of gravity $g = 9.8$ m/s^2 and H = height in meters of the water. For estimations, you may use $g = 10$ m/s^2.

For water, generally what is used is the volume, so the mass is obtained from density and volume.

$$\rho = m / V \text{ or } m = \rho * V, \text{ where } \rho = 1,000 \text{ kg/m}^3 \text{ for water}$$

Then, for water Equation 12.1 becomes

$$PE = \rho * g * H * V = 10,000 * H * V \tag{12.2}$$

Example 12.1

Find the potential energy for 2,000 m³ of water at a height of 20 m.

$$PE = 10,000 * 20 * 2,000 = 4 * 10^9 \, J = 4 \text{ GJ}$$

If the potential energy of a mass of water is converted into kinetic energy after falling from a height H, then the velocity can be calculated.

$$KE = PE$$

or

$$0.5 \, m * v^2 = m * g * H$$

Then, the velocity of the water is

$$v = \left(2 * g * H\right)^{0.5}, \text{ m/s} \tag{12.3}$$

Example 12.2

For data in Example 12.1, find the velocity of that water after falling through 20 m.

$$v = \left(2 * 10 * 20\right)^{0.5} = 20 \text{ m/s}$$

Instead of water at some height, there is a flow of water in a river or an ocean current, such as the Gulf Stream. The analysis for energy and power for moving water is similar to wind energy, except there is a large difference in density between water and air. Therefore, for the same amount of power, capture areas for water flow will be a lot smaller.

$$\frac{P}{A} = 0.5 * \rho * v^3, \text{W/m}^2 \tag{12.4}$$

Example 12.3

Find the power/area for an ocean current that is moving at 1.5 m/s.

$$\frac{P}{A} = 0.5 * \left(1.5\right)^3 = 1.7 \text{ k W/m}^2$$

Power is energy/time, and hydraulic power from water or for pumping water from some depth is generally defined in terms of water flow Q and the height. Of course, if you know the time and have either power or energy, then the energy or power can be calculated.

$$P = 10,000 * H * V/t = 10 * Q * H, \text{ kW} \tag{12.5}$$

where:

Q is the flow rate (m³/s)

In terms of pumping smaller volumes of water for residences, livestock, and villages, Q is generally noted as cubic meters per day, so be sure to note what units are used. There will be friction and other losses, so with efficiency ε the power is

$$P = 10,000 \ *ε*Q*H \tag{12.6}$$

Efficiencies from input to output (generally electric) range from 0.5 to 0.85. Small water turbines have efficiencies up to 80%, so when other losses are included (friction and generator), the overall efficiency is approximately 50%. Maximum efficiency is at the rated design flow and load, which is not always possible as the river flow fluctuates throughout the year or where daily load patterns vary.

The output from the turbine shaft can be used directly as mechanical power, or the turbine can be connected to an electric generator. For many rural industrial applications, shaft power is suitable for grinding grain or oil extraction, sawing wood, small-scale mining equipment, and so on.

12.2 WORLD RESOURCE

Around one-quarter of the solar energy incident on the Earth goes to the evaporation of water; however, as this water vapor condenses, most of the energy goes into the atmosphere as heat. Only 0.06% is rain and snow, and the power and energy of that water flow is the world resource, estimated at around 40,000 TWh/year. The technical potential (Table 12.1) is 15,000 TWh/year, and economic and environmental considerations reduce that potential.

The classification of hydropower differs by country, authors, and even over time. One classification is large (>30 MW), small (100–30 MW), and micro (≤100 kW). Some examples are as follows: In China, small hydro refers to capacities up to 50 MW, in India up to 15 MW, and in Sweden up to 1.5 MW. Now, in Europe, small hydro means a capacity of up to 10 MW. Today in China, the classifications are large (>50 MW), small (5–50 MW), mini (100 kW–5 MW), micro (5–100 kW), and pico (<5 kW). Others classify microhydro as 10–100 kW, so be sure to note the range when data are given for capacity and energy for hydropower.

12.3 HYDROELECTRIC

12.3.1 Large (≥30 MW)

In terms of renewable energy, large-scale hydropower (Figure 12.1) is a major contributor to electricity generation in the world, around 3,800 TWh in 2013. The world's installed capacity for large-scale hydroelectricity has increased around 2% per year, from 462 GW in 1980 to around 1,000 GW in 2014 with another 100 GW under construction. Increase in capacity since 1980 and proposed projects are in

TABLE 12.1

Technical Potential, Hydroelectric Production, and Capacity

	Potential, TWh/Year	Production, TWh/Year	Capacity, GW
Asia	5,090		
Asia and Oceanic		1,108	359
Central and South America	2,790	728	145
Europe	2,710	589	172
Eurasia		239	70
Middle East		20	12
Africa	1,890	109	26
North America	1,670	684	165
Oceania	230		
World	14,380	3,472	949

Source: Production and capacity data for 2011 from U.S. Energy Information Administration.

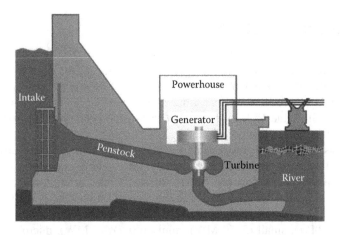

FIGURE 12.1 Diagram of hydroelectric plant. Height of water is level at dam to turbine generator. (From Tennessee Valley Authority.)

the developing world, mainly China. However, the hydroelectric percentage of electric power in the world has decreased from 21.5% in 1980 to 16% in 2012 as other sources of electrical energy have increased faster. China is now the leader in installed capacity and generation of electricity (Figure 12.2), with 20% of their electricity from hydroelectric sources. However, coal in China is still the major energy source for the production of electricity, and more coal power has been added than hydroelectric power. In Norway, 98% of the electrical energy is from hydro, and for several countries, it is over 50%; Paraguay sells most of its share of electricity from

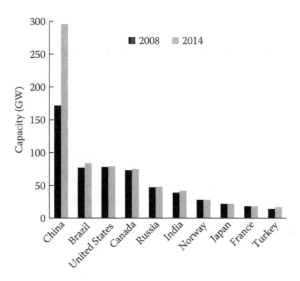

FIGURE 12.2 Installed hydroelectric capacity of top 10 countries. (Data from Energy Information Administration.)

the Itaipu Dam to Argentina. In the United States, the hydroelectric contribution is around 7%. The contribution from small or microhydro plants is difficult to estimate but could represent another 5%–10% in terms of world capacity. The capacity factor for hydroelectric power in the world has been fairly consistent at 40%–44% from 1980 to 2014. The capacity factor for hydroelectric power in the United States was 37% in 2013.

Large-scale hydroelectric plants have been constructed all across the world (Table 12.2). The Three Gorges Dam (Figure 12.3) on the Yangtze River is the largest power hydro plant in the world with 22.5 GW. Previously, the largest project was the Itaipu Dam on the Paraná River between Paraguay and Brazil. The series of dams is 7,744 m long and was built from 1975 to 1991. The Aswan High Dam, Egypt (2,100 MW), was completed in 1967 and produces more than 10 TWh/year, provides irrigation water for 3.2 million ha and produces 20,000 tons of fish per year. The entire Temple of Abu Simel had to be moved to higher ground, a major feat in archeology. One of the problems of the Aswan Dam was that farming practices on the banks of the river downstream had to be changed since no yearly floods meant no deposition of fertile silt.

The benefits or advantages of hydropower are as follows:

1. Renewable source, power on demand with reservoirs
2. Long life, 100 years
3. Flood control, water for irrigation and metropolitan areas
4. Low greenhouse gas emission
5. Reservoir for fishing, recreation

TABLE 12.2

Large Hydroelectric Plants in the World, Date Completed, and Capacity*

Country	Dam	Year	Capacity, MW
China	Three Gorges	2012	22,500
Brazil-Paraguay	Itaipu	1991	14,700
China	Xiluodu	2014	13,860
Venezuela	Guri	1986	10,055
Brazil	Tucurni	1984	8,370
United States	Grand Coulee	1941	6,809
China	Xiangjiaba	2014	6,488
Russia	Sayanao-Shushenskaya	1989	6,500
China	Longtan	2009	6,300
Russia	Krasnoyarsk	1972	6,000
Canada	Robert-Bourassa	1981	5,616
Canada	Churchill Falls	1971	5,429

* For photos, see Google images.

FIGURE 12.3 Three Gorges Dam, 22.5 GW, on the Yangtze River, China. (Courtesy of HydroChina.)

Some disadvantages or problems are the following:

1. There is a large initial cost and long construction time.
2. Displacement of population due to reservoir may occur. For example, 1.24 million people were relocated due to the Three Georges Dam.
3. On land downstream, there is loss of nutrients from floods.
4. Drought by season or year may restrict power output due to low water.

5. Lack of passage for fish to spawning areas, for example, salmon.
6. Rivers with high silt content may limit dam life.
7. Dam collapse means many problems downstream. There have been over 200 dam failures in the twentieth century, and it is estimated that 250,000 people died in a series of hydroelectric dam failures in China in 1975.
8. Resource allocation between countries can be a problem [4], especially if a series of dams that use a lot of water for irrigation are built upstream of a country.

In the United States, the first commercial hydroelectric plant (12.5 kW) was built in 1882 on the Fox River in Appleton, Wisconsin. Then, commercial power companies began to install a large number of small hydroelectric plants in mountainous regions near metropolitan areas. The creation of the Federal Power Commission in 1920 increased development of hydroelectric power with regulation and monetary support. The government supported projects for hydroelectric power and for flood control, navigation, and irrigation. The Tennessee Valley Authority was created in 1933 [5], and the Bonneville Power Administration was created in 1937 [6]. Construction of the Hoover Dam (Figure 12.4) started in 1931, and when completed in 1936, it was the largest hydroelectric project in the world with 2 GW [7]. Hoover Dam was then surpassed in 1941 by the Grand Coulee Dam (Figure 12.5) (6.8 GW) on the Columbia River [8]. The larger power output is due to the higher volume of water available.

A geographic information system (GIS) application, the Virtual Hydropower Prospector, provides maps and information for the United States [9]. The application allows the user to view the plants in the context of hydrography, power and

FIGURE 12.4 Hoover Dam, 2 GW, on the Colorado River, United States. (From U.S. Bureau of Reclamation.)

FIGURE 12.5   Grand Coulee Dam, 6.8 GW, on the Colombia River. (From U.S. Bureau of Reclamation.)

transportation infrastructure, cities and populated areas, and federal land use. Most of the possible sites will be for small hydro.

In the developed countries with significant hydroelectric capacity, many of the best sites for hydroelectric potential already have dams. Many more reservoirs have been built for irrigation, water supply, and flood control than for hydropower as only 3% of the 80,000 dams in the United States have hydropower. Also, the construction of dams has declined in the United States since 1980 (Figure 12.6). So, there

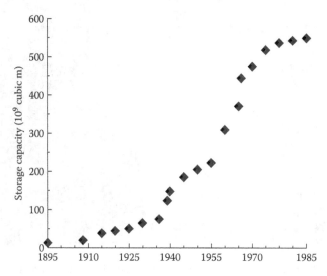

FIGURE 12.6   Reservoir capacity in the United States. (Data from U.S. Geological Survey.)

is a potential for hydropower by repowering some defunct hydroelectric installations or by new installations of small or medium hydropower at existing dams. The U.S. DOE estimates that existing dams not currently equipped for hydropower could provide an additional 12 GW of electric power (http://energy.gov/articles/powering-america-s-waterways).

12.3.2 SMALL HYDRO (100 kW TO 30 MW, 10 MW IN EUROPE)

The definition of small hydro differs by country, so the range in Europe is 100 kW to 10 MW and in other countries it is up to 25 or 30 MW. The World Energy Council estimated small hydro (up to 10 MW) at around 25.5 GW in 2006, energy production estimated at 60 TWh/year. Then in 2010, the installed capacity of small hydro was around 55 GW, with China having the largest capacity. In 2014, the estimated world capacity [10] was 75 GW with a total potential of 173 GW (Table 12.3). At 35%

TABLE 12.3
Installed and Potential Capacity for Small Hydropower

	Capacity, MW		Potential, MW	
Africa	561		7,901	
East		209		6,262
Middle		76		328
Northern		151		184
Southern		43		385
Western		82		743
Americas	10,301		22,982	
Caribbean		124		252
Central		599		4,166
South		1,735		9,465
North		7,843		9,099
Asia	46,345		112,706	
Central		186		4,880
Eastern		40,855		75,312
Southern		3,563		18,077
South eastern		1,252		6,683
Western		489		7,754
Europe	17,827		28,113	
Eastern		2,735		3,459
Northern		3,643		3,841
Southern		5,640		14,169
Western		5,809		6,644
Oceania	412		1,238	
Aus & NZ		310		932
Pac Island		102		306
Total	75,446		172,940	

capacity factor, small hydro would produce around 230 TWh/year. China is still the leader in installed capacity; however, China has a small hydropower definition of up to 50 MW (~45,000 plants, 65 GW capacity). The capacity of small hydro in Europe is around 17 GW from 22,000 plants which generate around 53 TWh/year (average capacity factor of 36%). The potential for small hydro in Europe is estimated at another 49 TWh/year, which consists mainly of low-head sites (below 30 m).

Hydroelectric plants in the United States are predominantly private (69%); however, 75% of the capacity is owned by federal and nonfederal public owners [11], primarily from large power plants. The percentage of low and small hydropower plants in terms of numbers is 86%. This indicates future expansion for hydroelectric power in the United States will be from distributed generation.

A resource assessment of hydropower for 49 states (no resource in Delaware) identified 5,667 sites (Figure 12.7) with a potential of around 30,000 MW [12]. The criteria were low power (<1 MW) or small hydro ($\geq$1 MW and $\leq$30 MW), and the working flow was restricted to half the stream flow rate at the site or sufficient flow to produce 30 MWa (megawatts average), whichever was less. Penstock lengths were limited by the lengths of penstocks of a majority of existing low-power or small hydroelectric plants in the region. The optimum penstock length and location on the stream was determined for the maximum hydraulic head with the minimum length. The number of sites studied was 500,000, with approximately 130,000 sites meeting the feasibility criteria. Then, application of the development model with the limits on working flow and penstock length resulted in a total hydropower potential of 30,000 MWa. The approximately 5,400 sites that could potentially be developed as small hydro plants have a total hydropower potential of 18,000 MWa. Idaho National Lab also developed a probability factor model, Hydropower Evaluation Software, to standardize the environmental assessment.

12.3.3 MICROHYDRO ($\leq$100 KW)

Estimation of the number of installations and capacity for microhydro is even more difficult. In general, microhydro does not need dams and a reservoir as water is diverted and then conducted in a penstock to a lower elevation and the water turbine. In most cases, the end production is the generation of electricity; however, in some parts of the world the watermills are for grinding grain.

There are thought to be tens of thousands of microhydro plants in China and significant numbers in Nepal, Sri Lanka, Pakistan, Vietnam, and Peru. The estimate for China was about 500 MW at the end of 2008. China started a program, SDDX [13], in 2003 that installed 146 hydro systems with a capacity of 113.8 MW in remote villages in the Western Provinces and Tibet. Hydropower was the predominant system in terms of capacity compared to wind and photovoltaics (PV), with 721 installations (15.5 MW) for villages and 15,458 installations (1.1 MW) for single households. The average size of the hydropower systems was 780 kW, which is much larger than average for the wind and PV systems (22 kW).

In Nepal, over 1,000 microhydro plants were installed in 52 districts with assistance from the World Bank [14]. Case studies are available for a number of countries [15], and case studies and software are available from Microhydro [16].

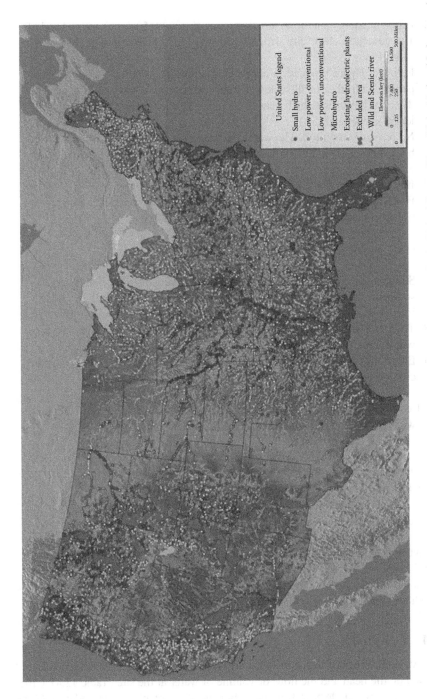

FIGURE 12.7 Present hydropower plants and possible sites for small and low hydropower in the United States. (From Idaho National Laboratory.)

The advantages of microhydro are the following:

1. Efficient energy source. A small amount of flow (0.5 L/min) with a head of 1 m generates electricity with micro hydro. Electricity can be delivered up to 1.5 km.
2. Reliable. Hydro produces a continuous supply of electrical energy in comparison with other small-scale renewable technologies. Also, backup, whether diesel or batteries (which causes operation and maintenance and cost problems), is not needed.
3. No reservoir required. The water passing through the generator is directed back into the stream with relatively little impact on the surrounding ecology.
4. It is a cost-effective energy solution for remote locations.
5. Power for developing countries. Besides providing power, developing countries can manufacture and implement the technology.

The disadvantages or problems are as follows:

1. Suitable site characteristics are required, including distance from the power source to the load and stream size (flow rate, output, and head).
2. Energy expansion may not be possible.
3. There is low power in the summer months. In many locations, stream size will fluctuate seasonally.
4. Environmental impact is minimal; however, environmental effects must be considered before construction begins.

Impulse turbines are generally more suitable for microhydro applications compared with reaction turbines because of

Greater tolerance of sand and other particles in the water
Better access to working parts
Lack of pressure seals around the shaft
Ease of fabrication and maintenance
Better part-flow efficiency

The major disadvantage of impulse turbines is that they are generally unsuitable for low-head sites. Pelton turbines can be used at heads down to about 10 m; however, they are not used at lower heads because their rotational speeds are too slow, and the runner required is too large. The cross-flow turbine is the best machine for construction by a user.

12.4 TURBINES

The two main types of hydro turbines are impulse and reaction. The type selected is based on the head and the flow, or volume of water at the site (Table 12.4). Other deciding factors include how deep the turbine must be set, efficiency, and cost. Many images are available on the Internet for the different types of turbines.

TABLE 12.4
Classification of Turbine Types

| Turbine | Head Pressure | | |
	High	Medium	Low
Impulse	Pelton	Cross flow	Cross flow
	Turgo	Turgo	
	Multijet Pelton	Multijet Pelton	
Reaction		Francis	Propeller
			Kaplan

12.4.1 IMPULSE TURBINES

The impulse turbine uses the velocity of the water to move the runner (rotating part) and discharges at atmospheric pressure. The water stream hits each bucket on the runner, and the water flows out the bottom of the turbine housing. An impulse turbine is generally used for high-head, low-flow applications.

A Pelton turbine (Figure 12.8) has one or more free jets of water impinging on the buckets of a runner. The jet is directed at the centerline of the two buckets. Draft tubes are not required for the impulse turbine since the runner must be located above the maximum tail water to permit operation at atmospheric pressure.

A cross-flow turbine (Figure 12.9) resembles a squirrel cage blower and uses an elongated, rectangular-section nozzle to direct a sheet of water to a limited portion of the runner, about midway on one side. The flow of water crosses through the empty center of the turbine and exits just below the center on the opposite side. A guide vane at the entrance to the turbine directs the flow to a limited portion of the runner.

FIGURE 12.8 Pelton runner (cast) showing bucket shape.

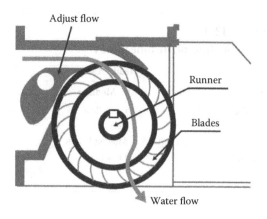

FIGURE 12.9 Diagram of cross-flow turbine.

The cross flow was developed to accommodate larger water flows and lower heads than the Pelton turbine.

12.4.2 REACTION TURBINES

A reaction turbine develops power from the combined action of pressure and moving water, as the pressure drop across the runner produces power. The runner is in the water stream flowing over the blades rather than striking each individually. Reaction turbines are generally used for sites with lower head and higher flows.

Francis turbines (Figure 12.10) are the most common for hydropower. They are an inward flow turbine that combines radial and axial components. The runner has fixed vanes, usually nine or more. The inlet is spiral shaped with guide vanes to direct the water tangentially to the runner. The guide vanes (or wicket gate) may be adjustable to allow efficient turbine operation for a range of water flow conditions. The other major components are the scroll case, wicket gates, and draft tube (as water speed is reduced, a larger area for the outflow is needed). However, the Francis turbine can be used for heads to 800 m.

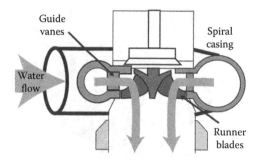

FIGURE 12.10 Diagram of Francis turbine.

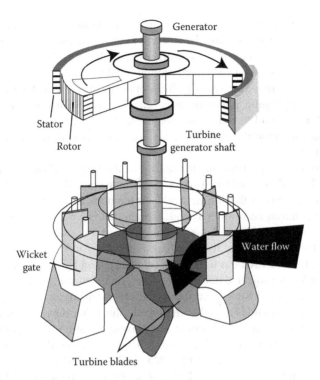

Generator

Stator

Rotor

Turbine
generator shaft

Wicket
gate

Water flow

Turbine blades

FIGURE 12.11   Diagram of propeller turbine. (From EERE.)

A propeller turbine (Figure 12.11) generally has a runner with three to six blades running in a pipe, where the pressure is constant. The pitch of the blades may be fixed or adjustable. The major components besides the runner are a scroll case, wicket gates, and a draft tube.

There are several different types of propeller turbines:

Bulb: The turbine and the generator are a sealed unit placed directly in the water stream.
Straflo: The generator is attached directly to the perimeter of the turbine.
Tube: The penstock bends just before or after the runner, allowing a straight-line connection to the generator.
Kaplan: Both the blades and the wicket gates are adjustable, allowing for a wider range of operation.

12.5   WATER FLOW

Kinetic energy turbines, also called *free-flow turbines*, generate electricity from the kinetic energy of the flowing water, similar to wind turbines, which generate energy from the flowing air. Systems are also referred to as hydrokinetic, tidal in-stream energy conversion, or river in-stream energy conversion. The systems may operate in rivers, tides, ocean currents, or even channels or conduits for water. Kinetic systems do not require large civil works, and they can be placed near existing structures such as bridges, tailraces, and channels that increase the natural flow of water. For tidal currents, unidirectional turbines are available; rotation is the same, even though current is from opposite directions. One hydrokinetic system has a hydraulic pump to drive an onshore electric generator. Kinetic energy turbines would have less environmental impact than dams, and like wind turbines, they are modular and can be installed in a short time compared to large civil structures.

The power/area is proportional to the cube of the velocity (Equation 12.4). Large rivers have large flows, and the Amazon, with an average flow of 210,000 m^3/s, has around 20% of the river flow of the world. At the narrows of Óbidos, 600 km from the sea, the Amazon narrows to a single stream that is 1.6 km wide and over 60 m deep and has a speed of 1.8–2.2 m/s. At New Orleans, the speed of the Mississippi is 1.3 m/s, and some sections of the river have flows of 2.2 m/s.

The U.S. DOE, EERE has a program on marine and hydrokinetic energy [17], which includes a project database of wave, tidal, current, and ocean thermal energy for the United States and the world [18]. An interactive map shows the project phase, from undeveloped to deployed, plus other details such as type and capacity. Under resource assessment and characterization are reports and U.S. resource maps for each type. The United States could produce 13,000 MW of power from hydrokinetic energy by 2025. In March 2010, the Federal Energy Regulatory Commission (FERC) had 134 preliminary permits for hydrokinetic projects (Figure 12.12) with a total capacity of 9,864 MW. Notice that many of the permits are on the Mississippi River. In March 2014, FERC [20] showed 18 permits for 1,516 MW at the pilot or pending stage (Table 12.5); however, there were no ocean thermal projects.

The technical recoverable resource from river flow is 120 TWh/year. At Hastings, Minnesota, a 250-kW hydrokinetic unit located below a dam (4.4 MW hydroelectric) was placed in operation in 2008. The ducted rotor is suspended from a barge with the generator on the barge (Figure 12.13). Free Flow Power is developing river projects at over 43 sites across the United States (http://free-flow-power.com/home).

The FERC requires consideration of any environmental effects of the proposed construction, installation, operation, and removal of the project. The description should include

1. Any physical disturbance (vessel collision or other project-related risks for fish, marine mammals, seabirds, and other wildlife as applicable)
2. Species-specific habitat creation or displacement
3. Increased vessel traffic
4. Exclusion or disturbance of recreational, commercial, industrial, or other uses of the waterway and changes in navigational safety
5. Any above- or below-water noise disturbance, including estimated decibel levels during project construction, installation, operation, and removal
6. Any electromagnetic field disturbance

7. Any changes in river or tidal flow, wave regime, or coastal or other geomorphic processes
8. Any accidental contamination from device failures, vessel collisions, and storm damage
9. Chemical toxicity of any component of, or biofouling coating on, the project devices or transmission line
10. Any socioeconomic effects on the commercial fishing industry from potential loss of harvest or effect on access routes to fishing grounds

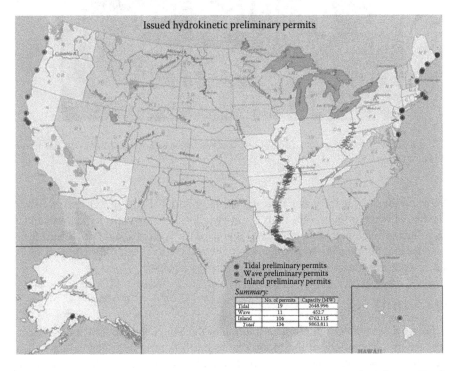

FIGURE 12.12 Proposed locations for hydrokinetic projects in the United States, March 2010. (From Federal Energy Regulatory Commission.)

TABLE 12.5
Hydrokinetic Projects in the United States

	Pilot		Issued		Preliminary		Pending		Total	
	#	MW	#	MW	#	MW	#	MW	#	MW
Wave			2	20.75	2	1300	1	20	5	1341
Tidal	3	2.35	3	165.14			2	5.1	8	173
Inland	1	0.1	4	2.64					5	3
									18	1516

FIGURE 12.13 In-river system, 250 kW, on Mississippi River, Hastings, Minnesota. (Courtesy of Hydro Green.)

An important factor for water flow is that, at good locations, power will not vary like that of wind turbines, especially for in-river locations, so capacity factors can be much higher. One manufacturer stated that capacity factors should be at least 30% for tides and 50% for in-river systems. As always, the final result for comparison is the cost per kilowatt hour, which should be life-cycle costs.

12.6 TIDES

Tides are due to the gravitational attraction of the Moon and the Sun at the surface of the Earth. The effect of the Moon on the Earth in terms of tides is larger than the effect of the Sun, even through the gravitational force of the Sun is larger. To find how the gravitational force of the Moon distorts any volume of the material body of the Earth, the gradient of the gravitational force of the Moon on that volume must be found (a gradient is how force changes with distance; in calculus, it is the differentiation with respect to length). The tidal effects (Figure 12.14) are superimposed on the near-spherical Earth, and there will be two tides per day due to the spin of Earth. When the tidal effects of the Sun and Moon are aligned, the tides are higher, spring tides. When the continents are added, the ocean bulges reflect from shorelines, which causes currents, resonant motions, and standing waves, so there are some places in the oceans where the tidal variations are nearly zero. In other locations, the coastal topography can intensify water heights with respect to the land. The largest tidal ranges in the world are the Bay of Fundy (11.7 m), Ungava Bay (9.75 m), Bristol Channel (9.6 m), and the Turnagain Arm of Cook Inlet, Alaska (9.2 m). The potential world tidal current energy is on about 2,200 TWh/year.

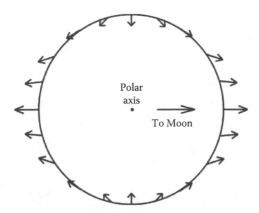

FIGURE 12.14   Tidal forces on the Earth due to the Moon.

Small mills were used on tidal sections of rivers in the Middle Ages for grinding grain. Today, there are only a few tidal systems installed in the world: the French installed a tidal system on the Rance Estuary (constructed from 1961 to 1967) with a power of 240 MW; an 18-MW rim generator at Annapolis Royal, Nova Scotia, Canada (1984); a 400-kW unit in the Bay of Kislaya, Russia (1968); a 500-kW unit at Jangxia Creek, East China Sea; and a 245 MW tidal barrage, Sihwa, South Korea (2012).

The simplest system for generation of electricity is an ebb system, which involves a dam, known as a barrage, across an estuary. Barrages make use of the potential energy in the difference in height between high and low tides. Sluice gates on the barrage allow the tidal basin to fill on high tides (flood tide) and to generate power on the outgoing tide (ebb tide). Flood generating systems generate power from both tides but are less favored than ebb generating systems. Barrages across the full width of a tidal estuary have high civil infrastructure costs, there is a worldwide shortage of viable sites, and there are more environmental issues.

Tidal lagoons are similar to barrages but can be constructed as self-contained structures, not fully across an estuary, and generally have much lower cost and environmental impact. Furthermore, they can be configured to generate continuously, which is not the case with barrages. Different tidal systems, installed and proposed plants, and prototype and demonstration projects are given in Marine and hydrokinetic technology database [18].

The potential for tidal in-stream systems was estimated at 250 TWh/year for the United States [20], with Alaska having the largest number of locations. The average power at the better locations exceeds 8,000 MW/m^2. A kinetic energy demonstration project (Figure 12.15) is installed in the East River, New York City, and consists of two 35-kW turbines, 5-m diameter, with passive yaw. In 9,000 h of operation, the system generated 70 MWh. A prototype was installed near the Jindo Island, South Korea, 5.3 m diameter, 110 kW at a current speed of 2.9 m/s. Then a larger unit was installed at the European Marine Energy Centre, 16 m diameter, 1 MW at current speed of 2.9 m/s. Notice the contrast in size and flow speed for a wind turbine, 60 m diameter, 1 MW at wind speed of 13 m/s.

FIGURE 12.15   Tidal turbine, 35 kW, being installed in East River, New York City. (Courtesy of Verdant Power.)

Another prototype, SeaGen, is installed in Strangford Narrows, Northern Ireland, with rated power of 1.2 MW at a current velocity of 2.4 m/s and with twin 16-m diameter rotors (Figure 12.16). The rotor blades can be pitched through 180° to generate power on both ebb and flood tides. The twin power units are mounted on wing-like extensions on a tubular steel monopole, and the system can be raised above sea

FIGURE 12.16   Tidal system, 1.2 MW, in Strangford Narrows, Northern Ireland. (Courtesy of Sea Generation).

level for maintenance. Hydrokinetic energy systems are being considered because of the lower cost, lower ecological impact, increased availability of sites compared to barrages, and shorter time for installation.

Advantages for tidal systems are as follows:

1. Renewable
2. Predictable

Disadvantages or problems are as follows:

1. A barrage across an estuary is expensive and affects a wide area.
2. The environment is changed upstream and downstream for some distance. Many birds rely on the tide uncovering the mudflats so that they can feed. Fish ladders are needed.
3. There is intermittent power as power is available for around 10 h each day when the tide is moving in or out.
4. There are few suitable sites for tidal barrages.

12.7 OCEAN

As with other renewable resources, the ocean energy is large [21]. The global technical resource exploitable with today's technology is estimated to be 20,000 TWh/year for ocean currents, 45,000 TWh/year for wave energy, 33,000 TWh/year for ocean thermal energy conversion (OTEC), and 20,000 TWh/year for salinity gradient energy. Of course, economics and other factors will greatly reduce the potential production, and future actual energy production will be even smaller.

Besides the environmental considerations mentioned, there are a number of technical challenges for ocean energy to be utilized on a commercial scale:

Avoidance of cavitations (bubble formation)
Prevention of marine growth buildup
Reliability (since maintenance costs may be high)
Corrosion resistance

12.7.1 CURRENTS

There are large currents in the ocean (Figure 12.17), and detailed information on surface currents by ocean is available [22]. For example, the Gulf Stream transports a significant amount of warm water toward the North Atlantic and the coast of Europe. The core of the Gulf Stream current is about 90 km wide and has peak velocities greater than 2 m/s. The relatively constant extractable energy density near the surface of the Gulf Stream, the Florida Straits Current, is about 1 kW/m^2. Although the volume and velocity are adequate for in-stream hydro-kinetic systems, an ocean current would need to be close to the shore.

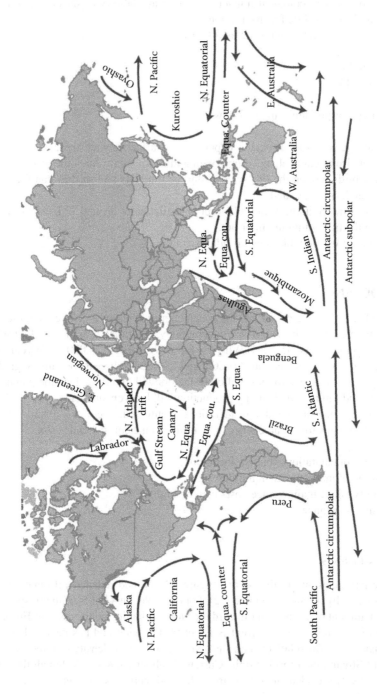

FIGURE 12.17 Major ocean currents in the world. (Courtesy of Michael Pidwirney.)

The total world power in ocean currents has been estimated to be about 5,000 GW, with power densities of up to 15 kW/m² [23]. The European Union, Japan, and China are interested in and pursuing the application of ocean current energy systems.

12.7.2 WAVES

Waves are created by the progressive transfer of energy from the wind as it blows over the surface of the water. Once created, waves can travel large distances without much reduction in energy. The energy in a wave is proportional to the height squared. In data for wave heights, be sure to note that height is for crest to trough, and amplitude, A, is midpoint to crest.

$$E = 0.5 * \rho * g * H^2 / 16, \text{ J} \tag{12.7}$$

where:
  H is wave height

This is for a single wave, but in the ocean, there is superposition of waves, and the energy transported is by group velocity. The speed of the wave, wave length, and frequency (or period, which is 1/frequency) are related by

$$\text{Speed} = \text{Wavelength} \ (\lambda) * \text{Frequency} \ (f)$$

In deep water where the water depth is larger than half the wavelength, the power per length (meter) of the wave front is given by

$$\frac{P}{L} = \rho * g^2 * H^2 * T / (64 * \pi) \sim 0.5 * H^2 * T, \text{ kW/m} \tag{12.8}$$

where:
  T is the period of the wave (time it takes for successive crests to pass one point)

In major storms, the largest waves offshore are about 15 m high and have a period of about 15 s, so the power is large, around 1.7 MW/m.

Example 12.4

Calculate power/length for waves off New Zealand if the average wave height is 7 m with a wave period of 8 s. The power/length is

$$\frac{P}{L} = 0.5 * 7^2 * 8 = 196 \text{ kW/m}$$

An effective wave energy system should capture as much energy as possible of the wave energy. As a result, the waves will be of lower height in the region behind the system. Offshore sites with water 25–40 m deep have more energy because waves have less energy as the depth of the ocean decreases toward the coast. Losses become significant as the depth becomes less than half a wavelength, and at 20 m deep, the wave energy is around one-third of that in deep water (depth greater than one-half

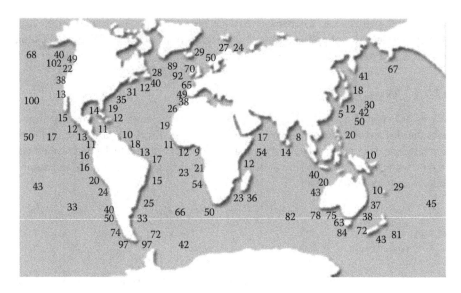

FIGURE 12.18 Average wave energy (kW/m) for coastlines around the world; values are for deep-water sites. (Courtesy of Pelamis Wave Energy.)

wavelength). The North Atlantic west of Ireland has wavelengths of around 180 m, and off the West Coast of the United States, the wavelengths can be 300 m.

The potential for wave energy (per meter of wave front) for the world is much larger than ocean currents due to the length of coastline (Figure 12.18). The potential for the United States is 240 GW, with an extractable energy of 2,100 TWh/year based on average wave power density of 10 kW/m. The technically and economically recoverable resource for the United Kingdom has been estimated to be 50–90 TWh of electricity per year or 15%–25% of total U.K. demand in 2010. The western coast of Europe and the Pacific coastlines of South America, Southern Africa, Australia, and New Zealand are also highly energetic. Any area with yearly averages of 15 kW/m has the potential to generate wave energy. Note that this threshold excludes areas such as the Mediterranean Sea and the Great Lakes of Northern America.

The resource or wave climate can be obtained from recorded data, and satellites now provide current worldwide data and are used for prediction of wave heights. For wave energy systems, it is also important to determine the statistical occurrence of the extreme waves that can be expected at the site over the lifetime of the system since the system should be designed to survive peak waves.

Once the general area of the wave farm site has been determined, more analysis is needed to pick the best site within that area, for example, by examining the mean wave direction, variability, and the possibility of local focusing of waves. Another essential task includes the calculation of calm periods that allow sufficient time for maintenance and other operations. However, as noted, large waves have lots of power and could damage or destroy the system, so design and construction must take these large waves into account.

The mechanisms for capture of wave energy are point absorber, reservoir, attenuator, oscillating water column, and other mechanisms. There are a number of prototypes and demonstration projects but few commercial projects. A point absorber has a small dimension in relation to the wavelength (Figure 12.19). A buoy system [24] in Australia is submerged below the ocean level and drives a seabed pump for electricity and high pressure water to supply a reverse osmosis desalination plant. In 2011 a prototype of 20 m diameter, 80 kW was tested at Garden Island and then larger units were built, 11 m diameter, 240 kW. The next design is the CETO 6 unit, 20 m diameter, 1 MW to be demonstrated in a 3 MW project with construction starting in 2016.

The reservoir system is where the waves are forced to higher heights by channels or ramps, and the water is captured in a reservoir (Figure 12.20). Locations for land installations for reservoir and oscillating water column systems will be much more limited than offshore systems; however, land installations are easier to construct and maintain. The Wave Dragon is a floating offshore platform (Figures 12.21 and 12.22).

The Pelamis Wave Energy Converter [25], an attenuator, is a semisubmerged, articulated cylindrical attenuator linked by hinged joints (Figure 12.23). The

FIGURE 12.19   PowerBuoy prototype, 40 kW, 14.6 m long, 3.5 m diameter; floats 4.25 m above surface of water. (Courtesy of Ocean Power Technologies, http://www.oceanpowertechnologies.com/home.)

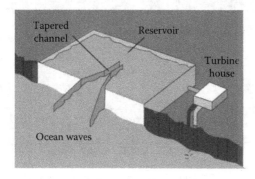

FIGURE 12.20   Diagram of a reservoir system on land.

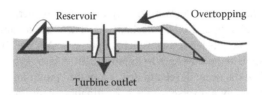

FIGURE 12.21 Diagram of floating reservoir system. (Courtesy of Wave Dragon, http://www.wavedragon.net.)

FIGURE 12.22 Prototype floating reservoir system, Nissum Bredning, Denmark. (Courtesy of Wave Dragon, http://www.wavedragon.net.)

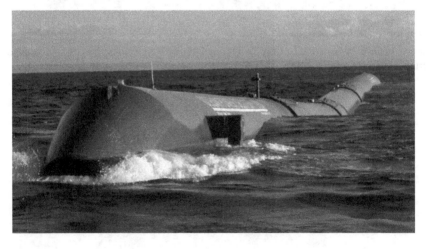

FIGURE 12.23 Sea trial of Pelamis Wave Energy Converter, 750 kW. (Courtesy of Pelamis Wave Energy.)

FIGURE 12.24 Installation of three units at Aguçadoura, Portugal, 2.25 MW total power. (Courtesy of Pelamis Wave Energy.)

wave-induced motion of these joints drives hydraulic rams, which pump high-pressure fluid through hydraulic motors via smoothing accumulators. The hydraulic motors drive an electrical generator, and the power from all the joints is fed down a single cable to a junction on the seabed. Several devices (Figure 12.24) can then be linked to shore through a single seabed cable. Current production machines have four power conversion modules: 750-kW rated power, 180 m long, 4-m diameter. The power table and the wave climate are combined to give the electrical power response over time and, from that, its average level and its variability. Depending on the wave resource, the capacity factor is 25%–40%. A three unit farm was installed at Aguçadoura, Portugal in 2008 (installed capacity 2.25 MW). Second generation systems are being tested at European Marine Energy Centre, Orkney, UK.

In an oscillating water column, as a wave enters the column, the air pressure within the column is increased, and as the wave retreats, air pressure is reduced (Figure 12.25). The Wells turbine turns in the same direction irrespective of the air-flow direction. The land-installed marine power energy transmitter (LIMPET) unit on Isle of Islay, Scotland [26], has an inclined oscillating water column, with an inlet width of 21 m (Figure 12.26) with the mean water depth at the entrance at 6 m. The system (rated power is 500 kW) has three water columns contained within concrete tubes, 6 m by 6 m, inclined at 40° to the horizontal, giving a total water surface area

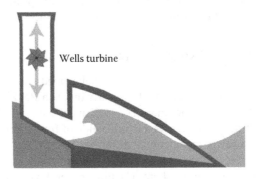

Wells turbine

FIGURE 12.25 Diagram of oscillating water column system.

FIGURE 12.26 LIMPET on Islay Island, Scotland, 500 kW; installed 2000. (Courtesy of Voith Hydro Wavgen.).

of 169 m². The upper parts of the tubes are connected to a single tube, which contains a Wells generator. Voith purchased Wavegen and in 2014 shut down the Wavegen business in Inverness, Scotland.

The design of the air chamber is important to maximize the conversion of wave energy to pneumatic power, and the turbines need to be matched to the air chamber. The performance has been optimized for annual average wave intensities of between 15 and 25 kW/m.

A wave power plant was constructed into the breakwater around the harbor at Mutriku, Spain [27]. The system has 16 Wells turbines, each rated at 18.5 kW. The plant produces around 600 GWh/year, which is fed into the grid.

In another system, waves drive a hinged flap connected (Figure 12.27) to the seabed at around 10 m depth, which then drives hydraulic pistons to deliver high-pressure water via a pipeline to an onshore electrical turbine.

12.7.3 OCEAN THERMAL ENERGY CONVERSION

OTEC for producing electricity is the same as solar ponds, for which the thermal difference between surface water and deep water drives a Rankine cycle. There is one major difference: the deep ocean water is rich in nutrients, which can be used for mariculture. In both systems, there is the production of freshwater.

An OTEC system needs a temperature difference of 20°C from cold water within 1,000 m of the surface, which occurs across vast areas of the world (Figure 12.28). The systems can be on or near the shore. The three general types of OTEC processes are closed cycle, open cycle, and hybrid cycle.

FIGURE 12.27 Oyster hydroelectric wave energy converter, 315 kW; unit installed at Billa Croo, Orkney, Scotland. (Courtesy of Aquamarine Power.)

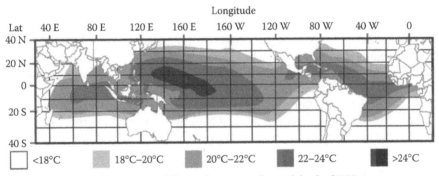

FIGURE 12.28 Ocean thermal differences, surface to depth of 1,000 m. (Data from NREL.)

In the closed-cycle system, heat transferred from the warm surface seawater causes a working fluid to turn to vapor, and the expanding vapor drives a turbine attached to an electric generator. Cold seawater passing through a condenser containing the vaporized working fluid turns the vapor back into a liquid, which is then recycled through the system.

An open-cycle system uses the warm surface water itself as the working fluid. The water vaporizes in a near vacuum at surface water temperatures. The expanding vapor drives a low-pressure turbine attached to an electrical generator. The vapor, which is almost pure freshwater, is condensed into a liquid by exposure to cold temperatures from deep ocean water. If the condenser keeps the vapor from direct contact with seawater, the water can be used for drinking water, irrigation, or aquaculture. A direct contact condenser produces more electricity, but the vapor is mixed with cold seawater, and the mixture is discharged into the ocean. Hybrid systems use parts of both open- and closed-cycle systems to optimize production of electricity and freshwater.

The first prototype OTEC project (22 kW) was installed at Matanzas Bay, Cuba, in 1930 [28]. Then, in the latter part of the twentieth century, experimental systems were installed in Hawaii and Japan. An experimental, open-cycle, onshore system was operated intermittently between 1992 and 1998 at the Keahole Point Facility, National Energy Laboratory, Hawaii. Surface water is 26°C, and the deep-water temperature is 6°C (depth of 823 m); the system produced a maximum power of 250 kW. However, the power requirements for pumping the surface (36.3 m³/min) and deep (24.6 m³/min) seawater were around 200 kW. A small fraction (10%) of the steam produced was diverted to a surface condenser for the production of freshwater, about 22 L/min. In 1981, Japan demonstrated a shore-based, 100-kWe closed-cycle plant in the Republic of Nauru in the Pacific Ocean. The cold-water pipe was laid on the seabed at a depth of 580 m. The plant produced 31.5 kWe of net power during continuous operating tests. In 2013, a 50-kW prototype was installed near Kumejima Island, Okimawa, and a 20-kW plant was installed in South Korea.

12.7.4 SALINITY GRADIENT

Salinity gradient energy is derived from the difference in the salt concentration between seawater and river water. Two practical methods for this are reverse electrodialysis and pressure-retarded osmosis; both rely on osmosis with ion-specific membranes. A small prototype (4 kW) started operation in 2009 in Tofte, Norway; however, development efforts by Statkraft were discontinued in 2014. The pressure generated is equal to a water column of 120 m, which is used to drive a turbine to generate electricity.

12.8 OTHERS

An application for water flow is ram pumps, where the pressure from water over a drop of a few meters is used to lift a small percentage of that water through a much greater height for water for people or irrigation. Ram pumps were developed over 200 years ago and can be made locally [29–31]. The operation of a ram pump (Figure 12.29) is as follows:

1. Water from a stream flows down the drive pipe and out of the waste valve.
2. As the flow of water accelerates, the waste valve is forced shut, causing a pressure surge (or water hammer) as the moving water is brought to a halt.
3. The pressure surge causes the check valve to open, allowing high-pressure water to enter the air chamber and delivery pipe. The pressurized air in the air chamber helps to smooth out the pressure surges to give a continuous flow through the delivery pipe.

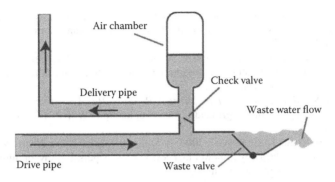

FIGURE 12.29 Diagram of ram pump.

4. As the pressure surge subsides, the pressurized air in the air chamber causes the check valve to close. The sudden closure of the check valve reduces the pressure in the drive pipe so that the waste valve opens, and the pump is returned to start the cycle again. Most ram pumps operate at 30–100 cycles a minute.

The Alternative Indigenous Development Foundation in the Philippines developed durable ram pumps, and the maintenance is done locally on the moving parts that need regular replacement. The five different size ram pumps can deliver between 1,500 and 72,000 L/day up to a height of 200 m. The 98 ram pumps installed by 2007 were delivering over 900 m³/day of water, serving over 15,000 people and irrigating large areas of land.

There have been proposals to use water flow in pipes and irrigations ditches where there is a height difference for the generation of electricity. A pilot project of in-conduit hydropower was tested in Riverside, California [32], and in 2015, four 1-m diameter turbines (total power 200 kW) were installed in water mains in Portland, Oregon. The power purchase agreement is for 20 yr.

REFERENCES

1. Water wheels of India. http://www.goodnewsindia.com/Pages/content/discovery/water-wheels.html.
2. Nepal Ghatta Project. Water mills in Nepal. http://alumni.media.mit.edu/~nathan/nepal/ghatta/ghattas.html.
3. S. P. Kapali. 2014. *Improved Watermill Development in Nepal.* Centre for Rural Technology, Nepal. http://www.crtnepal.org/docs/publications/IWM%20Status%20Review_Final_27%20Jan%202014.pdf. Go to Google for photos of Nepal watermills.
4. M. T. Klare. 2001. *Resource Wars: The New Landscape of Global Conflict.* Metropolitan Books, New York.
5. *From the New Deal to a New Century.* http://www.tva.com/abouttva/history.htm.
6. Booneville Power Administration, History. http://www.bpa.gov/news/AboutUs/History/Pages/default.aspx.

7. Hoover Dam. http://www.usbr.gov/lc/hooverdam/History/storymain.html, http://www.history.com/topics/hoover-dam, or http://www.arizona-leisure.com/hoover-dam-building.html.

8. Grand Coulee Dam. http://www.usbr.gov/pn/grandcoulee/history/index.html.

9. Idaho National Laboratory. Hydropower Evaluation Software. http://hydropower.inl.gov/resourceassessment/software/.

10. *World Small Hydropower Development Report 2013, Executive Summary.* http://www.smallhydroworld.org/fileadmin/user_upload/pdf/WSHPDR_2013_Executive_Summary.pdf.

11. D. G. Hall and K. S. Reeves. 2006. *A Study of United States Hydroelectric Plant Ownership.* http://hydropower.inel.gov/hydrofacts/pdfs/a_study_of_united_states_hydroelectric_plant_ownership.pdf.

12. *Feasibility Assessment of the Water Energy Resources of the United States for New Low Power and Small Hydro Classes of Hydroelectric Plants.* 2006. http://hydropower.inel.gov/resourceassessment/index.shtml.

13. S. Jingli, C. Dou, and R. Dongming. 2008. *Renewable Energy Based Chinese Un-Electrified Region Electrification.* Chemical Industry Press, Beijing, China, Chap. 6.

14. *Renewable Energy Powers Rural Nepal into the Future.* 2014. http://www.worldbank.org/en/news/feature/2014/02/05/renewable-energy-powers-rural-nepal-into-the-future.

15. S. Khennas and A. Barnett. 2000. *Best Practices for Sustainable Development of Micro Hydro Projects in Developing Countries.* http://www.microhydropower.net/download/bestpractsynthe.pdf.

16. Microhydro. http://www.microhydropower.net.

17. EERE. *Marine and Hydrokinetic Energy Research and Development.* http://energy.gov/eere/water/marine-and-hydrokinetic-energy-research-development.

18. *Marine and Hydrokinetic Technology Database.* http://en.openei.org/wiki/Marine_and_Hydrokinetic_Technology_Database.

19. EERE. *Marine and Hydrokinetic Resource Assessment and Characterization.* http://energy.gov/eere/water/marine-and-hydrokinetic-resource-assessment-and-characterization. Detailed report: Assessment of energy production potential from tidal streams in the United States. 2011. http://energy.gov/sites/prod/files/2013/12/f5/1023527.pdf.

20. FERC, hydrokinetic projects. 2014. http://www.ferc.gov/industries/hydropower/gen-info/licensing/hydrokinetics.asp.

21. World Ocean Review. http://worldoceanreview.com/en/wor-1/energy/renewable-energies/.

22. Cooperative Institute for Marine and Atmospheric Sciences, University of Miami. http://oceancurrents.rsmas.miami.edu.

23. *Technology White Paper on Ocean Current Energy Potential on U.S. Continental Shelf.* 2006. http://www.e-renewables.com/documents/Ocean/Ocean%20Current%20Energy%20Potential.pdf.

24. Carnegie Wave Energy. http://www.carnegiewave.com.

25. European Marine Energy Centre. Pelamis Wave Power. http://www.emec.org.uk/about-us/wave-clients/pelamis-wave-power/.

26. *Islay LIMPET Wave Power Plant.* 2002. http://cordis.europa.eu/documents/documentlibrary/66628981EN6.pdf.

27. Y. Torre-Enciso, I. Ortubia, L.I. Lópes de Aguileta, and J. Marqués. 2009. *Mutriku Wave Power Plant: From the Thinking Out to the Reality.* http://mhk.pnnl.gov/sites/default/files/publications/Torre-Enciso_et_al_2009.pdf.

28. International Renewable Energy Agency. 2014. *Ocean Thermal Energy Conversion, Technology Brief.* http://www.irena.org/DocumentDownloads/Publications/Ocean_Thermal_Energy_V4_web.pdf.

29. Clemson University. *Home-Made Hydraulic Ram Pump.* http://www.clemson.edu/irrig/Equip/ram.htm.

30. Case Study, Ashden Awards. http://www.ashden.org/winners/aidfoundation.

31. A. Idzenga. 2007. *Local Manufacture and Installation of Hydraulic Ram Pumps for Village Water Supply.* http://www.ashdenawards.org/files/reports/AIDFI_2007_technical_report.pdf.

32. Lucid Energy. http://www.lucidenergy.com. *Portland Water Bureau Conduit 3 Hydroelectric Project.* http://www.lucidenergy.com/wp-content/uploads/2015/01/Lucid-Energy-PDX-Fact-Sheet_20-Jan-2015.pdf.

RECOMMENDED RESOURCES

INTERACTIVE MAPS

International Renewable Energy Agency. Global atlas for renewable energy. http://globalatlas.irena.org/Default.aspx?tid=3.

Natural Resources Defense Council. Renewable energy map. http://www.nrdc.org/energy/renewables/map_hydro.asp#map.

Northern Great Plains Water Consortium. http://www.undeerc.org/Water/Interactive-Map.aspx.

U.S. DOE. Powering up America's waterways. http://energy.gov/articles/powering-america-s-waterways.

LINKS

EPRI Assessment and Mapping of the Riverine Hydrokinetic Resource in the Continental United States. 2012. http://www.epri.com/abstracts/Pages/ProductAbstract.aspx?ProductId=000000000001026880. Results of this study can be geo-spatially visualized, queried, and downloaded from the NREL website: http://maps.nrel.gov/river_atlas.

EPRI. Ocean energy web page. http://oceanenergy.epri.com/default.asp.

EPRI tidal in-stream energy conversion (TISEC) project. http://oceanenergy.epri.com/stream-energy.html.

European Ocean Energy Association. http://www.oceanenergy-europe.eu/index.php/en/.

Hydropower Research Foundation. http://www.hydrofoundation.org/index.html.

International Center on Small Hydro Power. http://www.inshp.org/main.asp.

International Energy Agency, Ocean Energy Systems. http://www.iea.org/topics/renewables/subtopics/ocean/. Small hydro. http://www.small-hydro.com/Home.aspx.

International Hydropower Association. http://www.hydropower.org.

International Renewable Energy Agency. Has technology briefs that highlights main features and current status for ocean energies; tidal, wave, OTEC, and salinity gradient. http://www.irena.org/menu/index.aspx?mnu=Subcat&PriMenuID=36&CatID=141&SubcatID=431.

Microhydro Power. http://practicalaction.org/energy/micro_hydro_expertise.

Microhydro Power. Links to case studies. http://www.microhydropower.net/index.php.

Microhydro projects, South Pacific. http://www.pelena.com.au/projects/index.html.

National Hydropower Association. http://www.hydro.org.

Neptune Wave Power. http://www.neptunewavepower.com/index.php.

Ocean Power Technologies. http://www.oceanpowertechnologies.com/mark3.html.

Oceanweather, current significant wave height and direction by regions of the world. http://www.oceanweather.com/data.

Perpetuwave Power. http://www.perpetuwavepower.com.

United States DOE, EERE. Water. http://energy.gov/eere/renewables/water.

Wave Dragon. http://www.wavedragon.net.

Wave power, point source.

PROBLEMS

12.1. What is the power/area for the current of the Amazon River at the narrows of Óbidos?

12.2. Suppose a village needs 15 m³/day for water; the dynamic head is 15 m. Calculate the hydraulic power.

12.3. What is the average flow rate of the Columbia River at the Grand Coulee Dam? Calculate the power available for the average water height between reservoir and discharge. How does that compare to generator capacity?

12.4. What is the capacity factor for the kinetic energy system (tidal) in the East River, New York City?

12.5. List the top three countries and capacity in the world for large hydroelectricity. Note source and date of information.

12.6. Find any case history for microhydro. Describe general specifications of the system. Note source and date of information.

12.7. Go to International Renewable Energy Agency (ocean technology briefs) or another source. What is the world installed capacity of tidal systems? Note source and date of information.

12.8. List two advantages and two disadvantages of onshore wave generation systems.

12.9. Calculate wave power/length (annual average) of the northwest coast of the United States.

12.10. Calculate power/area (annual average) of the Gulf Stream off the southern tip of Florida, the United States.

12.11. How much capture area, using an in-stream hydrokinetic system, would be needed for one 5-MW system in a good resource area? Be sure to note where you are locating your system: (a) tidal; (b) ocean current.

12.12. For a 50-MW point-absorber wave farm, how much area would be needed? Be sure to note where you are locating your system, the number of units, the rated power of the unit.

12.13. At rated power for the SeaGen tidal system, what is the power coefficient? Use data in text.

12.14. For the LIMPET wave system, annual average wave power input is 20 kW/m. What is the available input power? What is the rated power? Estimate the capacity factor.

12.15. What is the maximum theoretical efficiency for the OTEC plant in Hawaii? Remember to use degrees kelvin.

12.16. Briefly describe how power is obtained for an electric generator in a salt gradient system.

12.17. List the top two countries for ground source heat pumps with estimated capacity.

12.18. What is the productive use for watermills in Nepal?

12.19. What is approximate world capacity for ocean power (tidal, waves, OTEC)?

13 Storage

13.1 INTRODUCTION

Energy on demand means stored energy, and the most common storage is water in dams, batteries, and biomass. Of course, fossil fuels are stored solar energy from past geological ages. However, what means are available for storing renewable energy today? Storage is a billion-dollar idea, and anyone who comes up with cheap storage will be richer than Bill Gates. Economic storage would mean no new electric power plants would have to be constructed for many years, as energy could be stored from existing power plants during periods of low demand [1–3].

Energy cannot be created or destroyed, only transformed from one form to another, so in reality there are only two forms for storage: kinetic energy and potential energy. Storage as kinetic energy could be as flywheels, and thermal storage could be as heat (internal kinetic energy). For example, a passive solar home would use concrete or rock and maybe water for 2–4 days of thermal storage, and a ground source heat pump would use the earth as seasonal thermal storage. Compressed air is kind of a mixture; it is mechanical, but there is a thermal change. Super flywheels with high revolutions per minute and composite materials for strength have been designed and are used in uninterrupted power supplies and prototypes on buses. I kind of like flywheels as you would drive your car into the filling station and say "wind it up" rather than "fill it up."

Potential energy is due to the generalized interactions (gravitational, electromagnetic, nuclear weak, and strong), of which we consider gravitational and electromagnetic for storage systems. The gravitational potential energy is primarily water (Chapter 12): dams, tidal basins, and pumped storage. The electromagnetic interaction includes chemical, phase change, magnetic (superconductor magnetic energy storage [SMES]), electric (capacitors), and mechanical (springs) interactions. Chemical storage is by batteries, photosynthesis, production of methane and hydrogen, fertilizer, and other types. The storage of gas requires high pressure, converting to liquid, or as a chemical compound, or for example, storing hydrogen in metal hydrides. Of course, lots of solar energy is stored in chemical compounds as food and fiber (solar energy to sugar, starches, cellulose, etc.; liquid precursors for biofuels).

Thermal storage could be thermal mass, ice, molten salt (concentrating solar power [CSP] systems), cryogenic (liquid air, nitrogen, hydrogen), earth (heat pump), solar pond, and phase change. Molten salt for thermal storage is being used in combined heat and power systems. The main components for consideration of different storage systems are as follows:

Energy density
Efficiency and rate of charge/discharge
Duration or lifetime (number of cycles)
Economics

Different storage technologies (Table 13.1) can be for power or energy. Energy density and then size and weight for some applications are important factors (Figure 13.1). Liquid fuels have large energy density, while hydrogen gas has low energy density. In general, storage efficiencies range from 50% to 80%, and the lifetimes vary widely from 100 years for dams to 5–10 years for lead acid batteries

TABLE 13.1
Relative Rating of Storage Technologies

Type	Advantage	Disadvantage
Pumped hydro	High capacity	Site requirements
Compressed air	High capacity	Site requirements
Flywheels	High power	Low energy density
Superconductor magnetic	High power	Low energy density
Capacitors	Large number of cycles, high efficiency	Low energy density
Batteries		
Flow	High capacity	Low energy density
Metal-air	High energy density	Charging difficult
Pb-acid		Limited life when deep discharged
Li ion, Ni-Cd, Na-S	High power, energy density	
Hydrogen		Low energy density

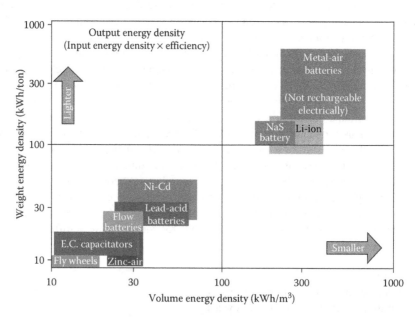

FIGURE 13.1 Energy density by weight and volume for electric storage. (Courtesy of Electricity Storage Association, Washington, DC.)

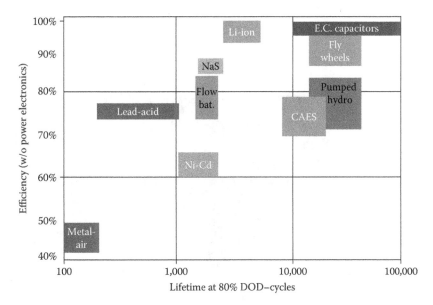

FIGURE 13.2 Efficiency and lifetime for different electric storage systems; CAES efficiency is for the storage only. (Courtesy of Electricity Storage Association, Washington, DC.)

in a photovoltaic (PV) system, to less than an hour for nonrechargeable batteries (Figure 13.2). The maximum rate and best rate of charging and discharging the storage is related to type and use of storage (Figure 13.3). Whether energy storage is included in an application and the type of storage is driven by economics (Figures 13.4 and 13.5) and specific power and energy (Figure 13.6). An electric car that takes 6 h to charge the batteries could not be used on a cross-country trip, even if charging stations were available. Note: Tesla has supercharger stations (80% charge in 30 min, free to Tesla owners) on major interstate highways and some cities, and in 2014, Teslas were driven from Las Angles to New York. Thus, you can see how versatile liquid fuels are for transportation. Also, the storage requirements for high power, short time is different from energy storage for a few days. For utilities (Table 13.2), the only large storage systems are pumped storage and compressed air energy systems (CAESs); however, battery systems for power shaving, conditioning, and reducing the variability from renewable energy sources have been installed. Both pumped hydro and compressed systems have long life and a large number of cycles. For remote village power systems and stand-alone systems, batteries are the most common storage.

The electricity produced by a generator cannot be stored; energy in minus energy losses is the demand, so the generation supplies that amount of demand, which varies by time of day and season. In addition, a utility system must meet peak demands and have spinning reserve for unforeseen conditions. If demand exceeds capacity, then users are taken off the grid, or as in some parts of the world, there are rolling blackouts, or electricity is only available for certain time periods. Finally, extreme events may force shutdown of the total grid.

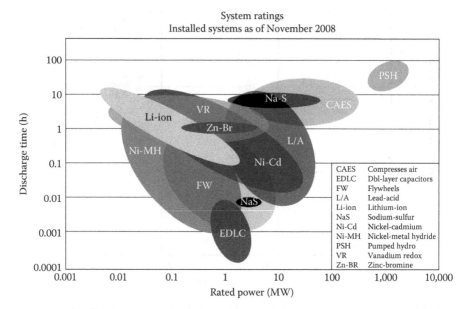

FIGURE 13.3 Discharge time and rated power of installed systems, electric storage. (Courtesy of Electricity Storage Association, Washington, DC.)

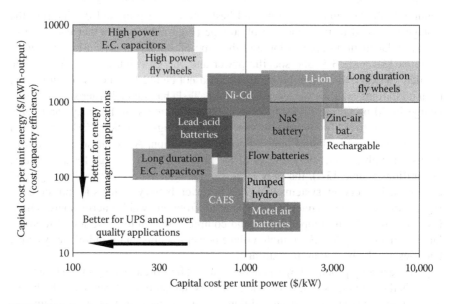

FIGURE 13.4 Capital costs for electric storage systems. (Courtesy of Electricity Storage Association, Washington, DC.)

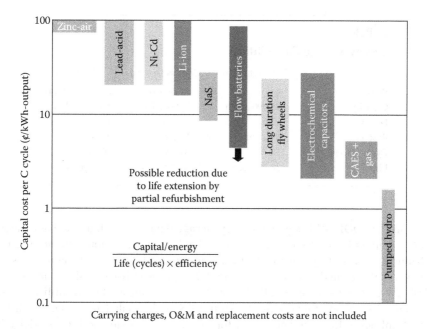

FIGURE 13.5 Cost per cycle for electric storage systems. (Courtesy of Electricity Storage Association, Washington, DC.)

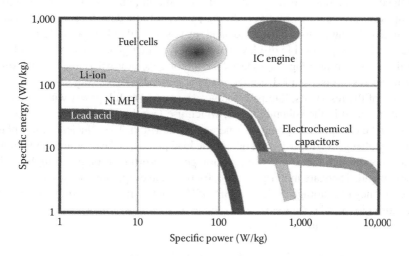

FIGURE 13.6 Energy and power by weight for different batteries.

TABLE 13.2
Energy Storage Capabilities

	Pumped Hydro	Compressed Air	Batteries
Power	4 GW	50–300 MW	50 kW to 50 MW
Capacity	25,000 MWh	2,500 MWh	50–250 MWh
Duration	12 h	4–24 h	1–8 h
Response	5–15 min	2–12 min	4 ms
Cycle efficiency (%)	70–85	70–85	60–90
Life	30–50 years	30–50 years	5–10 years

The U.S. DOE [4] has global energy storage database, with a searchable inter-active map and one of the filters is for technology type: compressed air, electro-chemical, flywheel, gravitational, hydrogen, pumped hydro, and thermal. As of 2014 for the United States, there were 1,242 projects listed with a capacity of 184 GW and an increase of 220 MW in 2015, with almost all in the utility sector. A source for energy storage case studies (CAES, molten salt, battery, and flywheel) is CleanEnergy [5]. Another source for information on storage for electricity is provided by the Energy Storage Association (http://energystorage.org), formerly the Electric Storage Association. Check out sections on technologies and technol-ogy applications.

13.2 PUMPED HYDRO

Pumped storage has two reservoirs, and the same motor/generator can be used to pump water to the upper reservoir during periods of low demand and then generate electricity during periods of high demand, just as any other hydroelectric plant with a reservoir. The pumped hydro levels the load for other generators on the grid. The need for a pump-priming head places the motor/generator below the water level of the lower reservoir. Pumped storage systems generally have a high head to reduce the size of the reservoirs. Pumped storage can respond to full power within minutes, and if operated in the spinning mode, which use less than 1% of their rated power, they can be changed to a pump or generator mode within 10 s. Nuclear power plants can only change load slowly, and other base load plants can be operated at maximum efficiency through the use of pumped storage to absorb their output at night during low demand. There are motor/generator, friction, and evaporation losses, so the over-all efficiency of a storage cycle is around 75%. The two main types of pumped hydro are closed loop and open loop.

The advantages are as follows:

Improved energy regulation and operation of the supply grid
Ancillary services such as standby and reserve, black-station start, frequency control, and flexible reactive loading

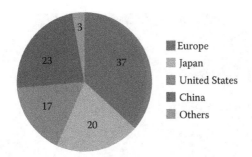

FIGURE 13.7   World pumped hydro storage (2012), percentage.

The disadvantages are as follows:

High capital cost due to fairly large reservoirs
As for other dams, land area needed, collapse could happen

There are over 142 GW of pumped storage in the world (Figure 13.7). Note that pumped storage is 77% of the world storage capacity and 99% of the world electricity storage capacity. In 2014, the United States had 26.6 GW of pumped storage, accounting for 2.5% of base load generating capacity.

In 2014, the Europe had around 50 GW net capacity of pumped storage out of a total of 173 GW of hydropower, and that represented 5% of total net electrical capacity in Europe. The potential for further pumped hydro is around double the present capacity. China had 20 GW of pumped storage from units 1 GW and larger, and Japan had 26 GW of pumped storage, which included a system that uses seawater for pumped storage.

13.2.1   CASE STUDY

The Taum Sauk plant in the United States is a pure pump-back operation in that there is no natural flow into the upper reservoir. Construction began in 1960, and operation began in 1963. The two pump-turbine units were each 175 MW and were upgraded in 1999 to 225 MW. The original upper reservoir had a capacity of 5,366,000 m³ and 244-m head to the hydroelectric plant and was connected by a 2,100-m tunnel through the mountain. The upper reservoir suffered a catastrophic failure [6] on December 14, 2005, due to a software error; water continued to be pumped to the reservoir when the reservoir was full, and then high winds from Hurricane Rita caused a breach in the reservoir walls (Figure 13.8). Approximately 4 million m³ of water was released in less than 0.5 h; luckily, no one was killed. The upper reservoir dam has been replaced with a roller-compacted concrete dam.

FIGURE 13.8 Upper reservoir for Taum Sauk pumped hydro plant, Missouri. Notice scouring of the creek due to wall failure. (Courtesy of U.S. Geological Survey, Reston, VA.)

13.3 COMPRESSED AIR

CAES is a peaking power plant that consumes 40% less natural gas than a conventional gas turbine, which uses about two-thirds of the input fuel to compressed air. In a CAES plant, air is compressed during off-peak periods and then is utilized during peak periods. The compressed air can be stored in underground mines or salt caverns, which take 1.5–2 years to create by dissolving the salt.

For an ideal gas, the amount of energy for an isothermal process from a pressure difference is

$$E = m\,R*T*\,\ln\!\left(\frac{P_A}{P_B}\right) = P*V*\,\ln\!\left(\frac{P_A}{P_B}\right),\ \mathrm{J} \tag{13.1}$$

where:
  P is the absolute pressure
  V is the volume
  m is the amount of gas
  R is the ideal gas constant
  T is the absolute temperature in degrees kelvin

The approximation is $100 * (P_A/P_B)$ kJ/m³ for gas at around atmospheric pressure.

A 290-MW CAES was built in Hundorf, Germany, in 1978, and a 110-MW CAES was built in McIntosh, Alabama, in 1991. The CAES in Germany provides 2 h storage from 300,000 m³ at a pressure of 48,000 Pa (1,000 psi). The CAES in Alabama provides 26 h of storage from a 540,000 m³ cavern at a pressure of 53,000 Pa (1,100 psi). The construction took 30 months and cost $65 million (about $591/kW). A 1.5-MW pilot project, modular isothermal CASE, where the heat from compression is stored in water is located in Seabrook, New Hampshire. Projects under construction are an underwater (air cavity at the bottom of a lake or ocean floor) CAES, 1 MW, 4 MWh in Toronto, Canada; an above ground CAES (80 kW for 30–60 min) in Promontory, Utah; an adiabatic CAES (200 MW) in Stassfurt, Germany; and a demonstration adiabatic CAES which uses an abandoned tunnel in the Swiss Alps. Texas Dispatchable Wind is a 1-MW wind with a 2-MW CAES storage system.

A CAES is relatively low efficiency, with a cost over $1,000/kW of storage. The input compressed air has to be cooled because the temperature would be too high for a salt cavern. The Electric Power Research Institute estimates that 80% of the United States has geology suitable for a CAES. Compressed air can be used for other applications, to power tools, pump water, and even power vehicles.

13.4   FLYWHEELS

Flywheels store energy due to rotational kinetic energy, which is proportional to the mass and the square of the rotational speed.

$$E = 0.5 * I * \omega^2, \text{J} \tag{13.2}$$

where:
I is the moment of inertia (kg m^2)
ω is the angular velocity (radian/s)

For a mass M and radius R, the moment of inertia for a ring is $I = M * R^2$, and for a homogeneous disk, $I = 0.5 * M * R^2$

Increasing the revolutions per minute increases the energy density, so high-speed flywheels have revolutions per minute in the tens of thousands. Low-speed flywheels are made from steel, and high-speed flywheels are made from carbon fiber or fiberglass. High-speed flywheels are housed in a low vacuum and use magnetic bearings to reduce or eliminate those frictional losses. Advances in power electronics, magnetic bearings, and materials have resulted in direct current (DC) flywheels. Note that if there is material failure, the container has to retain that energy inside. Cycle efficiency is around 80%.

Because of longer life, simpler maintenance, smaller footprint, and fast reaction time, flywheel systems are used in uninterruptible power supply (UPS) systems to provide backup power for the first 15 s until other backup generators come online. Installed cost depends on type, and ranges from $150 to $400/kW. Another application for flywheels is for cranes for ship and rail yards, where the flywheel provides the short period, high energy for lift, and then energy is returned to the flywheel as objects are lowered. One project has a 10 flywheel system (5 MW) to provide power quality support for a 20-MW wind farm [4].

Flywheels have been used in trains, cars, and buses; however, they were primarily experimental or prototypes. In the past, there were flywheels on tractors to smooth out the rotation of the crankshaft of two-cylinder engines. A hybrid vehicle could use flywheels for acceleration in conjunction with a smaller internal combustion engine, much like hybrid vehicles with batteries. Remember that, in a car, the flywheel is a gyroscope, which will change the handling.

13.5 BATTERIES

Batteries are common all over the world as lead-acid batteries are used for vehicles, and batteries are used for low-power and low-energy applications for lights, radio, TV, and electronic devices. PV or PV with rechargeable batteries has now replaced batteries for very low-power applications. Batteries convert chemical energy into electrical energy using electrodes immersed in a medium (liquids, gels, and even solids) that supports the transport of ions or electrolyte reactions at the two electrodes. Individual cells are placed in series for higher voltage and in parallel for higher current. Since there is an internal resistance and due to other factors, there are losses in charging and discharging a battery.

The power is the product of the voltage and current; however, batteries are generally specified by volts and storage capacity C_B, which is related to stored energy. C_B is the amount of charge that a battery can deliver to a load. It is not an exact number because it depends on the age of the battery (number of cycles), temperature, state of charge, and rate of discharge. If you discharge a lead-acid battery to essentially zero a few times, you have drastically reduced its lifetime. As a first approximation, the energy is

$$E = V * C_B, \text{ J} \tag{13.3}$$

where:
   V is the voltage
   C_B is the battery capacity (Ah)

Example 13.1

A 12-V battery is rated at 100 Ah. It could deliver 5 Ah for 20 h, $E = 12 * 20$ Wh or 1.2 kWh. However, at a faster discharge rate, the values would be lower: 85 Ah for 10 h, 70 Ah for 5 h.

Decreased temperature results in less battery capacity, and for a lead-acid battery, storage capacity decreases around 1% for every 1°C drop in temperature. Remember those cold mornings when the battery just had enough juice to start your car? Explosions by short circuit, generation of hydrogen, and disposal of used batteries and toxic chemicals are problems.

13.5.1 LEAD ACID

Lead-acid batteries are a low-cost and widely used technology for power quality, UPS, and some applications for spinning reserve. However, they are limited for large amounts of energy storage for utilities, primarily due to a short cycle life. As noted, the rate of charge/discharge (Figure 13.9) affects capacity (fast rate, fewer volts), and depth of discharge affects life (Figure 13.10). So, there is a trade-off between cost of more batteries and lifetime. Even with that disadvantage, lead-acid batteries are the most common energy storage for remote village power and stand-alone systems [7, Chapter 11, Energy Storage]. Examples of lead-acid systems are a 10-MW, 4-h system in Chino, California; a 20-MW, 40-min system in San Juan, Puerto Rico; and a 3.5-MW, 1-h system in Vernon, California.

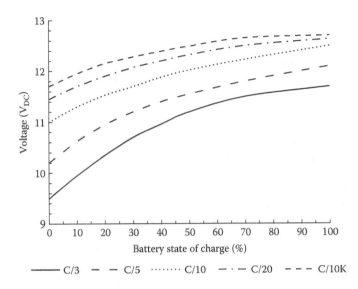

FIGURE 13.9 Battery voltage (12-V battery) and state of charge depend on rate of charge/ discharge (C/Number of hours).

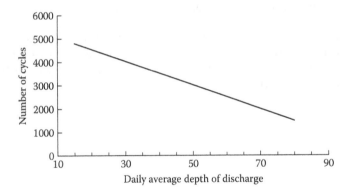

FIGURE 13.10 Life, number of cycles, for lead-acid battery versus depth of discharge.

Some battery practices for small renewable systems are as follows:

Do not add new batteries to old sets.
Avoid more than two (three at most) parallel strings.
Do not use different types of batteries in the same set.
Keep cable lengths the same.
Keep all components clean and connections tight.
Follow the manufacturer's recommendations for charging and equalization.
Do not wear jewelry when working on batteries.
Use insulated tools.
Wear protective clothing and eye protection when working on batteries.

There should be no smoking and no sparks around batteries.
Keep a battery log.

In general, car batteries are not suitable for storage in renewable energy systems, although they are available and have been used for some small systems.

13.5.2 LITHIUM ION

The main advantages of lithium ion batteries (Figure 13.11), compared to other advanced batteries, are as follows:

High energy density (300–400 kWh/m³, 130 kWh/ton)
High efficiency (near 100%)
Long cycle life (3,000 cycles at 80% depth of discharge [DOD])

Lithium ion batteries have captured 50% of the small portable market, and hybrid vehicles are one of the main drivers behind their increased use. However, they cost more. In 2015 Tesla announced a li-ion battery for the residential market, a 7 kWh daily cycle model that can be connected to the DC buss of a PV system. Preorders of 38,000 were reserved within 10 days after announcement, up to mid 2016 production.

There are 14 projects, 20 MW capacity in the world for storage using lithium ion batteries, and projects for renewable energy are as follows:

The Zhangbei wind/PV energy storage project consists of 100 MW wind, 40 MW
    PV and battery storage (lithium-ion iron-phosphate) of 36 MW, 4–6 h.
The AEI Laurel Mountain plant has 32 MW of batteries connected to a 98 MW
    wind farm for short-term smoothing.

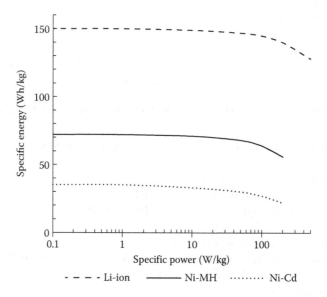

FIGURE 13.11 Energy and power by weight for advanced batteries.

A 1-MW/1-MWh battery is part of a smart grid demonstration project for wind integration at the Reese Technology Center, Lubbock, TX.

A 4-MW/2-MWh battery is integrated with 154 kW of renewable energy.

The Zhangbei National Wind and Solar Energy Storage and Transmission Project, China, will have 300 MW wind, 100 MW PV, and 110 MW of energy storage. Project currently has 14 MW of lithium-ion batteries and a 2-MW/8-MWh vanadium redox flow battery.

13.5.3 SODIUM SULFUR

The performance of the commercial sodium sulfur battery banks is as follows:

Capacity of 25–250 kW per bank.
Efficiency of 87%.
Lifetime of 2,500 cycles at 100% DOD or 4,500 cycles at 80% DOD.
Cost is around $2,500/kW.

Sodium sulfur batteries of over 270 MW at 6 h of storage have been demonstrated at numerous sites in Japan. The largest installation is a 34-MW, 245-MWh unit for a wind farm in northern Japan. U.S. utilities have deployed around 20 MW for peak shaving, backup power, wind farms, and other applications. Presidio, Texas, frequently experienced power outages due to their long connection to the main grid: a single, 60-year-old transmission line. A 4-MW, 32-MWh sodium sulfur battery was installed, which stores enough electricity for the whole town. Cost of the battery and substation was estimated at $23 million.

In the Great Plains of the United States, at 40 m and above, there is more wind energy at night than during the day, which is a poor load match to the demand of the grid. Xcel Energy installed a 1-MW battery storage next to an 11-MW wind farm [8]. The twenty 50-kW modules store about 8.2 MWh, which could power 500 homes for 7 h. GE Wind now offers batteries for short-term storage to provide ramp control, frequency regulation, and predictable power from 15 to 60 min. A wind farm in Japan has NaS batteries for load leveling; Rokasho-Futamata consists of 61 MW wind and a 34-MW, 238-MWh battery bank.

13.5.4 FLOW BATTERIES

A flow battery converts chemical energy to electricity; electrolytes containing one or more dissolved electroactive species flow through an electrochemical cell. Additional electrolyte is stored externally, generally in tanks, and is usually pumped through the reactor, although gravity feed systems are also known. The power and energy ratings are independent of each other. Flow batteries can be rapidly recharged by replacing the electrolyte liquid while simultaneously recovering the spent material for reenergization.

Vanadium can exist in four different oxidation states, so the vanadium redox battery [9] uses this to make a battery that has just one chemical electrolyte instead of two (Figure 13.12). H^+ ions (protons) are exchanged between the two electrolyte

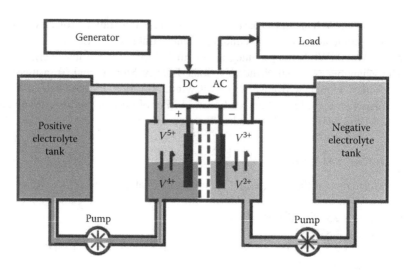

FIGURE 13.12   Diagram of a vanadium redox flow battery.

tanks through the permeable polymer membrane. The net efficiency of this battery can be as high as 85%. Current installations include

1.5-MW UPS system in a semiconductor fabrication plant in Japan
4-MW, 6-MWh output balancer, Tomaanae Wind, Hokkaido, Japan
250-kW, 2-MWh load leveler, Castle Valley, Utah
0.3 MW/1-MWh vanadium redox flow battery, 2-MWh lithium ion battery at community wind farm, Braderup, Germany.

13.5.4.1   Case Study

Hydro Tasmania, King Island Renewable Energy Integration Project [10].

Population around 1,800 provided electricity by diesel generators, capacity 6 MW which consumed around $4.5 * 10^6$ liters/year. Renewables supply around 65% of energy and during some periods provide 100%, so diesel fuel has been reduced to $2.6 * 10^6$ liters/year.

History of renewables integrated into the system.

Power station: Original, 2 MW firm capacity from two, 1,200 kW gensets, and one, 800 kW genset. Additional genset and upgrades provide 6 MW capacity, 4 MW firm capacity.

1998: Three wind turbines, 750 kW.

2003: Two wind turbines, 1,700 kW; vanadium flow battery, 200-kW, 800-kWh output leveler. Flow battery failed after a relatively short life and due to economics it was not put back into operation.

2018: Resistive frequency control.

2012: 2 diesel-UPS; flywheel, UPS to provide short term backup for the primary diesel generators. Excess wind energy is used to power flywheel.

2014: PV, 390 kW; biodiesel blending plant; smart grid, advanced lead acid battery, 3 MW, 1.6 MWh to store excess wind energy.

The smart grid provides demand side management to further save diesel. Wi-Fi provides access to users with App, KIREIP Dashboard, which provides real time data so participants in the smart grid can monitor their home energy usage. This may lead to differential tariffs to promote the use of renewable energy.

13.5.5 OTHER TYPES OF BATTERIES

Other types of batteries are made of metal-air, Ni-Cd, Zn-Br, Li-titanate, and even organic compounds. Lithium-titanate batteries have the advantage of faster charging time and estimated life of 20 years. The metal-air batteries could be less expensive; however, they are at the preproduction stage at which the consumed metal is mechanically replaced and processed separately. Recharge using electricity is under development, but they only have a life of a few hundred cycles and an efficiency of about 50%.

A Zn-Br battery has two different electrolytes that flow past carbon-plastic composite electrodes in two compartments separated by a microporous polyolefin membrane. Battery systems are available on transportable trailers (storage plus power electronics) with unit capacities of up to 1 MW, 3 MWh for utility-scale applications. Now, 5-kW, 20-kWh systems are being installed by electric utilities for community energy storage.

General Electric has a sodium halide battery for utility, telecommunication, and UPS applications.

A Ni-Cd battery system (40 MW, 7 min) was installed in Fairbanks, Alaska. Small rechargeable Ni-Cd and lithium ion batteries are available. Should you buy these or continue to buy one-time use batteries for your electronics?

13.6 OTHER STORAGE SYSTEMS

Inductors (from current in coils of wire) store magnetic fields, and capacitors store electric fields from charges, so both can be used to store energy. Their advantages are fast response time and large number of cycles; however, they are presently not used for storing a large amount of energy.

13.6.1 MAGNETIC SYSTEMS

SMES is the stored energy in the magnetic field due to flow of DC in a superconductor. When a superconductor is cooled below the critical temperature, there is no resistance, so there is essentially no energy loss over time. SMESs have long life, fast response, essentially no energy loss over time, and cycle efficiency of 95%; however, temperatures are very low, so there is the high cost for refrigeration and for the superconducting wire. High-temperature superconductors, defined as temperatures of liquid nitrogen (77 K) and above, are now commercially available, and there is much research on materials for very-high-temperature superconductors.

The amount of stored energy is

$$E = 0.5 * L * I^2, \text{J} \tag{13.4}$$

where:

  L is the inductance (henrys)
  I is the current (amperes)

The inductance depends on the shape of the coil of wire, which is solenoid or torus. See introductory physics books for information on the calculation of inductance.

There are commercial SMESs, 1-MW units, for power quality control in installations around the world, especially to provide clean power quality at manufacturing plants. There are several larger test projects to provide grid stability in distribution systems; for example, in northern Wisconsin, distributed SMES units enhance the stability of a transmission loop. Superconductors are now being considered for high-power transmission lines.

13.6.2 CAPACITORS

Capacitors are devices for storing charge on two plates separated by some distance, and the amount of stored charged can be increased by placing a dielectric between the plates. One problem with capacitors for energy storage is the leakage current. Electrochemical capacitors (ECs) store electrical energy in the electric double layer, which is formed between each of the electrodes and the electrolyte ions. The amount of stored energy is

$$E = 0.5 * Q^2/C \text{ or } 0.5 * C * V^2 \text{ or } 0.5 * Q * V, \text{J} \tag{13.5}$$

where:

  Q is the charge (coulombs)
  C is the capacitance (farads)
  V is the voltage (volts)

ECs have fast response, long life (up to 1 million cycles), and higher energy density than electrolytic capacitors but less than batteries. Aqueous capacitors have a lower energy density due to lower cell voltage; however, they cost less and have a wider temperature range. The asymmetrical capacitors that use metal for one of the electrodes have a significantly larger energy density than the symmetric ones and have lower leakage current.

ECs have been used in conjunction with batteries in hybrid vehicles.

13.6.3 PHASE CHANGE MATERIALS

In a phase change (gas–liquid–solid), there is heat absorption/release at almost constant temperature. For energy storage, the liquid–solid phase change is the only practical one, and the material should have large latent heat and high thermal conductivity [11]. Phase change materials store 5–14 times more thermal energy per unit volume than conventional storage materials (water, masonry, and rock). There are a large number of phase change materials in the temperature range from −5°C to 190°C. Organic materials such as paraffin and fatty acids have phase changes in this range. Some problems with using phase change materials are stability of thermal properties under extended cycling and sometimes phase segregation and subcooling.

The heat of fusion for water is 333.6 kJ/kg or 319.8 MJ/m^3. Remember, the density of ice is less than that for water, so containers have to have room for expansion for water in the solid phase. There are 86 ice and 16 chilled water thermal storage systems in the United States [4]. One example of phase change storage for an office building was ice made in the winter, stored in an underground reservoir, and then used for cooling during the summer.

13.7 HYDROGEN

There is a lot of information on hydrogen as the fuel of the future and on the storage of hydrogen [12, Chapter 11, Hydrogen Storage]. Even though hydrogen has a low environmental impact and can be produced by renewable energy sources, the major disadvantage is the low energy density per volume; hydrogen has one-third the energy content of methane. However, existing natural gas pipelines could carry about the same capacity because the hydrogen has lower viscosity.

Fuel cells are much more efficient than internal combustion engines and fuel cell cars are now on the market. However, the infrastructure for fuel cell cars is at the nascent stage and the energy source is natural gas or propane. So, if hydrogen is produced by renewable energy systems through electrolysis or through bioenergy, then transportation and storage become major factors. Hydrogen can be stored as compressed gas or liquid (must be cooled to 20 K in nonpressurized containers) or by an extraction process. The storage in materials could be adsorption, such as by activated carbon, chemical compounds, compounds that can be reversibly transformed into another substance of higher hydrogen content, and metal hydrides that change hydrogen content with temperature.

Metal hydrides could be used for storage of hydrogen for vehicles powered by fuel cells; for comparison, 100 kg of hydride could store around 500 MJ. However, 100 kg of gasoline provides 4,700 MJ of energy, which is a large difference. The efficiency of the hydrogen fuel cell is around 60%, and for the gasoline engine, it is around 20%. Hydrogen would have fewer emissions and is essentially nondepletable. The production of hydrogen from water using wind and sun would be a problem in arid and semiarid areas where there is already a lack of water. Some considerations for hydrogen storage systems are the following:

Ratio of mass of hydrogen to overall mass of the storage and retrieval system
Ratio of mass of hydrogen to total volume of the storage and retrieval system
Cycle efficiency
Retention (amount of hydrogen remaining over a long period of time)

Of course, as with other storage systems, economics (installed cost, operation and maintenance costs, and replacement costs) are paramount, and then there is safety, ease of use, and infrastructure for transportation.

REFERENCES

1. A. A. Akhil, G. Huff, A. B. Currier, B. C. Kaun, D. M. Rastler, S. B. Chen, A. L. Cotter, D. T. Bradshaw, W. D. Gauntlett. 2013. *DOE/EPRI 2013 Electricity Handbook in Collaboration with NRECA*. Sandia Report, SAND2013-5131. http://prod.sandia. gov/techlib/access-control.cgi/2004/046177.pdf.
2. I. Gyuk. 2009. *Energy Storage Status and Progress*. http://www.narucmeetings.org/ Presentations/Gyuk.pdf.
3. *Energy Storage System Projects*. http://www.sandia.gov/ess/projects_home.html.
4. U.S. DOE. *Global Energy Storage Database*. http://www.energystorageexchange.org/ projects.
5. CleanEnergy, *Energy Storage Case Studies*. http://www.cleanenergyactionproject. com/CleanEnergyActionProject/Energy_Storage_Case_Studies.html.
6. NOAA. December 14, 2005. *Taum Sauk Dam Failure at Johnson's Shut-In Park in Southeast Missouri*. http://www.crh.noaa.gov/lsx/?n=12_14_2005.
7. R. Foster, M. Ghassemi, and A. Cota. 2010. *Solar Energy, Renewable Energy and the Environment*. CRC Press, New York.
8. Wind-to-battery project. 2011. https://www.xcelenergy.com/staticfiles/xe/Corporate/ Renewable%20Energy%20Grants/Milestone%206%20Final%20Report%20PUBLIC. pdf.
9. *PowerPedia: Vanadium Redox Batteries*. http://peswiki.com/index.php/PowerPedia: Vanadium_redox_batteries.
10. Hydro Tasmania, *King Island Renewable Energy Integration Project*. http://www. kingislandrenewableenergy.com.au/project-information/overview.
11. M. M. Farid, A. M. Khudhair, S. A. K. Razack, and S. Al-Hallaj. 2004. A review on phase change energy storage: Materials and applications. *Energy Convers Manage*, 45(9–10), 1597–1615.
12. Ald Vieira da Rosa. 2009. *Fundamentals of Renewable Energy Processes*. 2nd Ed. Academic Press, New York.

RECOMMENDED RESOURCES

LINKS

Energy Storage Association. http://energystorage.org.
EPRI, electricity energy storage technology options, 1020676. 2010. http://www.epri.com/ abstracts/Pages/ProductAbstract.aspx?ProductId=000000000001020676.
Sandia National Laboratories, Energy storage systems program. http://www.sandia.gov/ess.
D. A. Scherson and A. Palencsár. Spring 2006. Batteries and electrochemical capacitors. *Electrochem Soc Interf*. https://www.electrochem.org/dl/interface/spr/spr06/spr06_p17-22.pdf.
A. Seki. 2007. Energy storage. http://www.heco.com/vcmcontent/IntegratedResource/IRP/ PDF/HECO_IRP4_AG13_101707_05_Energy_Storage.pdf.
U.S. FERC, pumped storage projects, map. http://www.ferc.gov/industries/hydropower/gen-info/licensing/pump-storage.asp.

PROBLEMS

13.1. How much energy could be delivered from a pumped hydro reservoir that is 5 million m^3 with a 250-m head? Reduce the amount due to an 80% efficiency.

13.2. Compare kinetic energy storage to batteries in terms of energy density and installed costs. Both systems have the same amount of energy storage.

13.3. Find any commercial SMES; note the source, company, and any other useful specifications. How much energy can it store?

13.4. Find any commercial flywheel for UPS; note the source, company, and any other useful specifications. How much energy can it store?

13.5. Find any commercial electrochemical capacitor storage system; note the source, company, and any other useful specifications. How much energy can it store?

13.6. High-speed flywheels are made from composite material. A flywheel is a ring: 3-m radius, mass = 50 kg, 20,000 rpm. How much energy is stored in the flywheel?

13.7. Estimate the energy stored in the CAES in McIntosh, Alabama. What is the cycle efficiency of that system?

13.8. Pick any 12-V car battery. How much energy does it store? Note the brand, specifications of the battery, cost, and expected lifetime.

13.9. What is the cost per kilowatt hour for a 12-V car battery used for lights and radio? Recharge once per week at a central charging station; life = 2,000 cycles. Cost of a recharge is around $1.00.

13.10. What are the advantages of a flow battery versus a deep-cycle lead-acid batteries for energy storage (load level) for a utility?

For Problems 13.11 through 13.13: A renewable energy village power system has 10-kW wind, 2-kW PV, 10-kW inverter, and battery bank (24 V, 2,000 Ah). The village is in a good solar and wind resource area, and energy consumption is around 300 kWh/month.

13.11. Estimate annual energy production and then daily energy production.

13.12. Estimate the installed cost for a remote area a 2-day drive from a large town over rough terrain. Get information from the Internet on equipment costs.

13.13. A Battery bank can be discharged to 50%. How many days can the battery provide power if there is no wind and no sun?

13.14. Calculate the cost per kilowatt hour for a D-cell battery for a flashlight (nonrechargeable; 1.5 V, 12 Ah, cost $4.00).

13.15. Compare cost per kilowatt hour for alkaline, nonrechargeable AA batteries with lithium ion rechargeable batteries. Alkaline batteries: 1.5 V, 3 Ah, $0.75. Lithium ion batteries: 1.2 V, 2.2 Ah, $6.50. Charger cost is $15, 50% efficient, grid electricity at $0.10/kWh; lithium ion life is 1,000 cycles.

13.16. A building uses the latent heat of the change of ice to water for cooling. How much energy can be absorbed in ice to a liquid if originally 500 m^3 of water was frozen?

13.17. A Greenhouse uses a phase change for storage; the tank contains 6,000 kg of paraffin, the melting temperature range is 48°C–60°C, and latent heat is 190 kJ/kg. How much energy was stored during the phase change?

13.18. Estimate the capital cost for pumped hydro storage today. Note specifications for reservoir size and head.

13.19. What are the major problems for using hydrogen for energy storage?

13.20. Find information on any car powered by fuel cells. Note brand, rated power. What is the fuel source?

13.21. Compare capital costs per kilowatt or capital costs per kilowatt hour for energy storage systems: pumped hydro, flywheels, lead-acid batteries, vanadium-redox flow battery, and hydrogen in metal hydrides.

13.22. Go to reference 4. For the United States, how many CAES are operational? Under construction? Note total installed capacity.

13.23. Go to reference 4. Pick any ice thermal system, state name, location, rated power, duration, and cost if available.

13.24. Go to reference 10, smart grid. Note date and values on the KIREIP Dashboard.

14 Institutional Issues

14.1 INTRODUCTION

The institutional issues (this is a noninclusive list as there are surely others) related to renewable energy include the following: legislation and regulation concerning the environment, incentives, externalities, world treaties and country responses to greenhouse gas emissions, connection to utility grids (large power generators such as wind farms, large photovoltaic [PV] arrays, concentrating solar power [CSP]; large numbers of small systems; distributed and community systems); incentives such as feed-in tariffs, renewable portfolio standards, rebates, tax credits; and certification standards for equipment and installation of systems. Of course, most of these issues are determined by politics and economics.

The following are some institutional and environmental issues by resource:

Solar, CSP: Large land use for collectors, but land generally is of low productivity for other uses

Wind: Visual, noise, and wildlife, primarily birds (avian) and bats

Bioenergy: Large land use for growth of biomass, erosion, burning of biomass releases nitrogen oxides, release of methane

Geothermal: Subsidence, seismic activity, reduction of hot springs, minerals in discharged water

Hydro: Visual, change in hydrology, impact on fish, large area and removal of people, silt, even dam collapse

Ocean: Reduction of tidal flats, marine life, shipping hazard for wave systems, and offshore wind

14.2 THE UNITED STATES

The U.S. National Energy Act of 1978 was a response to the 1973 energy crisis. The main purpose was to encourage conservation of energy and the efficient use of energy resources. The Public Utility Regulatory Policies Act (PURPA) covers small power producers and qualifying facilities (independent power producers), which are up to 80 MW. Sections 201 and 210 of PURPA encourage the use of renewable energy by mandating the purchase of power from qualifying facilities and exempting such facilities from much of the federal and state regulations. States had a large amount of flexibility in implementing PURPA. The main aspects of PURPA are as follows:

Utilities must offer to buy energy and capacity from small power producers at the marginal rate (avoided cost) the utility would pay to produce the same energy.
Utilities must sell power to these small power producers at nondiscriminatory rates.

333

Qualifying facilities are entitled to simultaneous purchase and sale. They have
the right to sell all their energy to the utility and purchase all the energy
needed.

Qualifying facilities are exempt from most federal and state regulations that
apply to utilities.

The implementation of PURPA was determined by public utility commissions,
utilities, independent power producers, and the courts. Determination of avoided
costs was the main point of contention between small power producers, independent
power producers, and utilities.

The National Energy Strategy Act of 1992 included the provision of wheeling
power over utility transmission lines. The Federal Energy Regulatory Commission
(FERC) can order the owner of transmission lines to wheel power at costs determined by FERC. The utilities are allowed to recover all legitimate, verifiable economic costs incurred in connection with the transmission services and necessary
associated services, including, but not limited to, an appropriate share, if any, of
the costs of any enlargement of transmission facilities. From the standpoint of wind
power, this legislation is important because the major source of wind energy is in
the Great Plains, which is far from the major load centers. In 1997, FERC opened
transmission access. The first wind farm (initial 35 MW, later expanded to 68 MW)
in Texas was in the far western part of the state, and power was wheeled to the Lower
Colorado River Authority area in central Texas. Those two wind farms were decommissioned in 2014.

The deregulation of the electric utility industry by some states has changed the
competition for renewable energy. Deregulation essentially means that the integrated
electric utility companies are split into three areas: generation of power, transmission, and distribution. Also, consumers can buy from different power producers. The
other aspect for increased use of renewable energy is green power and the push for
reduction of pollution and emissions from fossil fuel plants that generate electricity.

The goal of 20% electricity from wind energy by 2030 for the United States will
require major institutional changes, primarily in terms of transmission, carbon trading, and development of more load management. The location for much of those
wind farms would be in the wind corridor of the Great Plains.

14.2.1 Avoided Cost

The avoided costs were established by the public utility regulatory body in each
state. FERC defines avoided cost as the incremental or marginal cost to an electric
utility of energy or capacity, which the utility would have to generate or purchase
from another source if it did not buy power from the qualifying facility. Avoided
cost reflects the cost from new power plants, not the average cost from plants already
installed. However, many utilities said they did not need any new generation; therefore, avoided costs were only the fuel adjustment cost. The avoided cost includes not
only present but also future costs.

Utilities can set a standard purchase rate for qualifying facilities under 100-kW
capacity, and local public regulatory bodies can provide more information on small

power production. In the 1980s, the California Public Utilities Commission (PUC) set the avoided costs and types of contracts for qualifying facilities. Standard Offer Number 4 set the avoided costs for a period of 10 years, while Standard Offer Number 1 was variable depending on the cost of fuel. One of the reasons wind farms started in California was the high avoided costs set by the PUC.

Some utilities state that they have excess capacity, and therefore, the avoided cost is equal to the value of the cost of fuel at the power plant. The fuel adjustment cost for Southwestern Public Service in the Texas Panhandle in January 1994 was $0.02/kWh. The company was consolidated with a company in Colorado and Minnesota, now called Xcel Energy; however, in 2010 the avoided cost in this region was still the fuel adjustment cost, which was around $0.025/kWh, although when natural gas prices were high, the avoided cost was larger, depending on the percentage mix of natural gas for boiler fuel. However, the changing natural gas prices had utilities considering wind as a hedge against future volatility of natural gas costs, and some utilities were requesting proposals for wind farms for their systems. In 2014, the price of natural gas had declined due to increased production from shale formations by fracking, which then reduces the economics of new renewable installations.

14.2.2 Utility Concerns

For low penetration of renewable power generators, which produce variable power, on a large utility grid, there would be no problems with that small amount of power. It would be considered as a negative load: a conservation device, which is the same as turning off a load. For large penetration, 20% and greater, other factors such as the variability and dispatching become important. Also, operation and maintenance (O&M) costs could increase for conventional generators as they have more variable output, which could increase O&M costs due to variable renewable energy input to the grid. The utilities are concerned about safety and power quality due to any systems on their grid.

Safety is a primary consideration for any renewable energy project, just as with any other industrial enterprise. All energy industries have safety concerns, for example, underground coal mines or the drilling for natural gas in shale formations.

For large and small systems that generate electricity and are connected to the grid, primary concerns are energizing a dead utility line, grounding of equipment, and lightning. For wind turbines, high voltages, rotating blades and machinery, large weights, and working at hub heights up to 100 m make for a hazardous workplace. There is a summary and a list of accidents for the wind industry, including type of accident, turbine, date, and location [1]. In the survey, the longest recorded distance for blade failure was 400 m from the tower. The largest number of accidents was from blade failures, and the second largest was from fire; the most common fatality was from falls. Now, large wind turbine failures make news, and the videos appear on the web.

A mortality rate of 0.4 deaths/MWh is reported for the mid-1990s, which dropped to 0.15 deaths/MWh by the end of 2006 [2]. Some of the deaths in the database were associated with the transport of wind turbines. The mortality rate for any renewable energy industry needs to be compared to those of other conventional energy industries.

Quality of power refers to harmonics, power factor, voltage, and frequency control. A number of renewable energy systems on the end of a feeder line could require extra equipment to maintain quality of power. However, PV could increase the power quality and reliability on feeder lines as they supply power during peak air conditioning loads. Utility companies have to supply reactive power for induction generators for wind turbines, and they may require capacitors on the wind turbine or at the wind farm substation to maintain the power factor.

Connection of any renewable energy-generating system to the grid must be approved by the utility. The utility should be informed at the earliest possible stage of the intention to connect a renewable energy power source to its system. Information for the utility should include the following:

Specifications of the renewable energy system
Schematic (block diagram) of the electrical system
Description of machine controls when there is loss of load (utility power) and
    a lockable disconnect

The utility may require a meter that measures flow of energy in both directions, even if there is net energy billing. Smart meters for demand side management will be installed in more locations as the national grid is transformed and would make the energy from distributed renewable energy systems more valuable.

For large systems (10 MW and up), an interconnection study will cost from $30,000 to $120,000. This study determines the effect of the renewable energy system on the transmission lines and existing generators.

Liability for damage is another concern of utilities. The utilities would like to be insured against all damage due to the renewable energy system operation. Of course, the small power producer would like to be insured against damage to their system as a result of utility operation; however, that is impossible to obtain. For small systems, insurance should be available as part of a homeowner's policy or as part of a business policy. In the early period of wind power, some electric cooperatives were requiring proof of a $500,000 liability policy for connection of a wind turbine to their system.

Ancillary costs are those costs to the utility as renewable energy systems become a larger percentage of the generation capacity on the grid. The variability of renewable energy systems can increase operating costs, such as committing unneeded generation, scheduling unneeded generation, allocating extra load-following capability, violation of system performance criteria, and increasing cycling operation on other generators. Estimates of these costs range from $0.001 to $0.005/kWh [3]; however, one utility estimated the cost at $0.0185/kWh. In 2008, the Montana Public Service Commission set a rate up to $.00565/kWh for integration of wind power from the Two Dot wind farm into the utility grid.

A major storm occurred in Spain with winds above the cut-out wind speed, which resulted in a major drop in output of 7,000 MW from wind farms compared to the predicted input to the utility grid operation for that day. In another case, wind farms produced 53% of the total demand in Spain for 5 h (November 2009) when there was ample wind and low demand on the grid during early hours of the day. In the spring of 2014, The Electric Reliability Council of Texas recorded a penetration of 40%, a high capacity of 10,296 MW from wind farms, while the Southwest Power Pool recorded a capacity high of 6,448 MW from wind farms.

14.3 REGULATIONS

Regulations for renewable projects vary by country and by region or state, from simple (e.g., a review process from a single agency), to multiple (e.g., complex reviews for different agencies), and even to multiple levels of government. Sometimes, there seem to be competing regulations from different agencies, and the number of agencies can be large. National laws and policies may restrict connection of any renewable energy system to the utility grid. Most large projects require consideration of environmental impact, although actually enforcement may vary widely, depending on the country.

In the United States and other developed countries, permits for construction are generally required in residential areas and even in rural areas in some states. Zoning issues are esthetics (primarily visual), safety, and, for wind turbines, tower height, setbacks, and noise. Risks are accepted from other areas, such as cars and utility lines (electric and gas). Signs, trees, and even utility poles have failed in high winds or under conditions of icing.

Restriction of access and signs for high voltage are needed for renewable energy systems that generate electricity. One factor can never be dismissed: Anything that interferes with TV will be unacceptable to the public. Most locations do not have specific zoning regulations for renewable energy systems, so individuals must be prepared to educate public boards and their neighbors, although that is now changing as more systems are being sold due to availability of tax credits for small renewable energy systems.

In the United States, federal permitting requirements range from environmental to Federal Aviation Administration regulations on lights for tall towers and wind turbines (lights on towers or turbines taller than 200 ft). Those in the industry maintain that regulations are now a major portion of their cost of doing business. In most cases, those in the industry say that they cannot meet proposed regulations because it is uneconomical.

14.4 ENVIRONMENTAL ISSUES

There will be environmental issues for any large renewable energy project; the issues will vary by the renewable energy resource and the location, and there may even be environmental issues for small installations up to 100 kW. So, those developing projects should be prepared to have an environmental impact study or at least be able to obtain information as the Environmental Protection Agency has jurisdiction over many aspects. What is the biological impact on wildlife, plants, and habitat? Another common aspect is the visual impact, which can be detrimental, especially in those areas that are located close to scenic areas or parks. Some may like the view of large numbers of wind turbines, and others will be opposed. At the end of the project, decommissioning, recycling, and disposal, especially of any toxic components, have to be considered. The goal of the U.S. Fish and Wildlife Service is to protect wildlife resources, streamline site selection, and assist in avoiding environmental problems after construction. A U.S. Fish and Wildlife Service report addressed how to minimize the impacts of land-based wind farms on wildlife and its habitat [4].

There will be land areas that are excluded because of environmental considerations: national and state parks, wetlands, and wild life refuges. In addition, some states and even counties have regulations concerning the environment that will have to be met before a project can be considered. First, check with local officials before you install any system.

The developer should conduct an analysis of the environment regarding permits, licenses, and regulatory approvals; threatened or endangered species; wildlife habit; avian and bat species; wetlands and other protected areas; and location of known archeological and historical resources. Geographical information systems are an excellent tool for depicting environmental and land use constraints. Regulations on archeological sites differ by state, and in some states, private land is excluded. Even if it is not mandatory to check for archeological sites, it probably still should be done for a project of medium size and larger.

After the first analysis of environmental issues, a more detailed analysis should address possible impacts and possible mitigation of those impacts. After the project is operational, mitigation of the impacts has to be monitored on a scheduled basis. Biological concerns are habitat loss, alteration or fragmentation of habitat (e.g., prairie chickens), bird or bat collisions with wind turbines, electrocution of raptors, and effect on vegetation. Water, especially wetlands, soil erosion, and water quality have to be considered. For wind farms, the clearing of scrub brush for roads, sites, and even for laying underground wires is welcomed by ranchers; however, the cleared areas, such as shoulders of the roads, have to be seeded and monitored for growth, erosion, and noxious weeds. Another possibility in complex terrain and even range areas is that maintained roads may be welcome by the land owner or operator, and the roads can serve as firebreaks.

For protection against liability, a developer should perform a screening and environmental assessment prior to or early in the acquisition of the property. The American Society for Testing and Materials has screening tools and standards for environmental site assessment [5].

Some people are adamantly opposed to almost any large energy project in their vicinity, including those for renewable energy. In general, most individuals are neutral, and the rest are in favor, especially those who will receive some economic benefit. In the Great Plains of the United States, there is less opposition as wind farms are seen as rural economic development. Developers should provide community education at the planning and preconstruction phases of a project.

Wind farms are now a large renewable energy resource, so the environmental impacts of visual, noise, and birds and bats will be considered in more detail as an example.

The visual impact for wind farms is quite different because the number and height of the towers as wind turbines will be visible from 20 km. In the plains, they are visible from all angles (Figure 14.1), with only the curvature of the Earth limiting the distance from which they can be seen. In mountainous areas, the wind turbines will be in lines on the ridges, but in general they are not visible from all angles because

FIGURE 14.1 Visual aspect of wind turbines near Hitchland, Texas, looking south. Foreground: 3.2 km to one turbine, Vestas V 90, 3 MW, diameter = 90 m, 80-m tower; middle left, 4.4 km to row of eight Suzlon wind turbines, 1.25 MW, diameter = 64 m, 72-m tower; background, John Deere 4 wind farm, 9.5 km to first row, 14.5 km to back row, 38 Suzlon wind turbines, 2.1 MW, diameter = 88 m, 80-m tower.

most of the roads are in the valley. The moving rotors make wind turbines more noticeable, and flashing lights make them conspicuous at night, especially when the lights are synchronized to outline the wind farm. Shadow flicker happens, and the high impact is generally located within approximately 300 m of the turbine. In a pasture with no trees in the summertime, yearling calves and even horses line up in the shadows of the tower of the wind turbines, and the animals move to keep in the shade as the tower shadow moves.

Noise measurements have shown in general that wind turbine noise is below the level of ambient noise; however, the repetitive noise from the blades stands out. In general, one would not want a residence in the middle of a wind farm. The whine from gearboxes on some units is also noticeable. However, with larger wind turbines at higher hub heights and new airfoils, the noise has been much reduced. The farmers who live close to the wind turbines at the White Deer wind farm reported that noise was not a problem, especially since they were receiving money for the wind turbines on their land. Others near wind farms who are not receiving any economic benefit exaggerate the noise level, comparing it to jet engines on planes.

Birds and bats can be killed by wind turbines [6] as the tips of blades for both small and large wind turbines are moving around 70–80 m/s. The factors for mortality of birds and bats are number of fatalities, bird species, season, the threat to the population, and possible forms of mitigation. Collision rates for birds per turbine per year vary from 0.01 to 23 for a coastal site in Belgium, where the birds included gulls, ducks, and terns. Other coastal sites in northwest Europe had a rate of 0.01–1.2 birds per turbine per year. However, neither of these examples produced a significant decline in population. In general, migratory birds fly above the heights of wind turbines; however, overcast and ground clouds may lower flight paths. Two large wind farms, south of Corpus Christi, Texas, near the coast, have a radar for monitoring migration of birds, and the wind turbines are shut down if they pose a threat to the birds. Estimates of bird fatalities range from 140,000 to 325,000/year at monopole towers [7], while a higher estimate is 573,000/year for the installed wind capacity of 2012 [8]. Note that hundreds of thousands of birds are killed by communication towers, buildings (Geographica, photo in *National Geographic,* September 2003), hunters, and even cars.

After bats became a problem in West Virginia, information and guidelines became available for both bird and bat impacts [9]. The estimate of bat fatalities is 880,000/year [8]. One report stated that the air pressure differential of the passing blades could affect bats; it did not have to be a direct strike from the blade. Preconstruction data are used for predicting fatalities, and that projection needs to be compared with bird or bat fatalities after the wind turbines are operating.

Similar institutional and environmental problems will be associated with other renewable energy projects. Identification of problems, mitigation, and continued monitoring are essential.

14.5 POLITICS

As with any endeavor, politics enters the situation. To make a change in behavior of people and institutions, especially when the competition is an entrenched industry, you need *incentives, penalties,* and *education.* Someone estimated that the amount of each type of energy used is in direct proportion to the amount of subsidies in the past for that type of energy. Subsidies are in the form of taxes, tax breaks, and regulations, all of which generally require legislation: politics. What every entity (industry) wants are incentives for itself and penalties for its competitors. In addition, these entities want the government to fund research and development (R&D), demonstration or pilot projects, and even commercialization.

Incentives are tax breaks, subsidies, mandates, and regulations to promote R&D and commercialization. Public utility commissions are demanding that utilities use integrated resource planning (IRP), which means they have to consider renewable energy and conservation in the planning process. Can utilities make money for kilowatt hours saved? Who is supposed to take the risk, the consumers or the shareholders? Three Mile Island and the nuclear utility industry are good examples of politics, from the local to national level. The Price Anderson Act, a federal law, limited the amount of liability from a nuclear accident, and without that legislation, the nuclear industry could not have sold plants to utilities.

Penalties are generally in the form of taxes and regulations. Environmental groups have already indicated that utilities will be held accountable for the risk of a carbon tax if they plan on new coal plants. In other words, in their opinion the shareholders and not the consumers should take the risk.

Education is public awareness of the possibilities or options, a realistic cost-benefit comparison over the lifetime of the energy systems. Remember that you cannot fool Mother Nature, and you will pay one way or another; you will probably pay more later if the problem is not taken care of in the present.

Politics will continue to influence which and how much different energy sources are subsidized. Some incentives, or some may see them as penalties, include carbon trading or a carbon tax, rebates on equipment or incentives for electrical energy produced from renewable energy, renewable portfolio standards (renewable electric standards), feed-in tariff for renewable energy, and others. A policy table for renewable energy technologies around the world indicates that 138 countries have support policies at the state/provincial to national level [10]. In general, the United States has an unwritten policy of cheap energy and cheap food. I thought that President Ford and Congress would increase taxes on gasoline after the first oil crisis in 1973, however I was obviously wrong.

14.6 INCENTIVES

Energy subsidies have serious effects, generally in favor of conventional fossil fuels and established energy producers. Subsidies for renewable energy between 1974 and 1997 amounted to $20 billion worldwide. This amount can be compared with the much larger subsidies for conventional energy sources, which totaled *$300 billion per year,* and this number does not even take into account the expenditures for infrastructure, safeguards, and military actions [11] for continued flow of oil and natural gas. I estimated that the cost for Oil War I (Gulf War) and Oil War II (Iraq War) for the United States was approximately equal to a subsidy of $0.50/gallon for gasoline. What will be the future global cost for armed forces to keep fossil fuels available?

The International Energy Agency has an online database of energy subsidies, and their latest estimate for 2010 was that global fossil-fuel subsidies amounted to $409 billion and renewable energy subsidies amounted to $66 billion [12]. The World Watch Institute estimated fossil fuel subsidies between $775 billion and $1 trillion in 2012 [13]. For the United States from 2002 to 2008, fossil fuels receive 72.5 billion in subsidies, while renewable energy received $29 billion, divided into $12.2 billion for traditional renewable and $16.8 billion for corn ethanol. The privatization of the electric industry along with the restructuring into generation, transmission, and distribution has opened some doors for renewable energy.

In the support of national and state policies, the common tax incentives for renewable energy [14] are as follows:

Investment tax incentives: Large-scale applications provide income tax deductions or credits for some fraction of the capital investment.

Investment tax incentives: Residences and businesses receive tax deductions or credits for some fraction of the costs of renewable energy systems.

Production tax incentives: Provide income tax deductions or credits at a set rate per kilowatt hour.

Property tax reductions

Value-added tax (VAT) reductions: Exempt producers of renewable energy from taxes on up to 100% of the value added by an enterprise between purchase of inputs and sale of outputs.

Excise (sales) tax reductions: Exempt renewable energy equipment purchasers from up to 100% of excise (sales) tax for the purchase of renewable energy or related equipment.

Import duty reductions

Accelerated depreciation: Allows investors to depreciate plant and equipment at a faster rate than typically allowed, thereby reducing stated income for purposes of income taxes.

Research, development, demonstration, and equipment manufacturing tax credits

Tax holidays: Reduce or eliminate income, VAT, or property taxes for a temporary period of up to 10 years.

Taxes on conventional fuels: Some countries tax the consumption of nonrenewable energy (most often a fossil fuel or carbon tax).

Mandates for manufacturing: Percentage of components that must be made in country. In China, wind turbines installed in the country must have 70% of the components made in China. The American recovery and reinvestment act has a buy America provision.

14.6.1 THE UNITED STATES

The Database of State Incentives for Renewables and Efficiency (DSIRE) is a comprehensive source of information on state, local, utility, and selected federal incentives that promote renewable energy and energy efficiency [15]. Overview maps are also available by the type of incentives and policies. Check the database for detailed information.

The major impetus to the wind and the CSP industries was due to federal tax credits, the National Energy Act of 1978, and the avoided costs set by the California PUC. The credits for wind expired in 1985, while the solar credits were continued.

The second major impetus for the wind industry was the renewable energy production tax credit (PTC). The PTC was part of the National Energy Strategy Act of 1992, and it provides a $0.015/kWh incentive for production of electricity by renewable energy. A commercial or industrial entity can claim the PTC under Section 45 of the Internal Revenue Service code. The provisions are as follows:

The investor owns the wind facility, which was placed in service during the period December 31, 1993, to July 1, 1999.
The investor produces the electricity at the facility.
The investor sells the electricity to an unrelated party.

The credit applies to production through the first 10 years of the operation of the facility. The credit is intended to serve not only as a price incentive but also as a price support. The credit is phased out as the average national price exceeds $0.08/kWh, based on the average price paid during the previous year for contracts entered into after 1989. Both values are adjusted for inflation. The credit can be carried back for 3 years and carried forward for 15 years to offset taxes on income in the other years. The eligible technologies are landfill gas, wind, biomass, hydroelectric, geothermal electric, municipal solid waste, hydrokinetic, anaerobic digestion, small hydroelectric, tidal, wave, and ocean thermal.

The PTC has been extended a number of times and was available for projects under construction at the end of 2014. In 2014, the PTC was $0.023 for wind, geothermal, and closed-loop biomass technologies and $0.011 for the other technologies. Some technologies are only eligible for 5 years. Every year that the extension of the PTC was late, the next year saw a major decline in wind projects, so installations were small in 2013 and installations were large in 2014. As of spring 2015, planning is difficult for projects due to the uncertainty of the extension of the production tax credit.

Because of the problem of finding an entity with available tax liability, as part of the American Recovery and Reinvestment Act, the developer can choose to receive a 30% investment tax credit (ITC) instead of the PTC for facilities placed in service in 2009 and 2010 and for facilities placed in service before 2013 if construction began before the end of 2013, with the latest extension that projects had to be under construction by end of 2014. The ITC then qualifies to be converted to a grant from the Department of Treasury. The Treasury Department must pay the grant within 60 days of application submission.

The renewable energy production incentive was similar to the PTC, except the eligible entities were local government, state government, tribal government, municipal utility, rural electric cooperative, and native corporations. The problem was the amount of funding was capped, and Congress had to approve funding every year, so few projects were constructed that used this incentive.

Small renewable energy systems with 100 kW of capacity or less can receive a tax credit for 30% of the total installed cost of the system. This tax credit (Emergency Economic Stabilization Act of 2008) is available for equipment installed from October 3, 2008, through December 31, 2016. The value of the credit is now uncapped through the American Recovery and Reinvestment Act of 2009.

Federal and state incentives encourage ethanol production [16], for example, the mandates and incentives of the Energy Policy Act of 2005. Gasoline was mandated to contain 7.5 billion gallons of renewable fuel annually by 2012, and most of the requirement will be met with ethanol. The Energy Independence and Security Act of 2007 increases renewable fuel use to 36 billion gallons by 2022. The act requires advanced biofuels, which are defined as fuels that cut greenhouse gas emissions by at least 50%. The advanced biofuels could include ethanol derived from cellulosic biomass, biodiesel, butanol, and other fuels. The Volumetric Ethanol Excise Tax Credit amounts to around $0.45/gal subsidy for ethanol; phased out at end of 2013.

For PV, California had significant feed-in tariffs and investment subsidies:

Systems > 100 kW$_P$: $0.39/kWh
Systems < 100 kW$_P$: Can choose either $2.50/W$_P$ or $0.39/kWh
Contract duration: 5 years, constant remuneration
Net metering: Up to 2.5% of peak demand, rolls over month to month, granted
    to utility at end of 12-month billing cycle

Now the California Solar Initiative Programs pays an incentive based on system performance. The incentives are either an upfront lump-sum payment based on expected performance or a monthly payment based on actual performance over five years. The expected performance is available only for systems less than 30 kW.

Many states offer rebate programs for renewable energy systems, from a flat rate to $/W installed, with limits on systems size and maximum amount of rebate. Again go to Reference [15] for details on incentives.

14.6.1.1 Federal Support

The federal government continues to support renewable energy through the Department of Energy (DOE) budget for Energy Efficiency and Renewable Energy (EERE). For more detailed information, visit the EERE website (http://www.eere. energy.gov, then scroll to About EERE>Budget). Generally, the DOE budget for renewable energy is less than the budget for nuclear energy.

As an aside, every president from Nixon to today's and most politicians have touted energy independence. It is interesting to note that there are generally one to two energy acts per president, but neither any president nor Congress has taken the necessary steps to implement energy independence because it would require major sacrifices and some changes in lifestyle, and it is tough to get elected or reelected on those premises.

The tone or direction of energy policy is set by the administration, which changes with the president. The early direction was R&D plus demonstration projects, which was supposed to lead to commercialization. During the Reagan years, private industry was supposed to commercialize renewable energy, and federal funding was for generic R&D. During Reagan's term, the support for renewable energy was reduced every year.

Under President William Clinton, there was renewed interest in renewable energy, and the direction was commercialization. The Climate Change Action Plan moved the DOE from focusing primarily on technology development to playing an active role in renewable energy commercialization. This initiative was backed up with $72 million for fiscal year 1995 and a total of $432 million through the year 2000. For the emissions reductions from renewables, the DOE looked primarily to wind since it was the most economical renewable source at that time.

Under George W. Bush, the national energy plan first focused on increased production of oil and gas. With pressure from Congress, conservation, energy efficiency, and renewables were added to the package, and the PTC was extended in 2002. However, an increase in vehicle efficiency, CAFE (combined automobile fleet efficiency) standards, did not pass. Another national energy act, the Energy Policy Act of 2003, was passed, and finally, CAFE was increased in the last year of the Bush administration.

President Obama has changed the direction dramatically toward renewable energy, which means every university and national lab will be seeking that money by creating renewable institutes and centers. It would be interesting to count the number of new degree programs, institutes, and centers at universities since 2008.

14.6.1.2 State Support

States are also competing for renewable energy as a way to offset importation of energy and as a way to create jobs. Some states have mandated deregulation of the electric utility industry. Deregulation in some states gives the consumers choice of producers, and most of the states have a system benefits charge (SBC) that lets utilities recover stranded costs of power plants, primarily for nuclear plants. In some states, part of the SBC is set aside for renewable energy. For example, in California, funds from the SBC are available to offset part of the cost for small renewable energy systems.

The wind farm boom in Texas was fueled by a renewable portfolio standard (RPS) enacted in 1999, which was part of electric restructuring legislation. The mandate was for 2,000 MW of new renewables by 2009 in the following amounts by 2-year steps: 400 MW, 2003; 450 MW, 2005; 550 MW, 2007; and 600 MW, 2009. There was a rapid growth of wind farms in Texas, and now many states have a RPS. Texas expanded the RPS in 2000 with a new goal of 5,880 MW of renewables by 2015, with a carve-out of 500 MW to come from nonwind sources. Texas easily surpassed this goal as around 10,000 MW of wind power was installed by the end of 2010. The legislators and even utilities were surprised at the amount of wind capacity that was installed as a result of the RPS and the national PTC.

Another aspect of the electric restructuring in Texas is that electric retailers have to acquire renewable energy credits (RECs; 1 REC = 1 MWh) from renewable energy produced in Texas or face penalties of up to $50/MWh. Anybody may participate in the REC market: traders, environmental organizations, individuals, and so on. The market opened in January 2002, and early prices were around $5/REC, but due to the large amount of wind power installed, by 2015 the value was around $1/REC. The RECs are good for the year created and bankable for 2 years.

As always, industries seek tax breaks at every level. States and local entities give tax breaks for economic development, and renewable energy developers would like a tax break on installed costs as that is their major cost. Conventional power producers can deduct the cost of fuel, whereas for renewable energy these deductions are not available since the fuel is free. Legislators are now touting renewable energy, especially as rural economic development. States and development commissions are trying to lure businesses in renewable energy.

14.6.1.3 Green Power

For green power, the consumer pays a voluntary premium, which was around $3/month for a 100-kWh block in 2002, or contributes funds for the utility to invest in renewable energy development. Green power is an option in the policy in some states and has been driven by responses of utilities to customer surveys and town meetings. Green power represents a powerful market support for renewable energy development, which was mainly wind energy and ethanol. More than half of all U.S. electricity customers have an option to purchase green power from more than 850 utilities or about 28% of utilities nationally.

In 2012, retail sales of renewable energy from green power exceeded 48 TWh, approximately 1.6% of electricity sales [17]. Sources were wind (80.1%), landfill gas and biomass (12.8%), hydropower (6.2%), solar (0.6%), and geothermal (0.3%). When I lived in Canyon, Texas, I paid $3/month voluntary premium for green power for electricity, and after I moved to Round Rock, Texas, I had the option for retail purchase of electricity and I selected a company that was 100% green power (wind). In general, utilities have lowered the rate premium (around $1.50/kWh in 2014) on green power as traditional fossil fuel costs have increased or as green power becomes more competitive. If green power becomes cheaper than regular power, will those consumers who purchased green power pay below the regular rate? Other aspects of green power are community systems and crowd funding programs, which will increase in the future, and large corporations are purchasing and/or investing in green power. The U.S. DOE ranks the utility green power programs annually [18].

14.6.1.4 Net Metering

Forty-three states have net metering (Figure 14.2), which ranges from 10 to 1,000 kW, with most in the 10- to 100-kW range. If the renewable energy system produces more energy than is needed on site, the utility meter runs backward, and if the load on site

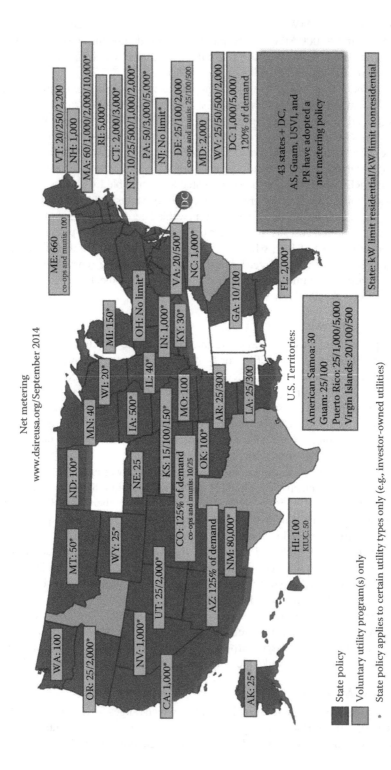

FIGURE 14.2 Map of net metering for the United States. (From the Database of State Incentives for Renewables and Efficiency [DSIRE, http://www. dsireusa.org].)

Net metering
www.dsireusa.org/September 2014

VT: 20/250/2,200
NH: 1,000
MA: 60/1,000/2,000/10,000*
RI: 5,000*
CT: 2,000/3,000*
NY: 10/25/500/1,000/2,000*
PA: 50/3,000/5,000*
NJ: No limit*
DE: 25/100/2,000
co-ops and munis: 25/100/500
MD: 2,000
WV: 25/50/500/2,000
DC: 1,000/5,000/
120% of demand

43 states + DC,
AS, Guam, USVI, and
PR have adopted a
net metering policy

State: kW limit residential/kW limit nonresidential

ME: 660
co-ops and munis: 100

DC

VA: 20/500*
NC: 1,000*
FL: 2,000*

OH: No limit*
IN: 1,000*
KY: 30*
GA: 10/100

MI: 150*
WI: 20*
IL: 40*
MO: 100
AR: 25/300
LA: 25/300

MN: 40
IA: 500*
KS: 15/100/150*
OK: 100*

ND: 100*
NE: 25

U.S. Territories:

American Samoa: 30
Guam: 25/100
Puerto Rico: 25/1,000/5,000
Virgin Islands: 20/100/500

MT: 50*
WY: 25*
UT: 25/2,000*
CO: 125% of demand
co-ops and munis: 10/25
AZ: 125% of demand
NM: 80,000*

WA: 100
OR: 25/2,000*
NV: 1,000*
CA: 1,000*

HI: 100
KIUC: 50

AK: 25*

State policy

Voluntary utility program(s) only

*   State policy applies to certain utility types only (e.g., investor-owned utilities)

is greater, then the meter runs forward. Then, the bill is determined at the end of the time period, which is generally 1 month. If the renewable energy system produced more energy over the billing period than was used on site, the utility company pays the avoided cost. In some states, any net to the consumer is carried over to the next month or the payment time period, and at the end of the year, any net production accrues to the utility with no payment to the consumer.

In general, net metering in the 10- to 50-kW range did not increase the sale of renewable energy systems since the electricity produced was still not cost competitive with retail electricity. If a small renewable system is installed, you want to use that energy on site as that is worth the retail rate. Also, if the time period is longer than 1 month or there is rollover for positive production, net metering is more useful to the consumer.

Of course, utility companies do not encourage net metering because it increases billing problems, and the utilities say that one group of customers would be subsidizing another group of customers. With electric restructuring, utilities are worried that large customers will find cheaper electricity, and then rates will rise for residential customers. Does that mean that many residential customers are subsidized today?

14.6.2   OTHER COUNTRIES

There have been incentives and goals for renewable energy around the world, with more announced every year. The amount of each type of renewable energy installed follows the legislation, regulations, and especially the amount of incentives provided. So far, the major impact has been large-scale installation for wind farms and ethanol plants, with PV and solar hot water making significant contributions. The Internet will have information on renewable energy incentives by country.

The European Union is a leader in the development and installation of renewable energy. The E.U. goal is 20% of energy consumption from renewables by 2020, with some countries setting higher goals. Those in the European Union are even discussing 100% from renewables by 2050. Japan, Spain, and Germany now are big markets for PV, and some countries are also requiring installation of solar hot water.

14.6.2.1   Wind

Several European countries started wind energy programs in the 1980s, with most emphasizing megawatt wind turbines; however, there was little success. The manufacturers in Denmark produced small units, then progressed to larger units in steps, and acquired around 50% of the early U.S. market and 66% of Europe's installed capacity in 1991. European manufacturers captured a major share of the world market; however, the large increase in installation of wind farms in China means that Chinese manufacturers are now major producers (Table 9.3).

The different policies for wind energy include tax incentives (tax credits, production incentives, accelerated depreciations, etc.), preferential tariff regimes, quota requirements and trading systems, with the two main mechanisms being feed-in tariff and renewable portfolio standards (RPS) [19]. Note that support by an incentive or subsidy is often a necessary, but never a sufficient condition because of other external elements in the implementation. Free trade in renewables in the E.U. market is complicated by the fact that renewables are supported by mandates or fixed prices at different levels by country and even state. This support could be regarded as a substitute for a pollution tax on fossil fuels.

Promotion of wind energy in Europe was based on two models: (1) price support for kilowatt hour production and (2) capacity based. In general, the minimum-based price (feed-in tariff) resulted in the most installations. Denmark, Germany, and Spain used the price support method, while the United Kingdom, Ireland, and France used the capacity-based method. In Denmark, Germany, Greece, and Scotland, the fact that local communities received benefits from local wind projects was important, for example, local farmers were encouraged to install wind turbines.

The main conditions for support for wind energy in Germany were (a) clear and long-term price stability through the feed-in tariff mechanism; (b) priority grid access; (c) local and regional banks made financing available; and (d) early and strong political commitment to renewable energy [20].

Germany accounted for half the European market for wind after 1995 due to the Electricity Feed Law (EFL, 1990), which was designed for climate protection, for saving fossil fuels, and for promoting renewable energy. The law obliges utilities to buy any renewable energy from independent power producers at a minimum price defined by the government, with the price based on the average revenue of all electricity sales in Germany. The initial value in 1991 was €160/MWh, and then it declined as more wind power was installed. Since the Renewable Energy Sources Act was enacted, electricity generation in Germany from renewable energy doubled from 6% in 2000 to 13% in 2007 [20], and most of that was generated by wind. The EFL was modified in 1998, which set a regional cap of 5% for renewable electricity. Some states also gave a 50% investment grant in the late 1980s and early 1990s. Special low-interest loans for environmental conservation measures were also available for financing projects. These factors contributed to the massive growth of wind in the 1990s in Germany. The law was changed in 2004 to €87/MWh on production for 5 years and €55/MWh for the next 15 years. For projects or wind turbines that came online in later years, there was a decrease of 2.5% per year. For example, for 2010, the payment was €79/MWh on production for 5 years and €50/MWh for the next 15 years. In 2012, the renewable energy sources act was amended to a regression rate to 1.5% per year for onshore wind, a repowering bonus and incentives for offshore wind, which included an offshore test site.

China now ranks first in the world in installed wind capacity (see Figure 9.12) due to favorable government policies for the installation of wind farms: mandatory market share, feed-in-tariff, and selection of seven areas of the country for the Wind Base Programme [19]. The wind industry has relied on financing through the Clean Development Mechanism (Kyoto Protocol), which has presented some problems for new projects. As in other countries, transmission needs to be upgraded from the windy areas to the load centers. Presently, China mandates that 70% of the components of wind turbines erected in China be produced by factories within the country. The feed-in tariff is now around €50/MWh. To meet the goal of 3% of electricity from nonhydro renewables by 2020, around 100 GW of wind capacity would need to be installed and 200 GW for the 2030 goal.

India ranks fourth in the world in installed wind capacity due to a favorable fiscal/policy environment. Wind power development in India has been promoted through R&D, demonstration projects and programs supported by government subsidies, fiscal incentives, and liberalized foreign investment procedures. The incentives are as follows:

Central government: Income tax holiday, accelerated depreciation, duty-free import, energy capital/interest subsidies.

State governments: Buyback, power wheeling, and banking; sales tax concession, electricity tax exemption; demand cut concession offered to industrial consumers who establish renewable power generation units; and capital subsidy. Tamil Nadu and several other state electric boards purchase wind energy at about $64/MWh.

India upgraded the Department of Non-conventional Energy Sources to a ministry in 1992. In 2003, the Electricity Act led to defined tariffs for wind energy, and in 2009, a generation-based incentive for grid-connected wind was implemented.

14.6.2.2 Photovoltaic Energy, Solar Hot Water

Japan, Spain, and Germany are the big markets for PV. The Japanese program started in 1994 with subsidies that were reduced by year to no subsidy in 2003 as more systems were installed. At the end of 2004, Japan led the world in installed PV capacity with over 1.1 GW. Then, in January 2009 Japan restored the subsidy system for domestic PV with grants of $760/kW.

Europe captured close to 80% of total annual installed PV capacity in 2009, with Germany and Spain leading the way due to substantial feed-in tariffs. In 2013, PV installations in Asia surpassed Europe, with China the top market in the world (see Figure 6.15 for installed PV capacities). Germany started a large-scale feed-in tariff system in 2000, and in 2007, Spain introduced a high feed-in tariff of 32–34 €/MWh. Both Germany and Spain are reducing or capping the amount of subsidy due to the large number of installations. In 2014, in Europe there were 13 support mechanisms for PV with feed-in-tariff, net metering, and energy sale on the electricity market being the most employed [21]. A degression rate applied at regular periods or rate depending on the capacity installed will reduce incentives over time. The full report has detailed analysis of current support schemes.

In October 2014, European Union (EU-28) governments agreed on a framework on climate and energy for 2030. The targets include a cut in greenhouse gas emissions by at least 40% by 2030 compared to 1990 levels, increase renewable energy to at least 27%, and reach an indicative energy efficiency target of at least 27%. The policy aims are for a more competitive, secure, and sustainable use of energy. There is a provision to reform and strengthen the EU emissions trading system.

A government regulation (1980) in Israel requires solar water heating in every new building. Now, 90% of the homes in Israel have solar hot water, and 80% of all the annual water heating requirements are met by solar hot water. This provides a saving of 3% of primary energy use. Now, other countries are mandating solar hot water for new construction. In British Columbia, Canada, 36 municipalities adopted a regulation that requires all new single homes to accommodate future installation of a solar hot water system. Other countries offer rebates for installation of solar hot water. China accounts for the largest amount of solar hot water collectors, primarily market driven. There are regulations even in the United States; starting in 2010, Hawaii has a law that every new single-family home must have solar hot water. Other states have solar ready regulations for new residential construction [22].

The production of ethanol has increased dramatically due to mandates and incentives. In the United States, there is a lot of political support as it is seen as improvement of rural economic income.

14.7   EXTERNALITIES (SOCIAL COSTS/BENEFITS)

Externalities are defined as social or external costs/benefits that are attributable to an activity that are not completely borne by the parties involved in that activity. Externalities are not paid by the producers or consumers and are not included in the market price, although someone at some time will pay for or be affected by them.

Social benefits, generally called *subsidies*, are paid by someone else and accrue to a group. An example is the Rural Electrification Act, which brought electricity to rural U.S. An example of a positive externality (social benefit) is the benefit everyone gets from cleaner air from installation of renewable energy systems. On the other side, a good example of a negative externality is the use of coal in China, as every city of 100,000 and over has terrible smog due to use of coal for heating, cooking, industry, and production of electricity. Smog alerts are common in the large cities and there is already a public health cost, which in 20 years will be a large public health cost for today's children.

External costs can be divided into the following categories:

Hidden costs borne by governments, including subsidies and support of R&D programs.

Costs associated with pollution: Health and environment damage, such as acid rain, destruction of ozone in the upper atmosphere, unclean air, and lost productivity.

Carbon dioxide (CO_2) emissions may have far-reaching effects, even though global warming is disputed by many in industry and by some scientists (http://cdiac. esd.ornl.gov/trends/co2/contents.htm).

Mechanisms for including externalities into the market are government regulation, pollution taxes, IRP, and subsidies for R&D and production.

Government regulation: The historical approach of regulation or mandates has led to inefficient and monopolistic industries and inflexibility, and it is highly resistant to change. The current vogue is for deregulation and privatization of energy industries. However, if external costs are not included, short-term interests prevail, and this generally distorts the economics toward conventional and entrenched suppliers of energy. Regulations can require a mix or minimum use of energy sources with lowest life-cycle cost, which include externalities.

Pollution taxes: Governments can impose taxes on the amount of pollution a company generates. European countries have such taxes. Another possibility is to give RECs for producing clean power. Pollution taxes and avoidance of pollution have the merit of simplicity and have only a marginal effect on energy costs but are not a true integration of external costs into market prices. The taxpayer pays, not the consumer. The pollution tax could be assessed in the consumer bill; therefore, it is paid on how much is used.

Integrated resource planning: This model combines the elements of a competitive market with long-term environmental responsibility. An IRP mandate from the government would require the selection of new generating capacity to include all factors, not just short-term economic ones.

Subsidies: Of course, subsidies for renewable energy promote that source and make that source more competitive with conventional fossil fuels. However, the recipients want the subsidy to continue, problems with timing of subsidy may contribute for difficult business decisions, and the subsidy may be harmful to an overall long-term, rational energy policy.

Many studies on externalities have been conducted. The European Union's six-volume *ExternE: Externalities of Energy* (http://www.externe.info/) is probably one of the most systematic and detailed studies to evaluate the external costs associated with a range of different fuel cycles. In their estimates, external costs for the production of electricity by coal can be as high as $0.10/kWh and external costs for nuclear power at $0.04/kWh.

Since 1995, companies in the United States have been trading sulfur dioxide (SOX) and nitrogen oxides (NOXs) emissions, which are precursors of acid rain and contributors to ground-level ozone and smog. Essentially, industries trade in units called *allowances*, which can be bought, sold, or banked for future use. CO_2 is not included in the United States; however, some states are now passing laws to reduce CO_2 production. It is difficult to predict whether and/or when the United States will have carbon trading or a tax on carbon CO_2 emissions. The United States and China lead the world in CO_2 emissions per year and the United States leads in total past emissions.

In the United States, emissions from the generation of electricity are primarily due to the burning of coal (Table 14.1). The average CO_2 emission is around 720 kg/MWh for all fuel types; of course, it is higher for coal, around 1,000 kg/MWh. Thus, wind turbines and PV reduce emissions of CO_2 by 1 metric ton per megawatt hour when displacing coal generation; in addition, they do not require water for the generation of electricity. The production of electricity by natural gas has increased in market share, which means that CO_2 emissions per megawatt hour are smaller, so the average has decreased since 1990. A Minnesota group (connected with the utility industry) estimated the external costs for CO_2 from coal as only $0.34 to $3.52/ton. In Europe, CO_2 emission reductions are worth $40/ton in some countries.

TABLE 14.1

U.S. Generation of Electricity and Air Emissions, 2013 Values

	2013	Carbon Dioxide		Sulfur Dioxide		Nitrogen Oxides	
	MWh 10^6	Kg/MWh	Tons 10^6	Kg/MWh	Tons 10^6	Kg/MWh	Tons 10^6
Coal	1,586	950	1,507	3.00	4.8	1.50	2.4
Oil	27	710	19	2.70	0.1	0.65	0.0
Natural gas	1,114	480	535	0.003	0.003	0.60	0.7
Total	2,727		2,061		4.8		3.1

The numbers for emission factors (Table 14.1) were adjusted somewhat to give total values in metric tons, the same as the numbers from the Energy Information Administration. The CO_2 emissions in 2013 were less than in 2009 due to the increase in natural gas and renewables and decreased use of coal. Emissions differ by source and plant operation, for example, the average value for sulfur dioxide from coal is 3.0 kg/MWh; however, the worst coal plant in the United States produces 18 kg/MWh. So, new coal plants have scrubbers, but nearly 40% of the coal plants do not have the same pollution control standards because they were online prior to the Clean Air Act of 1970.

14.8   TRANSMISSION

A major problem for renewable energy development, especially wind, is that many load centers are far away from the resource, and projects can be brought online much faster than new transmission lines can be constructed. For those states with electric restructuring, transmission is now by a separate company, and the questions regard jurisdiction, who pays for new lines, and if curtailment is needed because a project is producing too much power for the grid, who is curtailed, and the priority of curtailment. Even with new transmission lines, future development may be limited by transmission capacity. A large transmission investment of $12.6 billion would increase a retail bill of $70 by $1.

Texas has constructed high-voltage transmission lines (345 kV), primarily to bring wind power from West Texas and the Panhandle to the load centers within the Electric Reliability Council of Texas (ERCOT) [23]. The new lines (Figure 14.3) would have a capacity of 18.456 GW, which means that around 11 GW of new wind capacity will be added to the ERCOT system (9 GW installed when the law was passed, over 16 GW at end of 2015). The Panhandle of Texas is in the Southwest Power Pool (SPP), and SPP is also constructing new transmission lines, again primarily for wind power. Now, there is emphasis on a supranational transmission grid, similar to the interstate highway system, so planning and money are being spent on that program (see Links).

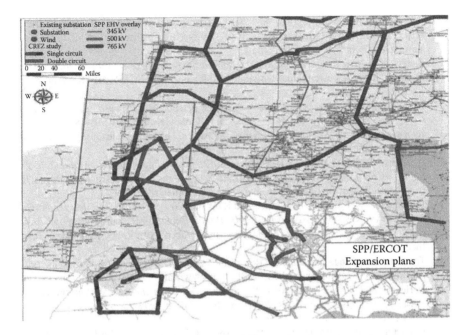

FIGURE 14.3 Transmission lines for ERCOT at planning stage for routes, construction completed in 2014. Transmission lines (northern set) for the Southwest Power Pool (SPP) at planning stage for some routes, others at study stage.

REFERENCES

1. Cathiness Windfarm Information Forum. *Summary of Wind Turbine Accident Data.* http://www.caithnesswindfarms.co.uk.
2. P. Gipe. 2012. *Wind Energy—The Breath of Life or the Kiss of Death: Contemporary Wind Mortality Rates.* http://www.wind-works.org/cms/index.php?id=43&tx_ttnews%5Btt_new s%5D=164&cHash=286b80c2e2af695492a9bb1d94513821.
3. B. Parsons, M. Milligan, B. Zavadil, D. Brooks, B. Kirby, K. Dragoon, and J. Caldwell. 2003. *Grid Impacts of Wind Power: A Summary of Recent Studies in the United States,* NREL/CP-500-34318, http://www.nrel.gov/docs/fy03osti/34318.pdf.
4. U.S. Fish and Wildlife Service. 2014. *Final Land-Based Wind Energy Guidelines.* http://www.fws.gov/windenergy/docs/WEG_final.pdf.
5. American Society for Testing and Materials. Screen process. http://www.astm.org/ Standards/E1528.htm; Phase I, http://www.astm.org/Standards/E1527.htm; Phase II, http://www.astm.org/Standards/E1903.htm.
6. NREL Wind-Wildlife impacts literature database (WILD). https://wild.nrel.gov.
7. S. R. Loss, T. Will, and P. P. Marra. 2013. Estimates of bird collision mortality at wind facilities in the contiguous United States. *Biol Conserv* 168, 201–209. http://www. sciencedirect.com/science/article/pii/S0006320713003522.
8. K. S. Smallwood. 2013. Comparing bird and bat fatality-rate estimates among North American wind energy projects. *Wildlife Soc Bull* 37(1), 19–33. http://onlinelibrary. wiley.com/doi/10.1002/wsb.260/abstract.

9. American Wind-Wildlife Institute. 2014. *Summary of Wind-Wildlife Interactions.* http://awwi.org/resources/summary-of-wind-wildlife-interactions-2/#section-summary-of-windwildlife-interactions.

10. *Renewables 2015 Global Status Report. Renewable Energy Policy Network for the 21st Century,* http://www.ren21.net/wp-content/uploads/2015/07/GSR2015_KeyFindings_lowres.pdf.

11. H. Scheer. 1998. Energy subsidies—A basic perspective, *Proceedings of the 2nd Conference on Financing Renewable Energies,* Bonn, Germany. Nov. 16–18.

12. International Energy Agency, Energy subsidies. http://www.iea.org/publications/worldenergyoutlook/resources/energysubsidies/

13. A. Ochs, E. Anderson, and R. Rogers. 2012. *Fossil Fuel and Renewable Energy Subsidies on the Rise.* Worldwatch Institute, Vital Signs. http://vitalsigns.worldwatch.org/vs-trend/fossil-fuel-and-renewable-energy-subsidies-rise.

14. D. Clement, M. Lehman, J. Hamrin, and R. Wiser. 2005. *International Tax Incentives for Renewable Energy: Lessons for Public Policy,* Center for Resource Solutions, draft report. http://www.resource-solutions.org/lib/librarypdfs/IntPolicy-Renewable_Tax_Incentives.pdf.

15. The Database of State Incentives for Renewable Energy (DSIRE). http://www.dsireusa.org.

16. Ethanol incentives and laws, EERE, DOE. http://www.afdc.energy.gov/fuels/laws/ETH.

17. J. Heeter and T. Nicholas. 2013. *Status and Trends in the U.S. Voluntary Green Power Market (2012 data).* NREL/TPl-6A20-60210. http://www.nrel.gov/docs/fy14osti/60210.pdf.

18. U.S. DOE, EERE. 2014. The Green Power Network. *Green Pricing, Top Ten Utility Green Power Programs.* http://apps3.eere.energy.gov/greenpower/markets/pricing.shtml?page=3.

19. International Renewable Energy Agency. 2012. *30 Years of Policies for Wind Energy, Lessons from 12 Wind Energy Markets.* http://www.irena.org/DocumentDownloads/Publications/IRENA_GWEC_WindReport_Full.pdf

20. German Wind Energy Association. Renewable Energy. http://www.wind-energie.de/en/policy/renewable-energy-act.

21. European Photovoltaic Industry Association. 2014. *European Support Schemes Overview-Short Public Version.* http://www.epia.org/fileadmin/user_upload/Policies/OSS-Q4_2014_short_version.pdf.

22. A. Watson et.al. 2012. *Solar Ready: An Overview of Implementation Practices.* NRELTP-7A40-51296. http://www.nrel.gov/docs/fy12osti/51296.pdf. (See Appendix C.)

23. Public Utility Commission of Texas, CREZ Transmission Program Information Center. http://www.texascrezprojects.com. Maps with transmission lines and substations are provided for the five CREZ zones.

RECOMMENDED RESOURCES

Links

T.D. Couture, K. Cory, C. Kreycik, and E. Williams. 2010. *A Policymaker's Guide to Feed-in-Tariff Policy Design.* NREL/TP-62A-44849. http://www.nrel.gov/docs/fy10osti/44849.pdf.

Database of State Incentives for Renewables and Efficiency. www.dsireusa.org. Excellent site.

DOE Energy Efficiency and Renewable Energy. www.eere.energy.gov/.

Energy Information Administration. Interactive map for all power plants plus resource potential. http://www.eia.gov/state/maps.cfm?v=Renewable.

IRENA. *Evaluating Renewable Energy Policy: Review of Criteria and Indicators for Assessment.* http://www.irena.org/DocumentDownloads/Publications/Evaluating_RE_Policy.pdf.

D. Koplow. *Fueling Global Warming, Federal Subsidies to Oil in the United States*, http://www.greenpeace.org/usa/Global/usa/report/2007/7/fueling-global-warming.pdf.

National Wind Coordinating Committee. www.nationalwind.org/. Comprehensive guide to studying wind energy/wildlife interactions. https://nationalwind.org/research/publications/comprehensive-guide/.

NET METERING, GREEN POWER, AIR EMISSIONS, TRANSMISSION

Carbon trading. http://www.carbontradewatch.org/issues.html.

Climate Change. http://www.epa.gov/climatechange/index.html.

EERE. http://energy.gov/eere/office-energy-efficiency-renewable-energy.

EWEA. 2005. *Large Scale Integration of Wind Energy in the European Power Supply; Analysis, Issues and Recommendations*. http://www.ewea.org/fileadmin/ewea_documents/documents/publications/grid/051215_Grid_report.pdf.

EWEA. 2008. *Integration of Wind Energy into the Grid*. http://www.ewea.org/index.php?id=196.

Fresh Energy. http://www.fresh-energy.org/.

Green Power Markets, Net Metering Policies. http://apps3.eere.energy.gov/greenpower/markets/netmetering.shtml.

Green Power Network. http://apps3.eere.energy.gov/greenpower/.

High Plains Express Transmission Project. 2007. http://www.rmao.com/wtpp/HPX/HPX_Stakeholder_111407.pdf.

National Electric Transmission Corridor. http://energy.gov/oe/services/electricity-policy-coordination-and-implementation/transmission-planning/national-2.

Sustainable Development. http://sustainabledevelopment.un.org.

PROBLEMS

14.1. List your choice of top three incentives for renewable energy? Give brief explanation for your choices

14.2. How much support should the U.S. government provide for renewable energy? Why?

14.3. What type of projects should the federal government support? (Some examples are R&D, prototypes, demonstration projects, commercialization projects.) Give reasons for your answer.

14.4. Should state and local governments provide incentives for renewable energy? If the answer is yes, list your top three choices and explain why?

14.5. What type of education would be most effective for promoting renewable energy? At what level and to whom?

14.6. What are the major environmental concerns if a renewable energy system is planned for your area?

14.7. List two environmental concerns if a wind farm were to be located in your area.

14.8. At what dollar level should the federal government fund renewable energy? Fossil fuel? Nuclear energy? Compare your numbers to the federal budget for this fiscal year.

14.9. How many states have net energy metering of 100 kW or greater?

14.10. What is the longest period for net energy billing?

14.11. What state incentives are there for residential size systems in your state?

14.12. Go to http://www.dsireusa.org. How many states have renewable portfolio standards?

14.13. Does your utility offer green power? If yes, what are the costs?

14.14. Wikipedia has financial incentives for PV. Choose any country except the United States. What are the PV incentives?

14.15. Why is there a large amount of PV installed in Germany, Spain, and Japan?

14.16. How many states have rebate programs for renewable energy systems? Give range of dollar values.

14.17. Which states have mandates for ethanol in gasoline?

14.18. How many countries have mandates for solar hot water?

14.19. You want to install a renewable energy system for your home or residence. Choose any system and then determine if permits are needed and if so what kind you would need.

14.20. Discuss the pros and cons of trading versus tax on CO_2 emissions.

14.21. For your country for the period of 2005 to 2015, estimate total subsidy for fossil fuels, for renewable energy.

14.22. If you were in charge of the national energy policy for your country, what incentive would you choose to promote renewable energy? Be specific about cost and length of time.

15 Economics

15.1 INTRODUCTION

The critical factors in determining whether it is financially worthwhile to install renewable energy systems are (1) initial cost of the installation and (2) the net annual energy production. If the renewable energy system produces electrical energy and is connected to the grid, the next important factor is the value of that energy. For large systems, it is the value of the electricity sold to the utility company; for systems using energy on site, it is generally the value of the electricity displaced, the retail value. In determining economic feasibility, renewable energy must compete with the energy available from competing technologies. Natural gas and oil prices have had large fluctuations in the past years, and the future prices for fossil fuels are uncertain, especially when carbon emissions are included. For the United States, if the military costs for ensuring the flow of oil from the Middle East were included, that would probably add $0.15–0.30/L ($0.50–1.00/gal) to the cost of gasoline. To increase market penetration of renewable energy, the return from the energy generated should exceed all costs in a reasonable time. For remote locations where there is no electricity, high values for electricity from renewable energy are probably cost competitive with other sources of energy. Of course, all values for electricity produced by renewable energy systems depend on the resource, so there is a range of values.

The general uncertainty regarding future energy costs, dependence on imported oil, reduction of pollution and emissions, and to some extent availability have provided the driving force for development of renewable sources. The prediction of energy cost escalation is a hazardous endeavor as the cost of energy (COE) is driven primarily by the cost of oil. The price of oil was $12 to $25/barrel in the 1990s, and predictions at that time for 2003 by the U.S. Energy Information Administration (EIA) were for a gradual increase to $30/bbl by the year 2020. However, actual values were $99/bbl in 2007, with a peak of $140/bbl in April 2008; the price of oil was $80/bbl in June 2010, back to $110/bbl in June 2014, with a reduction to $50/bbl in March 2015; however, that value is still larger than the predicted value (made in 2003) for 2030. In 2010, the EIA predicted that oil will be around $120/bbl in 2020 and $140/bbl in 2035; however, in 2014 EIA predictions were $95/bbl in 2020 and $135/bbl in 2035 for the reference case. This dramatically demonstrates that oil prices have not been and will not be uniform, in terms of either time or geography. At the point of time when demand exceeds production, there will be a sharp increase in the price of oil. Some experts predicted that the peak of world oil production (http://www.oilposter.org) would be in 2007 to 2010, while others predicted it will be around 2015, and others even predicted peak oil in 2040. The EIA has a wide range of predictions for peak oil from 2021 to 2112, depending on growth in demand [1] and implementation of advanced technology, such as fracking for shale oil. The most important factors are the estimated total reserves and what amount is recoverable. As price increases, it becomes economic to recover more from existing reservoirs and from deep sea, polar, and tar sands. The EIA predictions for oil production, reference case, are 43 * 10^9 bbls/years in 2040 from the base of 33 * 10^9 bbl/years in 2014, so peak oil would be beyond that year.

As stated in the previous chapter (Chapter 14), economics is intertwined with incentives and penalties, so actual life-cycle costs (LCCs) are hard to determine, especially when externalities of pollution and government support for research and development for competing energy sources are not included. Incentives for large and small renewable systems have driven and will drive the world market.

15.2 FACTORS AFFECTING ECONOMICS

The following list includes most of the factors that should be considered when purchasing a renewable energy system for residence, business/commercial, farm and ranch, and industry uses:

Load (power) and energy
COE from competing energy sources to meet need
Initial installed cost (purchase price, shipping, installation [foundation, utility intertie, labor, etc.], and cost of land [if needed])
Production of energy
Types and sizes of systems
   Warranty
   Company (reputation, past history, number of years in business, and future prospects)
Renewable energy resource
   Variations within a year
   Variations from year to year
Reliability
Selling price of energy produced or unit worth of energy displaced and anticipated energy cost changes (escalation) of competing sources
Operation and maintenance (O&M)
General operation and ease of service
Emergency services and repairs
Major replacement cost over lifetime (e.g., batteries 5 to 7 years)
Insurance
Infrastructure (are service personnel available locally?)
Cost of money (fixed or variable interest rate)
Inflation (estimated for future years)
Legal fees (negotiation of contracts, titles, easements, and permits)
Depreciation if system is a business expense
Any national or state incentives

Every effort should be made to benefit from all incentives, and the difference in incentives may determine type and size of renewable system. The cost of land is a real cost, even to those using their own land. This cost is often obscured because it occurs as unidentified lost income. Reliability, or availability, is important in determining the quantity of energy produced. For optimum return, the system must be kept in operation as much time as possible, consistent with safety considerations. Background information on system performance, including failures, should be sought and used to estimate the downtime. The distribution of the energy production throughout the year can affect the value of the energy. If most of the energy comes during a time of increased demand on the utility system or during the time energy is needed on the site, then that energy is clearly of more value. For example, photovoltaic (PV) systems produce energy that matches the load for air conditioning.

Renewable energy systems can produce electricity (1) for consumption on site, (2) to sell to a utility, or (3) both. The electricity used on site displaces electricity at the retail rate. If net energy billing is available, even the energy fed back to the utility is worth the retail rate, up to the point of positive feedback (dependent on period or rollover). If more energy was produced than was used during the billing period, then that energy is sold for avoided cost, may roll over to the next billing period, or may accrue to the utility the end of the year. The price paid by the utility is either negotiated with the utility or decided by a public regulatory agency.

Example 15.1

A wind turbine produces 2,000 kWh in a month. There are two meters: one measures energy purchased (3,000 kWh) from the utility company, and the second measures energy fed back to the grid (1,200 kWh). The energy displaced by the wind turbine is 800 kWh (2,000–1,200), the on-site use. The retail rate from the grid is $0.09/kWh. The value of the excess energy sold to the grid is $0.03/kWh, which is the avoided cost and in many cases is the fuel adjustment cost of the utility.

This is the billing if two meters are used:

Meter	kWh	Rate ($)	Bill ($)
One	3,000	0.09	270
Two	1,200	0.03	−36
Monthly charge for meter 2			15
Total			249
In net energy billing, one meter runs forward and backward:			
Meter	1,800[a]	0.09	162

[a] 3,000–1,200 kWh

Clearly, net energy billing is preferable because all the energy produced by the renewable energy system is worth the retail rate, up to the point at which the meter reads no difference for the billing period. Notice that energy displaced, 800 kWh, is worth, $72, the retail rate.

The costs of routine O&M for individuals represent the time and parts costs. Information on system reliability and durability for long time periods may be difficult to obtain, so the cost of repairs will be difficult to estimate. It is important that the owner has a clear understanding of the manufacturer's warranty and that the manufacturer has a good reputation. Estimates should be made on costs of repairing the most probable failures. Insurance costs may be complicated by companies that are uncertain about the risks involved in a comparatively new technology. However, the risks are less than operating a car.

Inflation will have its principal impact on expenses incurred over the lifetime of the product. The costs of O&M and especially the unanticipated repairs fall into this category. On the other hand, generally with inflation, cheaper dollars would be used to repay borrowed money (for fixed-rate loans).

15.3 ECONOMIC ANALYSES

Economic analyses, both simple and complicated, provide guidelines, and simple calculations should be made first. Commonly calculated quantities are (1) simple payback, (2) COE, and (3) cash flow.

A renewable energy system is economically feasible only if its overall earnings exceed its overall costs within a time period up to the lifetime of the system. The time when earnings equal the cost is called the *payback time*. The relatively large initial cost means that this period could be a number of years, and in some cases, earnings would never exceed the costs. Of course, a short payback is preferred, and a payback of 5–7 years is acceptable. Longer paybacks should be viewed with caution.

How do you calculate the overall earnings or value of energy? If you did not have any source of energy for lights, radio, and maybe a TV, a cost of $0.50 to $1.00/ kWh may be acceptable for the benefits received. Many people are willing to pay more for green power because they know it produces less pollution. Finally, a few people want to be completely independent from the utility grid, no matter if the system would never meet costs when compared to COE from the utility. As noted, independence from the grid means efficient use and conservation (low energy use).

15.3.1 SIMPLE PAYBACK

A simple payback calculation can provide a preliminary judgment of economic feasibility for a renewable energy system. The difference is usually around 5%–7% between borrowing money for a system and lost interest if you use your money to pay for the system. Note the low interest rates since the recession of 2008, which has continued into 2015. Since the lost interest rate was very low, paying for the system and counting on future escalation in competing energy cost made renewable systems more economic. The easiest calculation is cost of the system divided by cost displaced per year and assuming that O&M are minimal and will be done by the owner.

$$SP = \frac{IC}{(AEP * \$ / kWh)} \tag{15.1}$$

where:
SP is simple payback (years)
IC is the initial cost of installation (dollars)

AEP is the annual energy production in (energy units)/year if comparing to electricity (kWh/year)

$/(energy unit) is the price of energy displaced; if electricity, that value is cost per kilowatt hour, the rate paid to the utility company

Note that the price for PV modules has declined dramatically since 2008.

Example 15.2

You purchased a 300-W solar system for electricity for lights, radio, and television.

Installed cost = $2,000, produces 500 kWh/year at $0.50/kWh
$0.50/kWh is the estimated cost for remote electricity.
SP = $2,000/(500 kWh/year * 0.50 $/kWh)
SP = 2,000/250 = 8 years

A more complex calculation would include the value of money, borrowed or lost interest, and annual O&M costs.

$$SP = \frac{IC}{AEP * \$/kWh - IC * FCR - AOM} \qquad (15.2)$$

where:
$/kWh is the price of energy displaced or the price obtained for energy generated
FCR is the fixed charge rate (per year)
AOM is the annual O&M cost (dollars/year)

The FCR could be the interest paid on a loan or the value of interest received if you had not displaced money from savings. An average value for a number of years (5 years) will have to be assumed for cost per kilowatt hour for electricity displaced. Equation 15.2 involves several assumptions; the same number of kilowatt hours is produced each year, the value of the electricity is constant, and there is no inflation. More sophisticated analysis would include details such as escalating fuel costs of conventional electricity and depreciation. In general, these factors might reduce the payback.

Example 15.3

You purchase a 2-kW wind turbine with inverter to connect to the grid, IC = $11,000. The unit produces 5,000 kWh/year. You are losing interest at 4% on the installed cost, but that value is reduced as you generate electricity, so assume half. The retail rate of electricity is $0.09/kWh.

$$SP = \frac{11,000}{(5,000*0.09-5,500*0.04)} = \frac{11,000}{(450-220)} = \frac{11,000}{230} = 48 \text{ years}$$

You would think twice before purchasing this system on an economic basis, and no O&M was included. A cash flow would give a better idea of payback time.

15.3.2 Cost of Energy

The COE (value of the energy produced by the renewable energy system) gives a levelized value over the life of the system. The lifetime depends on the type of system and is assumed to be 30 years for PV and 20 years for wind turbines. Lifetimes for other renewable energy systems will probably fall within this range, except for large hydro, which will be much longer. The COE is primarily driven by the installed cost and the annual energy production.

$$COE = \frac{(IC * FCR + AOM)}{AEP} \tag{15.3}$$

where:
AEP is the annual energy production (net)

The COE is one measure of economic feasibility, and it is compared to the price of electricity from other sources (primarily the utility company), the price at which generated energy can be sold, or the price of energy from other sources.

Example 15.4

A renewable energy system has the following costs and production:

IC = $250,000, FCR = 8% = 0.08, AOM = 1% of IC = $2,500 / year

AEP = 120,000 kWh/year

COE = (250,000 * 0.08 + 2,500) / 120,000 = $0.19 / kWh

The COE should be compared with an estimated average cost of electricity from the utility over the next 10 years.

The Electric Power Research Institute (EPRI; tag-supply method, Equation 15.4) includes the addition of levelized replacement costs (major repairs) and fuel costs [2]. The cost of fuel for most renewable energy systems is zero, so that term would not be included. However, some of the concentrating solar power (CSP) systems have backup heat from fossil fuels to provide dispatchable power, and the combustion of biomass in some cases would have a fuel cost. In Equation 15.3, the major replacement costs are included in the annual O&M costs, so that the AOM should be larger than the AOM used in Equation 15.4.

$$COE = (IC * FCR + LRC + AOM + AFC)/AEP \tag{15.4}$$

where:
LRC is the levelized replacement cost (dollars/year)
AFC is the annual fuel cost (dollars/year)

The COE can be calculated for cost per kilowatt hour or cost per megawatt hour, and the last term could be separated as AOM/AEP, again in terms of cost per kilowatt hour or cost per megawatt hour. It may be difficult to obtain good numbers for the LRC since repair costs are generally proprietary. One method is to use a 20-year lifetime and estimate the LRC as IC/20. That means the major repairs are equal to the initial cost spread over the lifetime.

Example 15.5

For a 1-MW wind turbine, IC = $2 million, FCR = 0.07, AEP = 3,000 MWh/year, LRC = $100,000/year, AOM/AEP = $8/MWh.

$$COE = \frac{(\$2,000,000 * 0.07 + \$100,000)}{3,000 + 8 = 80 + 8} = \$88 / MWh$$

The COE needs to be compared to all expected incomes, any incentives, accelerated depreciation, and so on.

The LRC distributes the costs for major overhauls and replacements over the life of the system. For example, for a system with batteries, they will need to be replaced every 5 to 7 or up to 10 years. The LRC is an estimate for future replacement costs in terms of today's costs of components.

1. Year in which the replacement is required n
2. Replacement cost, including parts, supplies, and labor RC
3. Present value of each year's replacement cost PV

$$PV(n) = PVF(n) * RC(n) \tag{15.5}$$

where:
  PVF(n) is the present value factor for year n and is equal to $(1 + I)^{-n}$
  I is the discount rate and is equal to 0.069
  RC(n) is the replacement cost in year n

The LRC is the sum of the present values multiplied by the capital recovery factor (CRF):

$$LRC = CRF + \sum_{n=1}^{20} PV(n) \tag{15.6}$$

where:
  CRF is equal to 0.093

A good source of information is the interactive transparent cost database, U.S. DOE EERE (http://en.openei.org/apps/TCDB/). Historical values and projections are provided for generation (renewables, fossil fuels, and nuclear), fuels, and vehicles, and costs are available by categories: levelized cost of energy (LCOE), overnight capital cost, fixed operating cost, variable operation cost, and capacity factor.

15.4 LIFE-CYCLE COSTS

An LCC analysis gives the total cost of the system, including all expenses incurred over the life of the system and salvage value, if any [3]. There are two reasons to do an LCC analysis: (1) to compare different power options and (2) to determine the most cost-effective system designs. The competing options to small renewable energy systems are batteries or small diesel generators. For these applications, the initial cost of the system, the infrastructure to operate and maintain the system, and the price people pay for the energy are the main concerns. However, even if small renewable systems are the only option, an LCC analysis can be helpful for comparing costs of different designs or determining whether a hybrid system would be a cost-effective option. An LCC analysis allows the designer to study the effect of using different components with different reliabilities and lifetimes. For instance, a less-expensive battery might be expected to last 4 years, while a more expensive battery might last 7 years. Which battery is the best buy? This type of question can be answered with an LCC analysis.

$$\text{LCC} = \text{IC} + M_{pw} + E_{pw} + R_{pw} - S_{pw} \tag{15.7}$$

where:
  LCC is the life-cycle cost
  IC is the initial cost of installation
  M_{pw} is the sum of all yearly O&M costs
  E_{pw} is the energy cost (sum of all yearly fuel costs)
  R_{pw} is the sum of all yearly replacement costs
  S_{pw} is the salvage value (net worth at the end of the final year; 20% for mechanical
      equipment)

Future costs must be discounted because of the time value of money, so the present worth is calculated for costs for each year. Life spans for renewable energy systems are assumed to be 20–40 years; however, replacement costs for components need to be calculated. Present worth factors are given in tables or can be calculated.

LCCs are the best way of making purchasing decisions. On this basis, many renewable energy systems are economical. The financial evaluation can be done on a yearly basis to obtain cash flow, break-even point, and payback time. A cash flow analysis will be different in each situation. Cash flow for a business will be different from a residential application because of depreciation and tax implications. The payback time is easily seen if the data are graphed.

Example 15.6

For a residential application (tax credit available), IC = $20,000, loan rate = 6%, payment = $2500/year, value of energy saved is $2,500 (first year).

Year	0	1	2	3	4	5	6	7	8	9	10
Down payment	5,000										
Principal	15,000	13,400	11,704	9,906	8,001	5,981	3,839	1,570	0	0	0
Toward principal		1,600	1,696	1,798	1,906	2,020	2,141	2,270	1,570	0	0
Interest		900	804	702	594	480	359	230	94	0	0
Maintenance		500	500	500	500	500	500	500	500	500	500
Insurance, tax		115	115	115	115	115	115	115	115	115	115
Costs		3,115	3,115	3,115	3,115	3,115	3,115	3,115	2,279	615	615
$ energy saved		2,500	2,500	2,500	2,500	2,600	2,600	2,600	2,600	2,600	2,700
Tax credit	6,000										
Income	6,000	2,500	2,500	2,500	2,500	2,600	2,600	2,600	2,600	2,600	2,700
Cash flow	1,000	–615	–615	–615	–615	–515	–515	–515	321	1,985	2,085
Cumulative	1,000	385	–230	–845	–1,460	–1,975	–2,490	–3,005	–2,684	–699	1,386

Note: $ energy saved = amount that would have been paid for electricity purchased from utility or the value of electricity generated by residential system and used on site. Income = value of electricity fed back to utility.

Notice that positive cash flow occurs in year 8 after the loan is paid off, and payback time is 10 years. This cash flow analysis did not take into account the present value of money.

There are a number of assumptions about the future in such an analysis. A more detailed analysis would include inflation and increases of costs for O&M as the equipment gets older.

A cash flow analysis for a business with a $0.023/kWh tax credit on electric production and depreciation of the installed costs would give a different answer. Also, all operating expenses are business expenses. The economic utilization factor is calculated from the ratio of the costs of electricity used at the site and electricity sold to the utility.

The core of the RETScreen tools consists of a standardized and integrated renewable energy project analysis software that can be used to evaluate the energy production, LCCs, and greenhouse gas emission reductions for the following renewable energy technologies: wind, small hydro, PV, passive solar heating, solar air heating, solar water heating, biomass heating, and ground-source heat pumps (http://retscreen.gc.ca/ang/menu.html).

15.5   PRESENT WORTH AND LEVELIZED COSTS

Money increases or decreases with time depending on interest rates for borrowing or saving and inflation. Many people assume that energy costs in the future will increase faster than inflation. The same mechanism of determining future value of a given amount of money can be used to move money backward in time. If each cost and benefit over the lifetime of the system were brought back to the present and then summed, the present worth can be determined.

$$PW = \frac{(\text{cost total for year } S) - (\text{financial benefit total for year } S)}{(1+d)^M} \qquad (15.8)$$

where:
  cost total is the negative cash flow
  S is the specific year in the system lifetime
  M is the years from the present to year S
  d is the discount rate

The discount rate determines how the money increases or decreases with time. Therefore, the proper discount rate for any LCC calculation must be chosen with care. Sometimes, the cost of capital (interest paid to the bank or, alternately, lost opportunity cost) is appropriate. Possibly, the rate of return on a given investment perceived as desirable by an individual may be used as the discount rate. Adoption of unrealistically high discount rates can lead to unrealistic LCCs. The cost of capital can be calculated from

$$CC = \frac{1 + \text{loan interest rate}}{1 - \text{inflation rate}} - 1$$

If the total dollars are spread uniformly over the lifetime of the system, this operation is called *levelizing*.

$$\text{Annualized cost} = \frac{\text{PW}\,d(1+d)^P}{(1+d)^P - 1} \tag{15.9}$$

where:

P is the number of years in the lifetime

One further step has been utilized in assessing renewable energy systems versus other sources of energy, such as electricity. This is the calculation of the annualized COE from each alternative. The annualized cost is divided by the net annual energy production of that alternative source.

$$\text{COE} = \frac{\text{annualized cost}}{\text{AEP}}$$

It is important that annualized costs of energy calculated for renewable energy systems are compared to annualized costs of energy from the other sources. Direct comparison of the annualized COE to the current COE is not rational. Costs of energy calculated in the above manner provide a better basis for the selection of the sources of energy. This type of calculation also shows that renewable energy systems are economical today.

15.6 EXTERNALITIES

Externalities are now playing a role in integrated resource planning (IRP) as future costs for pollution, carbon dioxide, and so on are added to the LCCs. Values for externalities range from zero (past and present value assigned by many utilities) to as high as $0.10/kWh for steam plants fired with dirty coal. Again, values are assigned by legislation and regulation (public utility commissions).

As always, there is and will be litigation by all sides as providers of energy do not want externalities included in their costs. The Lignite Energy Council petitioned the Minnesota Public Utilities Commission to reconsider its interim externality values. The council represented major producers of lignite, investor-owned utilities, rural electric cooperatives, and others. It focused the protest on values assigned to CO_2 emissions because from their standpoint there is an acknowledged lack of reliable science that CO_2 emissions are harmful to society. In Europe, different values have been assigned to CO_2 emissions, which makes renewable energy more cost competitive.

15.7 PROJECT DEVELOPMENT

The three most important considerations for development of large projects are the following:

1. Land (surface and offshore) with good-to-excellent resource
2. Contract to sell electricity produced
3. Access to transmission lines (proximity and carrying capacity)

Again, a good source of information is RETScreen International (http://www.retscreen.net/ang/home.php).

The RETScreen Clean Energy Project Analysis Software is a unique decision support tool developed with the contribution of numerous experts from government, industry, and academia. The software, provided free-of-charge, can be used worldwide to evaluate the energy production and savings, costs, emission reductions, financial viability, and risk for various types of Renewable-energy and Energy-efficient Technologies (RETs). The software (available in multiple languages) also includes product, project, hydrology and climate databases, a detailed user manual, and a case study-based college/university-level training course, including an engineering e-textbook.

National governments and trade associations will have information on project development and costs; for example, National Renewable Energy Laboratory (NREL) has Cost of Renewable Energy Spreadsheet Tool (CREST), a cash flow model (https://financere.nrel.gov/finance/content/crest-cost-energy-models), and Windustry has a Wind Project Calculator, under Community Wind Toolbox, Chapter 3 (http://www.windustry.org/CommunityWindToolbox). In any case, there is the need for public involvement at an early stage for any project. However, in a competitive environment for land or rights offshore, how much information at what time period is up to the developer. Once the area is under contract, community involvement and education are highly recommended before any construction begins.

The following list provides more details on project development:

1.0 Site selection
    1.1 Evidence of significant resource
    1.2 Proximity to transmission lines (note possibility of future high-voltage lines)
    1.3 Reasonable access
    1.4 Few environmental concerns
    1.5 Receptive community
2.0 Land or surface
    2.1 Term: Expected life of the project
    2.2 Rights: Ingress/egress, transmission
    2.3 Compensation: Percentage of revenue, per megawatt, or combination
    2.4 Assignable: Financial requirements
    2.5 Indemnification
    2.6 Reclamation provision: Bond to remove equipment at end of the project
    2.7 Project life after resource assessment
3.0 Resource assessment
    3.1 Lease, cost/acre: 1 to 3 years, or flat fee
    3.2 Corollary or existing data
    3.3 Install measurement systems, number needed
    3.4 Collect 10-min or hour data; minimum time period is generally 1 year
    3.5 Report on resource
    3.6 Estimated energy output of project; may be for different manufacturers' equipment

    3.7 Data, report, and output projections to landowner if developer does not exercise option for installation of project

4.0 Environmental
    4.1 Review for endangered species
    4.2 Biological
        4.2.1 Wildlife habitat
        4.2.2 Fragmentation of habitat
        4.2.3 Required studies and reports
    4.3 Archeological studies
    4.4 Noise, visual
    4.5 Erosion, water quality
    4.6 Solid and hazardous wastes
    4.7 Construction material (gravel) and water from landowner

5.0 Economic modeling
    5.1 Output projections
    5.2 Equipment costs
    5.3 Installation costs
    5.4 Communication and control
    5.5 Taxes: Sales, income, property (depreciation schedule), and tax abatement (payment in lieu of taxes)
    5.6 O&M estimates
    5.7 Finance assumptions: Tax credits, equity rate of return, incentives, debt rate and term (coverage ratios), and debt/equity ratio
    5.8 Other: Insurance, legal

6.0 Interconnection studies
    6.1 Interconnection request (electric reliability council)
    6.2 Capacity limitation
    6.3 Load flow analysis
    6.4 Voltage controls
    6.5 System protection

7.0 Permits
    7.1 Local, state, federal, public land, private land
    7.2 Land, marine use permit
    7.3 Building permit

8.0 Sale of energy/power
    8.1 Energy/power purchase agreement
        8.1.1 Long-term contract with utility
        8.1.2 Green power market
        8.1.3 Market (merchant) and avoided cost
        8.1.4 Renewable energy credits
        8.1.5 Pollution/emission credits
    8.2 Kilowatt hour: Real or nominal levelized
    8.3 Capacity, power
    8.4 Term
    8.5 Credit-worthy buyer
    8.6 Facility sales agreement

8.7 Turnkey price, complete project
9.0 Financing
    9.1 Source of equity: Rate of return 15–18%
    9.2 Source of debt
        9.2.1 Market rates
        9.2.2 Term of debt
    9.3 Assignable documents
    9.4 Third-party due diligence
10.0 Equipment purchase
    10.1 Power curve (output projection)
    10.2 Renewable equipment cost
    10.3 Turnkey construction cost
    10.4 Warranties: Equipment and maintenance
    10.5 Construction financing
    10.6 Past history of manufacturer
    10.7 Availability of equity, down payment, delivery date
11.0 Construction
    11.1 Access: Land, roads; marine, docks, transportation
    11.2 Foundations (excavation, concrete, and others)
    11.3 Interconnection to utility
    11.4 Equipment assembly and installation
    11.5 Commissioning
    11.6 Environmental
        11.6.1 Continued monitoring of impact on wildlife
        11.6.2 Continued control of liquid and solid wastes
12.0 Operation and maintenance
    12.1 Fixed cost per unit per year
    12.2 Fixed price per kilowatt hour produced
    12.3 Availability of warranties
    12.4 Penalties for nonperformance
    12.5 Types of costs
        12.5.1 Labor
        12.5.2 Management
        12.5.3 Insurance, taxes
        12.5.4 Maintenance equipment
        12.5.5 Parts on hand
        12.5.6 Nonrecurring costs: Major repairs
        12.5.7 Roads, maintenance and access for landowner
13.0 Public information

Information for the public ranges from visitor centers at the O&M offices, kiosks, brochures with general information at the O&M office, to no public information and avoidance of the general public at the wind farm. One wind farm operator near an interstate highway removed an outdoor kiosk as too many visitors came into the office area seeking more information and disrupting the workforce at the site. So, the developer or operator needs to have an idea of how much access to provide and where it is located. They especially do not want unaccompanied visitors in the operational area. Of course, public roads through a renewable energy project are accessible to the public. How is access controlled for offshore projects? Some projects have websites, and again the question concerns what information to display or provide. Do you provide updates on a web page (e.g., about energy production), and how often?

15.7.1 Landowner Considerations

For some large projects, such as hydro, the land may be purchased, the nation or state might use eminent domain, or the land may be owned or controlled by the nation. In general, offshore areas are controlled by the nation; however, some distance from shore or tidal areas may be under local or state control. The considerations are as follows:

1. Lease resource assessment, 1–3 years (2-years extension may be requested due to financial and other problems).

   Flat rate or cost/acre per year ($1.75–2.00/acre).

   Access to land and installation of met stations.

   If option not exercised for project installation, collected data become the property of landowner.

   Data have an estimated value of $20,000 to $25,000 for the first year of data and $10,000/year thereafter.

   Also, if construction is not started, be sure that all rights revert to landowner at that point in time.

2. Project 20–30 to how many years? Option for extension at end, generally 10 years. Payment quarterly or yearly.

   A. Royalty on production, 2%–5%; escalation clause after which year?

   B. Per system, based on rated power, $4,000 to $6,000/MW/year.

   C. A or B, whichever is larger for that year.

   D. If there is additional future revenue, for example, pollution credits, landowner should share in that return.

3. Land consideration.

   A. Fee/turbine during construction; $3,500 for $5,000/MW.

   B. Laydown, assembly area.

   C. Substation area, transmission lines (underground on site and overhead substation to utility).

   D. Road easement and material (value).

   E. Water.

   F. Gates and cattle guards.

   G. Hunting rights (none during construction phase). After commissioned, restrictions, workers in area, locations?

   H. Renewable energy rights (if in windy area and close to transmission line, keep, sell, or sell with land?).

   I. Easement issues (http://www.windustry.org/opportunities/easements. htm).

   J. Insurance during construction.

   K. Bond for removal of turbines after project life or developer defaults (salvage value).

TABLE 15.1

Representative Timeline for Wind Farm Project

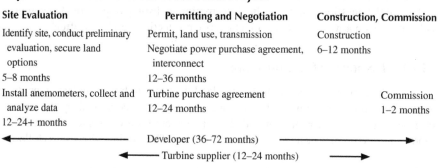

Site Evaluation	Permitting and Negotiation	Construction, Commission
Identify site, conduct preliminary evaluation, secure land options	Permit, land use, transmission	Construction
	Negotiate power purchase agreement, interconnect	6–12 months
5–8 months	12–36 months	
Install anemometers, collect and analyze data	Turbine purchase agreement	Commission
12–24+ months	12–24 months	1–2 months
◄──────────	Developer (36–72 months)	──────────►
	◄──── Turbine supplier (12–24 months) ───►	

Contracts and leases will differ by region and type of surface (offshore), developer, resource, and access to transmission. The landowner should have an attorney to read and advice on the proposed contract before signing the contract.

The landowner may receive one or more offers, and some landowners are forming cooperatives to deal with developers as in most cases the amount of contiguous or adjacent land will encompass multiple landowners. A developer trying to tie up land may say that, within a project, whatever the best offer is for one landowner, then all will receive that offer. For multiple land owners, for wind projects, there is generally no guarantee that there will be wind turbines located on their land; however, for cooperatives, everyone may share in the revenue, depending on the agreement among the landowners.

In countries where national or state governments control the land or where it is communal land, the questions for the present occupants are how much is fair value for the land removed from production and who receives payment and when (once or annual). What is the fair value to the population for relocation due to large dam projects? For offshore projects, who controls the surface, and who receives money on energy produced?

The total time from land/surface acquisition to an operating large project may take 3–6 years (Table 15.1) for PV, CSP, wind, bioenergy, geothermal, and water projects. However, construction time for large dams will be much longer. The construction phase for most renewable energy projects can take from 6 months to 2 years, and projects can be installed much faster than transmission lines.

15.8   COST (VALUE) OF ENERGY AND DIFFERENT SOURCES

The installed costs for renewable energy systems (Figure 15.1) have decreased from the 1980s; however, the COE for most renewable energy systems increased after 2003 and peaked around 2008, with some decrease thereafter and a large decrease for PV as costs for modules declined. The increase was due primarily to the increased cost for steel, copper, and cement. Sometimes, project costs are noted in press releases, and for the generation of electricity, project information may be

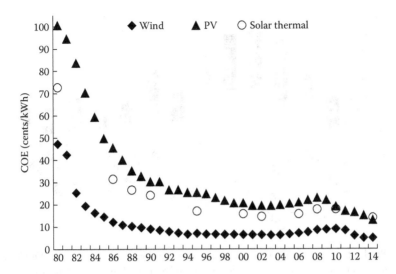

FIGURE 15.1 Cost of energy for renewable energy systems has declined dramatically since 1980. There is actually a range of values depending on resource, so plotted values are averages for large systems and locations with good-to-excellent resource.

available from regulatory agencies, the Federal Energy Regulatory Commission (FERC) for electric plants (http://eqrdds.ferc.gov/eqr2/frame-summary-report.asp) and state public utility commissions. Renewable energy projects are included in FERC quarterly reports, which include type of plant, megawatt hours generated, income, and the rate paid to the project for energy sold. The problem is that the reporting of the name of the company has to be known.

Comprehensive reports for installed cost and COE for renewable energy technologies (wind, hydro, PV, CSP, biomass for power generation, and geothermal) by region and some countries are available from the International Renewable Energy Agency [4]. Then more detailed reports, Renewable Energy Cost Analysis Technology Briefs, are available for each technology, except geothermal, and then technology briefs are also available.

The COE can be estimated using Equation 15.4. Most COE values will be for projects installed within the past few years. Note that some reports use COE and others use LCOE. For large installed capacity for the generation of electricity by PV, CSP, wind, geothermal power, and hydropower, the numbers are fairly good. Of course, there will be a fairly wide range of values (Figure 15.2) for each type of system [4], and the predicted range for LCOE for 2020 decreases for all technologies except for hydropower and geothermal power. For prototypes and a few installations, the proponents tend to have optimistic numbers for the COE, generally based on proposed and future reduction of costs for utility-scale systems and not on the COE of the prototype system. A sensitivity analysis (Figure 15.3) shows how the different factors in Equation 15.4 affect the COE. The most important factor is the renewable energy production, and the second is installed cost. Note that this COE had zero fuel cost.

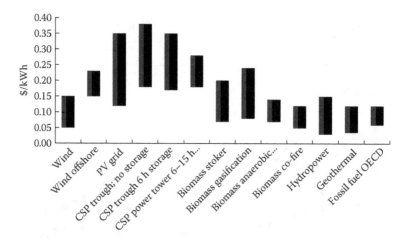

FIGURE 15.2 Typical LCOE cost ranges for renewable power generation. OECD is the Organization for Economic Co-operation and Development (developed countries).

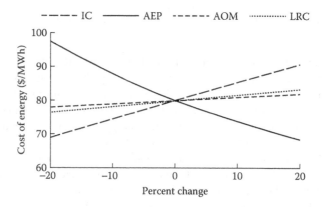

FIGURE 15.3 Sensitivity analysis for COE (base or reference COE = $80/MWh).

The COE for small systems such as solar hot water, PV, wind, and geothermal heat pumps will vary significantly due to resources, local installation costs, incentives, and value for energy displaced. The value of energy displaced in the future, fuel escalation, will vary over time, generally an increase. What value do you use: zero, general inflation, less than inflation? In general, the COE from fossil fuels will probably continue to increase at historical rates. The fuel escalation chosen for small systems will influence payback and LCCs. Cost of energy can be calculated for small systems; however, use conservative estimates for annual energy production.

Systems up to 1 kW are not cost effective when connected in parallel to the utility grid, even for single residences. Residences connected to the utility grid need 5 to 10 kW, and farms, ranches, and businesses need a minimum size of 25 kW or larger. The positive aspect of PV and solar hot water is that they are modular.

15.8.1 PASSIVE SOLAR

The COE for passive solar is the most difficult to estimate since building costs vary so much by location. In general, passive solar adds around 10% to the cost of the building and can reduce heating and cooling costs by 40%–80%. In most areas, there are no builders of passive solar homes, and if an architect is hired, that is an additional cost. However, in New Mexico there are a number of passive solar homes, so finding a builder is not a big problem. The best method to estimate the value of energy is to use simple payback or cash flow.

Example 15.7

A new 150-m² home cost is $1,000/m². The additional cost for passive solar is $15,000, which is added to the 6% interest loan. The value of energy saved is estimated at $1,100/year (10,000 kWh per year at $0.11/kWh). The estimated payment for the addition of the passive solar is $97/month or $1,164/year for a 25-year loan, calculated with a free loan calculator on the web. It is easy to calculate the payment for other loan periods. If the COE from an outside source increases at the rate of inflation, then the value of the passive solar increases of that year, and the additional cost is paid by reduced utility costs. To improve the results, try to reduce the cost of the passive solar and have the passive solar provide more of the energy.

15.8.2 ACTIVE SOLAR HEAT

Systems are available for industry and commercial applications; however, there are essentially few or no commercial systems for the home market. There is a market for solar hot water heaters and solar hot water for swimming pools. For a new home, the price for solar hot water is around $15–20/month on a 30-year loan. Installed costs for solar hot water systems for the home range from $1,500 for do-it-yourself systems to $6,000 for a 3- to 4-m² system. The system should provide around 60%–80% of the hot water needed. If you need to replace a hot water heater, you might want to check on the economics of solar hot water, especially if there are incentives.

Energy Efficiency and Renewable Energy (EERE) has a web page for Energy Saver>Water Heating, which includes estimating the cost of solar hot water (http://energy.gov/energysaver/articles/estimating-cost-and-energy-efficiency-solar-water-heater). The factors are system solar energy factor and the fuel type for the auxiliary tank. Once the cost is determined, then that cost can be compared to costs of a conventional hot water heater. A similar procedure is used for solar hot water systems for swimming pools (http://energy.gov/energysaver/articles/solar-swimming-pool-heaters).

15.8.3 PHOTOVOLTAICS

The major change is that the price of PV modules has declined dramatically from $3.80/kW$_{DC}$ in 2008 to around $0.75/kW$_{DC}$ in 2014. PV use to represent around 50% of the cost of the large systems; however, with the decrease in price, it is around 35%. However, for residential systems it remains at 50% and larger, depending on the country. If prices for modules are around $1.00/W_{DC}$, then installed system costs are around $4,000–5,000/kW for residential systems. COE for PV in Figure 15.2 provides residential and commercial in the 10–100 kW range and smaller residential is slightly higher as that was the predominate installations in the early years. Also note that the system prices are substantially lower in Germany than in the United States, which means that the balance-of-system costs reductions is possible in the United States. Of course, there are no fuel costs, and O&M costs are low.

Example 15.8

This example involves a 4-kW PV system for a home in Albuquerque, New Mexico. From the program PVWatts, the system would produce 6,700 kWh/year. Installed costs are $5/kW, the discount rate is 6%, the AOM is $0.001/kWh, and the LRC is 0. Use Equation 15.4.

$$COE = \frac{(20,000*0.6)}{6,700} + 0.001 = \$0.18 / kWh$$

If the 30% credit is available, then the COE = (16,800 * 0.06)/6,700 + 0.001 = $0.13/kWh.

There are economies of scale for utility systems and the COE is becoming competitive to wind power. An installed system cost of $2,800/kW [5] corresponds to a COE of around $0.09/kWh or $90/MWh. PV has an advantage that, in general, it is a better load match. However, PV does not match the late afternoon, early evening load; therefore, economic storage is needed.

15.8.4 CONCENTRATING SOLAR POWER

The COE for CSP is around $140/MWh (Figure 15.1), without incentives. A COE was estimated in 2008 at $150/MWh for a 100-MW parabolic trough plant with 6 h of storage in California and use of 30% investment tax credit [6]. That was based on an installed cost of $4.58 million/MW, an AOM of $6 million, and a discount rate of 6.8%. An International Energy Agency report, *Energy Technology Perspectives, 2008*, indicates that the COE was from $125 to $225/MWh, depending on location. Future costs in 2008 were predicted to be $43–62/MWh for trough plants and $35–55/MWh for tower plants, numbers that are competitive with fossil fuel for the generation of electricity. However, in 2014 those predicted future costs are still in the future.

Since 2008, a number of CSP plants have been built (Table 7.1), and the largest is the Ivanpah plant in the Mojave Desert, California. The plant consists of three towers [7] and has predicted energy production of 1.08 TWh/year, with an estimated COE of $0.15/kWh. Plant has a dry air condenser to reduce water use.

Current COE from CSP plants ranges from $0.20–0.33/kWh for parabolic trough and $0.16–0.27/kWh for power tower [8]. Projected COE for 2020 is $0.10–0.14/kWh for parabolic trough and $0.09–0.16/kWh for power tower. Many of the CSP plants have thermal storage to provide power during intermittent solar during the day and to meet the load demand of late afternoon and early evening. If a CSP plant has auxiliary energy from fossil fuels for dispatchable power, those fuel costs have to be included in the calculation of COE. Note that the annual operation and maintenance cost for CSP is larger than for PV.

15.8.5 WIND

The COE for wind farms installed in 2014 is around $60/MWh (Figure 15.1), without incentives. There are economies of scale for wind, and the COE for wind farms is less than the COE for small wind systems for residences, agribusiness, and even for community and distributed wind (Table 15.2). The annual energy production is estimated by the generator size method in a good wind regime, and capacity factors of 25%–35% are used.

The installed costs for wind farms in the United States increased from around $1.2 million/MW in 2003 to over $2.0 million/MW in 2009 (both in 2009 dollars). The increase was due primarily to the increased cost for steel, copper, and cement and the availability of wind turbines. However, after the recession, installed costs declined and the weighted average was around $1.75 million/MW for 2013 and 2014 when compared to the weighted average of $1.9 million/MW for 2012 [9]. Of course, the range is large, and the low value for 2012 was $1.4 million/MW and the high value was $3 million/MW. There were only 11 projects in 2013 compared to a larger number in 2012; however, there were a large number of projects for 2014. Power purchase agreements range from $20 to $97/MWh, with average values for 2013 less than $30/MWh. Remember that there is a production tax credit of $23/MWh for 10 years for those projects that qualified and other incentives such as accelerated depreciation; that is why power purchase agreements are less than the estimated COEs.

For the United States, an estimated COE for wind farm projects in 2012 is $47/MWh; FCR = 0.06, capacity factor = 0.30, availability = 98%, AOM = 10/MWh. The capacity factor for Wildorado Wind Ranch in the Texas Panhandle is 0.45, so that would result in an estimated COE of $31/MWh.

$$COE = \frac{(1,900 * 0.6)}{2,575 + 0.01} = \$0.054 / kWh \text{ or } \$54 / MWh$$

TABLE 15.2
COE for Wind Turbines in Good Wind Resource Area

Power (kW)	IC ($/kW)	COE ($/kWh)
1–5	6,000	0.20–0.25
10	5,000	0.15–0.20
50	4,500	0.13–0.18
100	4,000	0.12–0.18
1,000	1,800	0.06–0.08

TABLE 15.3

Percentage Cost for Wind Farm Installation

Component	Percentage
Turbine	70–82
Foundation	1–6
Electric	2–9
Connections to grid	2–9
Land	1–3
Roads, ditching	1–5
Consultants, resource assessment, other	1–3

$$COE = \frac{(1,900 * 0.6)}{3,863 + 0.01} = \$0.039 \,/\, kWh \text{ or } \$39 \,/\, MWh$$

The weighted averages for LCOE (2012 data) for wind projects across the world are similar [10]: Europe: $120/MWh, North America: $80/MWh, Africa: $95/MWh, China: $75/MWh, India: $80/MWh, and Latin America: $95/MWh.

The capital cost is the major cost for a project, and of that the wind turbine is the major component (Table 15.3). A detailed breakdown of the capital costs for the Amayo wind farm (39.9 MW) in Nicaragua shows the turbine accounts for 69% [10]. Most renewable energy projects will be similar in that equipment costs are the major item. The installed cost for offshore wind farms is over twice that of wind farms on land; however, the turbine cost share is 30% to 50% because of the increased cost of installing offshore. The largest offshore wind farm, the London Array (630 MW, commissioned in spring 2013), costs an estimated 4.4 million per MW.

The value to the landowner can be estimated from the annual energy production or megawatts installed and the type of contract with the landowner. Examples are as follows:

A. Royalty on production, 4%–6% with escalation, generally at 10-year periods
B. $4,000–$6,000/MW installed per year
C. Use A or B, whichever is larger

Example 15.9

Suppose a 50-MW project has a contract to sell electricity at $35/MWh (the landowner will not receive any royalty on the production tax credit [PTC]). The income of the wind farm is

$$1.35 * 10^5 \, MWh \,/\, year * \$35 \,/\, MWh = \$4,730,000 \,/\, year$$

Option A: Landowner would be paid $189,000/year at 4% royalty.
Option B: Landowner would be paid $4,000 * 50 = $200,000/year.

At 2 acres per turbine taken out of production, then 100 acres are lost to production. Return value per acre to the landowner is then $1,890/acre/year. This is a much greater return per acre than a farmer can make from crops or livestock.

The FERC has information for electric power generation. The type of sale is shown by the rate: power purchase at fixed value, power purchase with peak and off-peak values, or if sold at market the high and low values plus the average are given. As an example, for 2011, the Wildorado Wind Ranch west of Amarillo, Texas, received $18.1 million for 644 GWh from a power purchase agreement at $28.12/MWh. The wind farm has an installed capacity of 161 MW, so the calculated capacity factor was 45.7% for that year. The capacity factor will be less during the quarters when the winds are lower.

NREL has a COE model for economic analysis of renewable energy projects: solar, wind, geothermal, anaerobic digestion, and fuel cell (https://financere.nrel.gov/finance/content/crest-cost-energy-models). The output includes minimum annual energy to meet the financial criteria, levelized cost, payback period, internal rate of return, detailed cash flows, and summary. Check NREL and other sites for energy calculators for other renewable energy sources.

15.8.6 BIOENERGY

There are many different types of bioenergy systems, from those for the generation of electricity from biomass, biowaste, and biogas to those that produce liquid fuels for transportation. In the generation of electricity from biomass, there will be fuel costs, while there may be a negative fuel cost for the generation of electricity from biowaste and biogas as a result of payment for disposal of the waste. Installed cost would be for conventional boiler plants or conventional combustion engines connected to generators, with an installed cost of $1,200–2,000/kW. For biomass power generation, capital costs range from less than $1,000/kW for co-firing to near $7,000 for stoker combined heat and power and the LCOEs range from $0.05 to 025/kWh (Figure 15.2). However, in non-OECD countries capital costs are lower, range from $600 to 1,400/kW. Capacity factors are 50%–60%.

For liquid fuels for transportation from bioenergy, there will be cost for the crops and transporting biomass to the conversion plant. The cost per volume for that fuel (Table 15.4) can be compared to the cost per volume for gasoline and diesel. It is important to consider the energetics of these systems; for example, the energetics for maize (corn) is 2.3 when distillers grains are included and is probably near 1 for ethanol produced from irrigated corn (well depth of 100 m). For bioenergy systems, the decommissioning costs should be covered by the scrap value.

15.8.7 GEOTHERMAL SYSTEMS

The Geysers California plant electricity is sold for $0.03–0.035/kWh; however, for a plant built today, the cost would be $0.05–0.07/kWh. Installed costs are around $3,000–4,000/kW with an AOM of 0.01–0.03/kWh. Most geothermal power plants can run at 90% capacity factor and LCOEs range from $0.078 to 0.10/kWh. The value for

TABLE 15.4

Average Production Costs by Feedstock for Liquid Fuels and Biomethane (2012 USD)

Biodiesel	$/liter	$/gal
U.S. soybean oil	1.25	4.75
Malaysian palm oil	1.05	3.99
Malaysian jatropha oil	0.85	3.23
Europe rapeseed oil	1.30	4.94
Ethanol		
U.S. grain	0.95	3.20
Brazil sugar cane	0.78	2.95
	$/Nm³	
Biomethane	0.50–1.00	

Source: IRENA, Road Transport: The Cost of Renewable Solutions (http://www.irena.org/DocumentDownloads/Publications/Road_Transport.pdf)

Note: Nm, Normal cubic meter.

direct use of geothermal energy is compared to the value of energy displaced from conventional sources: electricity, natural gas, oil, and coal-fired boilers.

The general rule for geothermal heat pumps is $2,500 per ton of capacity, and a typical home would use a 3- to 4-ton unit. However, the installed cost will depend on the ground source (drilled vertically or horizontal loops in ground or water). The additional cost for the geothermal aspect ranges from $4,000 to $11,000 for a 3-ton system. Again, the COE will be compared to the value of energy displaced.

15.8.8 WATER

The cost of large hydro is site specific, and most of the cost is for the structure; installed costs are $1,200–3,500/kW. The construction period is long, the same as for nuclear power plants, so the cost of capital is high, and payment of interest will start before the project is completed. However, equipment has a 25- to 50-year life, and the structure has a 50- to 100-year life, so the COE is low. Installed costs of small hydro plants (100-kW to 30-MW) are $800–1,500 in China to $1,100–7,800 in Europe [11]. In China, small hydro for remote villages was the cheapest source of renewable energy, larger rated power, and easiest to increase power output; however, it is also dependent on suitable locations.

A tidal system capacity factor is 20%–25% and current demonstration projects have LCOE in the range of $0.32–0.75/kWh. For commercial-scale tidal energy, the COE is estimated at $0.10–0.20/kWh. For the proposed 8.6-MW Severn Barrage in the United Kingdom, the COE was estimated at $0.10/kWh. An EPRI report estimated that tidal kinetic systems would have a COE of $0.06–0.013/kWh for utility-scale systems. COE estimates based on demonstration and pilot projects and then scaled up to larger sizes are in general optimistic.

Demonstration projects LCOEs for wave energy to be $4–8/kWh, and predictions are estimated at $2.00–2.50/kWh. Earlier predicted COEs were estimated at $0.06–0.08/kWh in the United Kingdom, and projected COEs were $0.06/kWh for utility-scale systems. Marine current system COEs were estimated at $0.10–0.14/kWh, again with projected values at $0.06/kWh. Again the early predictions were optimistic.

Microhydro installations are generally remote or village systems, so the COEs will be higher as a minigrid would be installed. However, the cost in developing countries can be reduced by in-kind labor. The installed costs ranged from $900 to $6,000 for five developing countries [12].

15.8.9 VILLAGE POWER

The economics vary widely for village power due to components from different manufacturers and difficulty of reaching remote locations. The source of energy chosen for village power depends on the renewable resource and how much storage (1 to 3 days) is needed. Wind, solar, minihydro, and maybe even geothermal systems are considered, and then LCCs are used for one or more components of the system, which may also include fossil fuels, generally diesel. LCCs will help determine the ratio of different renewable energies in a hybrid system.

The China SDDX project (2002–2005) consisted of 866 village power systems in the western provinces of China. There were 146 minihydro systems (113,765 kW) and 721 systems (15,540 kW) powered by PV, wind, or a wind/PV hybrid. The average cost was $4,370/kW, which is remarkable considering the remote locations. For the China SDDX project, minihydro was the cheapest source of energy, and for good-to-excellent wind regimes, wind was the next-lowest-cost system. Notice that the average size of the minihydro was 780 kW, compared to 22 kW for PV and wind. The advantages of PV are that there are no moving parts, and everything is at ground level.

An example of remote village power is the system at Subashi (Figure 15.4), Xinjiang Province, China. The hybrid system has wind/PV/diesel (54 kW: two 10-kW wind turbines, 4-kW PV, 30-kW diesel, 1,000 Ah battery bank, and a 38-kVA inverter). The cost was $178,000 (2003 dollars), which included the minigrid. The wind/PV produces around 150 kWh/day. To estimate the COE, you need to include fuel cost (percentage of system generation not known), major replacement cost (one for sure is the battery bank every 5–7 years), and O&M. Since none of these are known, only a rough estimate can be made.

$$IC = \$178,000, AEP = 75,000 \, kWh/year \, (25\% \text{ generated by diesel})$$

$$FCR = 0.04, LRC = \$2,000 \, / \, year, AOM = \$0.01 \, / \, kWh, \text{ fuel costs} = \$1.50 \, / \, L$$

One liter will generate 4 kWh, so 18,750 kWh/year uses 4,700 L; at $1.50/L, this equals $7,000/year.

$$COE = (0.04 * 178,000) + 2,000 + 7,000)/75,000 + 0.01 = 0.215 + 0.01 = \$0.23/kWh$$

FIGURE 15.4 Hybrid system (54 kW) for village power, Subashi, Xingiang Province, China. (Courtesy of Charlie Dou.)

If they can really generate electricity for $0.23/kWh for 20 years, that would be very good. Remember that major problems will be O&M, replacement costs, and load growth, so village power systems need to be modular.

For example, if 20 kW is needed for a village power system, and the local resources for both wind and solar are good, do you choose wind, PV, or a hybrid system? First, try wind alone and then PV alone for the 20 kW. The capacity factor for wind is 25%, and for solar, the average is 4 h/day at peak power, 80% sunshine. The estimated yearly production for a 20-kW wind turbine is 43,000 kWh, and for a 20-kW PV array, the yearly production is 23,000 kWh. Also, the installed cost for wind is cheaper than the installed cost for PV, so the wind system is the obvious choice. However, a hybrid system with a small portion of PV may be a better choice for more consistent power or a smaller battery bank. For the reasons stated, for a hybrid system the ratio of wind power to PV power would be around five to one.

Small hybrid systems (Table 15.5) are available, which usually can be purchased as modular systems. Most manufacturers do not supply prices on their websites, so you have to get quotes from the manufacturer or a dealer. Shipping and installation in remote locations will increase the cost; sometimes, that will double the cost of the energy components, and for overseas, import taxes will also increase the cost. From the initial cost and estimated energy production, the COE can be estimated.

15.8.10 Wind Diesel

Wind turbines or PV added to an existing diesel generation plant are fuel savers, and the economics depend primarily on the cost of diesel fuel (Table 15.6), which

TABLE 15.5
Wind/PV Hybrid Systems for Producing AC Power

Company	Size (kW)	Wind (kW)	PV (kW)	Battery (kWh)	Inverter (kW)	Energy (kWh/year)	$ (2009)
Bergey	10.1	7.5	2.6	84	6	12,000	78,300
Bergey	1.2	1.0	0.18	10.6	1.5	1,200	8,100

TABLE 15.6
Fuel Prices Drive the Percentage Cost for Electricity from Diesel Plants

	2004 (%)	2007 (%)
Fuel	46	77
Operation and maintenance	21	9
Major repairs, replacement	19	8
General and administration	14	6

depends on remote locations; of course, the economics also fluctuate with the price of oil.

At Ascension Island, four 225-kW wind turbines were connected to a grid powered by two 1,900-kW diesel generators for a low-penetration system (14%–24%). Then, in 2003, two 900-kW wind turbines, controllable electric boiler, and a synchronous condenser were installed for a high-penetration system (43–64%). This saved an additional 2.4 million liter of diesel fuel per year for a saving of over $3 million per year with the cost of diesel at $1.50/L. The simple payback was estimated at 7 years, and with the increased cost of diesel fuel, simple payback became less than 5 years.

Three 100-kW wind turbines at Toksook Bay, Alaska, produce around 675,000 kWh/year for a wind-diesel system. The wind turbines displace 196,000 L of diesel per year, and with a cost of diesel at $1.50/L, there is a saving of $300,000/year. If the installed cost for wind turbines is $10,000/kW, then the simple payback is 7 years.

15.9 SUMMARY

National, state, and local entities are promoting renewable energy as a source of economic development, especially rural economic development. Proponents for renewable energy systems will tout COEs that will compete with present fossil fuels and how much economic development they will bring. For example, wind projects provide 150 construction jobs (6–12 months) and 10–14 full-time jobs per 100 MW for O&M, administration, and clerical jobs. The wind farm will also pay property taxes, but in most cases, the project developers try to obtain tax reductions for the economic development.

In 2009, there were over 6,000 MW of wind turbines installed in the corridor from Abilene, Roscoe, Big Spring, and then north from Roscoe to Sweetwater, Texas. In the middle of that is Nolan County, with over 3,000 MW of wind power; the economic impact was estimated at $360 million (14/1 multiplier) just from direct jobs [13]. Taxable property increased from $500 million in 1999 to $3 billion in 2009, and royalty payments to landowners were estimated at over $17 million in 2009.

The Colorado Green Mountain wind farm near Lamar, Colorado, started construction in the summer of 2003. The 162-MW project is located on 4,790 ha (11,840 acres) with 14 landowners. During construction, there were 200 to 300 jobs and after completion around 15 local jobs. The project owner receives an income from electricity sold of around $18 million/year, of which the landowners receive around $800,000/year (only 2% of land taken out of production), and the wind farm will pay around $2 million per year in property taxes. The project was purchased for $212 million by Shell and PPM from GE Wind.

Wind farms are the cheapest renewable energy source for generating electricity as COEs are $50–$70/MWh, and these numbers are starting to compete with new plants powered by fossil fuels. Note that future predictions are for trends (Figure 15.5), not actual values. Of course, this graph from 2000 did not predict the increase in COE after 2003 or the effect of the recession in 2008–2009. The COE numbers are for good-to-excellent wind regimes and for 30-MW and larger wind farms for economy of scale on construction and O&M. Actual values for wind turbine in 2015 will probably be around $0.02/kWh higher than the trends shown in Figure 15.5.

Levelized costs of energy in the United States for different sources (Figure 15.6) show wind as competitive [14] with other sources of energy and even with combined cycle gas turbines, which produce electricity from natural gas. If the external cost of CO_2 was added, then coal would be quite a bit higher and natural gas a little higher. Natural gas in 2008 was at $5.00/mcf (1000 cubic feet). However, with the increased amount of natural gas from shale formations, the spot price was $3.60–$4.50/mcf in 2014;

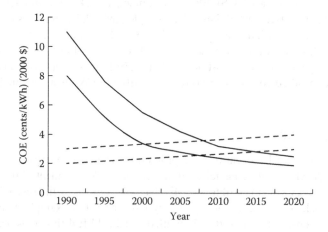

FIGURE 15.5 Range of cost of electricity (solid lines for high and low wind speed sites) from wind turbines compared to bulk power (dashed lines). (From the National Renewable Energy Laboratory.)

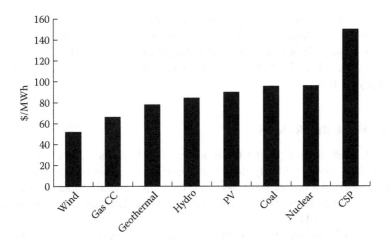

FIGURE 15.6 Estimated cost of electricity from new power plants, 2014.

however, the spot price peaked at $7.50/mcf in January. This shows that there is and will be volatility in the price of oil and natural gas. However, with the price in the $4/mcf range, it is more difficult for new renewable sources to obtain financing to compete in the generation of electricity. The values in the graph are averages, and there will be a range of values for each source. Another source is the work of Lazard [15] for comparing COEs for the generation of electricity. Cost of energy from nuclear plants is somewhat difficult to estimate since none has been built in the United States in a number of years; however, the Energy Administration Agency estimate for installed cost is $5,500/kW for a 2,000 MW plant.

From economics, from mandates (legislation or regulation), or on a voluntary basis, there will be more use of renewable energy. Traditional energy sources have an advantage in that the fuel costs are not taxed, while for renewable energy the fuel costs are free. The problem is the high initial costs for renewable energy, and most people would rather pay as they go for the fuel. Even with the increased production of small wind turbines (100 kW and smaller), in 2009 they are still not cost competitive with electricity from the grid. However, if LCCs are used, then many renewable energy systems are competitive in many situations.

Green pricing is now available in many locations, and the number is increasing. The premium has decreased from the initial value of $0.03/kWh for a block of 100 kWh/month. Pacific Gas and Electric estimated that up to 40% of their power could come from renewables without adding storage. Another major driving force for renewable energy is economic development and jobs at the local or state level. That is because renewable energy is local; it does not have to be shipped from another state or country.

The capacity of existing transmission lines and curtailment of wind farms is a major problem. The other major problem is that the wind and solar resources are generally quite distant from major loads, and geothermal is localized. Enhanced geothermal and geothermal heat pumps can be used over wide areas. However, new transmission lines will have to be built. The questions are with deregulation, who will finance the construction and overcome the right-of-way problems?

The value of externalities ranges from zero (past and present value assigned by many utilities) to as high as $0.10/kWh for steam plants fired with dirty coal. Again, values are being assigned by legislation and regulation (public utility commissions). As always, there is and will be litigation by both sides on external costs and who should pay for them.

15.10   FUTURE TRENDS

As stated, predictions about the future are risky and generally wrong on specifics, but sometimes trends are fairly clear. For example, a prediction for the price of oil at $200/bbl by 2020 is questionable; however, I am fairly confident that the price of oil will increase over the next decade. With that in mind, here are some comments on the future of renewable energy.

At some point in time, there will be a distributed renewable energy market, very similar to the farm implement business today. A farmer, rancher, or agribusiness will go to the bank and obtain a loan for a renewable energy system (size range from 25 to 500 kW). This system will expect to provide a payback of around 5 years, and it will make money for the next 15 years. The nice thing about dollars from renewable energy is that the COE will not fluctuate like for other agriculture commodities. Also for utilities, renewable energy generation of electricity can serve as a hedge against the volatility of cost for natural gas.

High- and ultrahigh-voltage transmission lines will be built from the plain areas in the United States to load centers. The same will be done in other countries, which will install large renewable energy projects for production of electricity: European Union, China, India, and others. Income (dollars per kilowatt hour) that landowners receive as royalties for new large, renewable energy systems will decrease as installed costs decrease. Within 5–8 years, renewable energy power will compete with fuel adjustment cost without PTCs but with carbon credits.

There will be trading in carbon dioxide in the United States, much as there is now trading in nitrogen oxides (NO_x) and sulfur dioxide (SO_2); however, implementation is very dubious by 2020 because of the political and industrial opposition. At that point, renewable energy, especially wind energy, becomes the cheapest source of electricity. Renewable energy systems are being installed in the world with part of the income derived from carbon trading. It is the same as European countries buying forests in South America to reduce carbon dioxide emissions. Cooperative or community systems, from 1 to 10 units, will become common because of the economies of scale. Near Luverne, Minnesota, 66 farmers formed a limited liability corporation to purchase four 950-kW wind turbines. They raised 30% of the $3.6 million and borrowed the rest through local banks. With the PTC, they expect a 17% return on investment.

The world faces a tremendous energy problem in terms of supply and in terms of emissions from the use of fossil fuels. The first priority is conservation and energy efficiency, and the second is a shift to renewable energy for a sustainable energy future. This shift has started to occur, and the renewable energy market will grow rapidly over the next 30 years.

REFERENCES

1. J. H. Wood, G. R. Long, and D. F. Morehouse. 2004. *Long-Term World Oil Supply Scenarios, the Future Is Neither as Bleak or Rosy as Some Assert.* http://www.eia.gov/pub/oil_gas/petroleum/feature_articles/2004/worldoilsupply/oilsupply04.html.
2. J. M. Cohen, T. C. Schweizer, S. M. Hock, and J. B. Cadogan. 1989. A methodology for computing wind turbine cost of electricity using utility economic assumptions. *Windpower'89 Proceedings,* American Wind Energy Association.
3. R. J. Brown and R. R. Yanuck. 1980. *Life Cycle Costing: A Practical Guide for Energy Managers.* Fairmont Press, Atlanta, GA.
4. International Renewable Energy Agency. 2012. *Renewable Power Generation Costs in 2012: An Overview.* http://www.irena.org/DocumentDownloads/Publications/Overview_Renewable%20Power%20Generation%20Costs%20in%202012.pdf.
5. D. Feldman et.al. 2012. *Photovoltaic (PV) Pricing Trends: Historical, Recent, and Neat-Term Projections.* SunShot, DOE/GO-102012-3839. http://www.nrel.gov/docs/fy13osti/56776.pdf.
6. L. Stoddard, J. Abiecunas, and R. O'Connell. 2006. *Economic, Energy, and Environmental Benefits of Concentrating Solar Power in California.* http://www.nrel.gov/docs/fy06osti/39291.pdf.
7. Ivanpah Solar Electric Generating System. 2014. http://www.brightsourceenergy.com/stuff/contentmgr/files/0/3eac1a9fed7f13fe4006aaab8c088277/attachment/ivanpah_white_paper_0414.pdf.
8. *Renewable Energy Technologies: Cost Analysis Series, Concentration Solar Power.* 2012. IRENA. http://costing.irena.org/media/2794/re_technologies_cost_analysis-csp.pdf.
9. R. Wiser and M. Bolinger. 2013. *Wind Technologies Market Report.* U.S. DOE, EERE. http://emp.lbl.gov/sites/all/files/2013_Wind_Technologies_Market_Report_Final3.pdf.
10. *Renewable Energy Costs, Technologies and Markets.* 2012. IRENA. http://costing.irena.org/irena-costing.aspx. A number of charts are available.
11. *Micro-Hydro Power.* Practical Action. http://practicalaction.org/micro-hydro-power.
12. S. Khennas and A. Barnett. 1999. *Micro-Hydro Power: An Option for Socio-Economic Development.* http://www.afghaneic.net/renewable/16%20smail.pdf.
13. *Nolan County: Texas Wind Energy Economics Case Study.* 2008. http://cleanenergyfortexas.org/downloads/Nolan_County_case_study_070908.pdf.
14. NREL. *Energy Technology and Cost Performance Data.* http://www.nrel.gov/analysis/tech_cost_data.html.
15. Lazard. June 2008. *Levelized Cost of Energy—Version 2.0.* http://www.narucmeetings.org/Presentations/2008%20EMP%20Levelized%20Cost%20of%20Energy%20-%20Master%20June%202008%20(2).pdf.

RECOMMENDED RESOURCES

LINKS

Long-Term World Oil Supply

Fueling global warming subsidies for oil. 2005. http://www.greenpeace.org/usa/Global/usa/report/2007/7/fueling-global-warming.pdf.

International Energy Outlook 2014, U.S. Energy Information Administration. http://www.eia.gov/forecasts/ieo/pdf/0484(2014).pdf.

The Oil Age Poster. http://www.oilposter.org.

Cost Modeling

NREL, Energy Analysis, Models and Tools. http://www.nrel.gov/analysis/models_tools.
html; Energy technology cost and performance data for distributed generation. http://
www.nrel.gov/analysis/tech_cost_data.html.

P. W. Petersik. Modeling the costs of U.S. wind supply. 1999. http://www.eia.doe.gov/oiaf/
issues/wind_supply.html#rwgt. For historical perspective.

U.S. DOE EERE. Interactive: Transparent cost database. http://en.openei.org/wiki/
Transparent_Cost_Database. Very informative as it gives values, box and whisker or
scatter.

U.S. Federal Budget

U.S. Department of Energy. http://energy.gov/eere/about-us/office-budget.

Other

Biomass for power generation. http://www.irena.org/DocumentDownloads/Publications/
RE_Technologies_Cost_Analysis-BIOMASS.pdf.

The cost of wind energy in the U.S. http://www.awea.org/Resources/Content.aspx?ItemNum
ber=5547#CostofWindEnergy.

CSP. http://costing.irena.org/media/2794/re_technologies_cost_analysis-csp.pdf.

Hydropower http://www.irena.org/DocumentDownloads/Publications/RE_Technologies_
Cost_Analysis-HYDROPOWER.pdf.

International Renewable Energy Agency.

NREL. Energy technology cost and performance data for distributed generation. Has information on capital cost, operation and maintenance, utility scale capacity factors, useful
life, land use by system technology, and LCOE calculator. http://www.nrel.gov/analysis/
tech_cost_data.html.

PV. http://www.irena.org/DocumentDownloads/Publications/RE_Technologies_Cost_
Analysis-SOLAR_PV.pdf.

Technology briefs: Ocean Energy, OTEC, Salinity Gradient, Tidal Energy, Wave Energy.
http://www.irena.org/menu/index.aspx?mnu=Subcat&PriMenuID=36&CatID=141&S
ubcatID=445.

U.S. Environmental Protection Agency. Renewable energy cost database. http://www.epa.
gov/cleanenergy/energy-resources/renewabledatabase.html.

Wind. http://www.irena.org/DocumentDownloads/Publications/RE_Technologies_Cost_
Analysis-WIND_POWER.pdf.

PROBLEMS

15.1. What are the two most important factors (the factors that influence COE
the most) in the COE formula?

15.2. Estimate simple payback for 10 kW PV system for your home. Be sure
and show values you used in the calculation.

15.3. Calculate the simple payback for solar hot water (four modules) for a
swimming pool in northern Florida. Choose any manufacturer, note the
type, specifications, installed cost, energy production, and value of energy
displaced. You will have to calculate or estimate energy production.

15.4. Use the EERE web page to estimate the cost of a solar hot water system
for your home. http://energy.gov/energysaver/articles/solar-water-heaters.

15.5. Calculate the COE (use Equation 15.3) for a Bergey 10-kW wind turbine (grid connected) on an 80-ft (24 m) tower for a good wind regime. You can use a simple method for estimating the annual kilowatt hours.

15.6. Calculate the COE (use Equation 12.4) for a 50-kW wind turbine that produces 110,000 kWh/years. The installed cost is $200,000, the fixed charge rate is 6%, O&M is 1% of installed cost, and levelized replacement costs are $2,000/years.

15.7. Explain LCCs for a renewable energy system.

15.8. Do an LCC analysis for a 5-kW PV system installed in Amarillo, Texas. Use tilt angle = latitude.

15.9. Calculate the COE for a proposed tidal system: 8 GW, 20% capacity factor, installed costs = $2,500/kW, annual O&M = $120 million, FCR = 6%.

15.10. The COE from a wind farm is around $0.06/kWh. Make a comparison to nuclear power plants. What is the retail rate for the latest nuclear power plants installed in the United States? (Do not calculate; find an estimate from any source.) For nuclear plant, you may use installed cost of $5,500/kW, small AOM, and capacity factor of 90%.

15.11. What are today's values for inflation, discount rate, interest rate? What is your estimate of fuel escalation, average/year between now and the year 2030?

15.12. An 80-MW wind farm (80 wind turbines, 1 MW) was installed in White Deer, Texas. The utility company is paying an estimated $0.026/kWh for the electricity produced. Estimate the yearly income from the wind farm. You could find actual income from the FERC site. If the landowners get 4% royalty, how much money do they receive per year?

15.13. For wind farm in problem 12, installed costs were $1,000/kW, FCR = 9%, capacity factor (annual efficiency) = 35%, AOM = 0.01/kWh. Calculate the COE using Equation 15.4. You will need to estimate the levelized replacement costs or calculate using Equations 12.5 through 12.7. Compare your answer to the $0.026/kWh, which is the estimated price the utility company is paying the wind farm. How can the wind farm make money?

15.14. Estimate the COE for a hydrokinetic tidal system: IC = 2,500 kW, capacity factor = 25%, FCR = 8%.

15.15. What is the price of oil ($/bbl) today? Estimate the price for oil ($/bbl) for the years 2020 and 2030. Estimate the price for oil when the costs for the military to keep the oil flowing from the Middle East are added. Place results in a table.

15.16. For a remote 2-kW microhydro system, estimate the COE. Installed cost = $12,000 and it produces 40 kWh/day; FCR = 6%; neglect AOM.

15.17. Go to the FERC website for power plant reporting. Pick any wind farm. What were the data reported for the latest quarter, energy production, cost, cost per megawatt hour? Calculate the capacity factor for that quarter. (http://eqrdds.ferc.gov/eqr2/frame-summary-report.asp)

15.18. A village power system is to have 20-kW wind, 5-kW PV, a battery bank for 2 days, a 20-kW inverter. Estimate the annual energy production.

15.19. For Problem 15.18, estimate the installed cost. You may use components from any manufacturer. Also, include the cost for a minigrid for 200 households, 1 clinic, 1 school, 1 government building.

15.20. Estimate the cost of energy for a wave system with a 50-MW plant. You may use any type or manufacturer. Be sure to note specifications, energy production, capacity factor, and so on.

15.21. Why is present worth used in estimating future costs/benefits?

15.22. Estimate the COE for any CSP system. Note all input data, specifications, and so on.

15.23. In your opinion, what do you foresee for the cost ($/kWh) of externalities for electricity generated from coal for 2020? For 2030? You might want to write this number down and see how it compares to the actual value.

15.24. What is the U.S. DOE budget for renewables, nuclear, fossil fuels? Note the year you are using.

15.25. For this year, what two renewable sources have the lowest COE. Note the year, values, source of information.

15.26. For those regions or countries that have CO_2 trading, what is the value/ton? Note year, region/country, value.

15.27. Go to Transparent Cost Database (see Links). What is the medium, max, and minimum LCOE for PV for 2014?

16 Observations

This chapter presents observations, comments, predictions, and some opinions by the authors, primarily for students. Of course, others may find the observations useful. Hopefully, there will be some new insights, and there will be disagreements from mild to outrage about some of the comments and opinions. Predictions can only delineate trends and different possible paths or alternate futures (sea level rise of 1 m vs. 10 m). Novel synergetic technologies from today's science are hard to predict, and new breakthroughs or transformations are completely unknown. We will use the authors' initials to indicate our separate specific opinions and comments.

Today's students were raised in the age of instant access to information at any time and almost anywhere: smart phones, tablets, laptops, PCs, Wi-Fi, and 4G networks. The amount of information on the Internet is astounding, located in the cloud or some nebulous place (servers across the world). Lecture classes with some professors using the blackboard (math, science, and engineering) are almost ancient history, as students are accustomed to visual media, power point, and other apps. How many students in the classroom are looking at their smart phones or sending and receiving information (texts, emails, songs, photos, videos, and ?), especially in the large introductory classes. Teachers are admonished that their class presentations need to be entertaining so the students will pay attention. Notice at the next conference luncheon how many people are using their smart phones and tablets, rather than listening to the speaker.

My career (VN) spans the gamut from traditional lecture classes and labs on the receiving end, Kansas State Teachers College 1953–1957 and summer courses 1959–1960, Harvard University 1960–1961, and University of Kansas 1961–1967, and then on the delivery end, Universidad de Oriente, Cumaná, Venezuela, 1967–1969 and West Texas A&M University (WTAMU), 1969–2003 and then 2009–2010. In the beginning at WTAMU, I taught physics courses in the traditional lecture and lab method. I developed a quantum mechanics course that was self-paced; however, most students did not have the discipline for self-paced, leaving units until the end of the semester, and thus, it was too late, so I only did it one semester. Then because of small enrollments, WTAMU joined other regional institutions to deliver physics courses using interactive TV [1]. The problem with interactive TV is the need for a classroom and equipment at sending and receiving locations, constraint on number of students that can be handled by the instructor, and limitation on number of classes that can be offered from one or two special interactive TV classrooms. Plus you need the technical help to operate the systems. How many interactive classrooms can a university afford? Some have characterized interactive TV as the talking head and the writing hand. As Dean of the Graduate School, Research, and Information Technology, in collaboration with Dr. LeAnn Thomason, we were instrumental in starting WTOnline, with the first online class in 1997.

I developed and taught two courses, solar energy and wind energy, and then taught those courses online. That material was then used for web-based CDs, along with CDs for Wind Water Pumping and Bombeo de Agua con Energía Eólica (www. windenergy.org for information and purchasing). The wind energy material was expanded into a conventional introductory textbook titled *Wind Energy* published in 2009 and subsequently second edition in 2013, and in 2010, I developed and taught an online course on renewable energy, which was also expanded into a conventional textbook titled *Introduction to Renewable Energy* (this book is the second edition). Ken and Vaughn are working on *Introduction to Bioenergy*, which should be published in 2016.

The advantages of online courses versus interactive TV were less cost, no restriction on number of students and location, better interaction, flexibility for students in terms of time (except for chat sessions) and for students that commute, and a cost saving. However, there is a limit on number of students (30–40) that one instructor can handle without assistance, grading problems and tests, and interaction using chat (can handle 10–12 students in 1 h, so need multiple chat times). The first time I taught wind energy and wind turbines online I had students from cities too far away for commuting, from other states, and even one from a country in the Middle East. Now even students on campus take online courses because of the flexibility of scheduling. However, some students do not care for online courses and some students need more structure. I had essentially the same group of students for a fall traditional course and then in the spring, an online course. Anecdotal comments from the students were that they enjoyed my lectures but they learned more in the online course, as they had to read the material and do the homework, although chat was available for questions and comments. From my experience then there are three major types of universities: research, traditional or teaching (which can also be called the *social university*), and the learning university (online).

Now every university and community college has online courses and there are public and for-profit universities that have changed their delivery method to online or only online, for example, Phoenix University and Western Governor's University in the United States, and The Open University, United Kingdom. Now online classes have progressed to the massive open online course (MOOC), unlimited participation and open access via the web [2], best universities and entities. Then how do regional universities compete? Major universities have placed their courses online for anybody to audit, not for credit.

My career (KS) brackets the traditional learning method and ends in the current state of learning. I grew up in a small Texas town of 1,300 people and a graduating class of 28. This led to focused one-on-one education with my teachers, and when I went to school at West Texas State University in the 1970s, I expected the same attention. Dr. Nelson was one of my first instructor (would have been the first, but I went to the wrong room the first day of classes), and all the instructors at WT were committed to working with the students. My work with renewable energy research began at the time I started school, and my involvement was at the installation and operations level. But that led to the development of skills in data collection and analysis, archiving information, photographic records of projects, video capture of experiments, creating test articles and demonstration vehicles, wiring and operating data loggers, computer programming for data reduction, report generation, and long distance data collection by telecommunication.

This hands on and theoretical overview led to a better understanding of many of the subcomponents of the renewable systems: aerodynamics, mechanical power transfer, electrical power conversion and grid connection, control systems, and meteorological resource assessment. But this information was not useful unless applied or passed along, so I also taught training programs at the national and international levels. It means a small town boy got to see every continent (except Antarctica). I was able to compare how things were done and how energy and resources were used throughout our world, which showed how important that renewables were needed to fill the gap for the potential loss of fossil fuel resources in the future. We have to see that recognizing trends and predicting future trends is needed and should be part of the toolbag of every student. I have taken over the online instruction in the Renewable Energy Program at WTAMU which includes courses in introduction to renewable energy, wind, solar, bioenergy, and geothermal energy. In each class, I try to stress that if I can get the student to recognize the light approaching them in the tunnel is a train (finite amount of fossil fuels), then they should have time to step off the tracks, get out of the tunnel, or face it head on. But at least they were aware that it is there.

The following comments (VN) are about state supported universities in the United States. Regional universities face future competition from community colleges offering more junior–senior level courses (less expensive) and the large research universities claiming that regional universities do not deserve much funding for research (fewer graduate courses). All universities acquire a significant amount of their funding from enrollment in introductory courses (many are required courses), so there will be the competition from textbook publishers and companies such as Microsoft and Apple. For-profit entities will develop online courses and will tell state legislators that they will provide courses that are cheaper than state supported institutions and they will use comparative criteria for knowledge acquired. In general, state budgets for education are discretionary, so there is always pressure to reduce or keep that budget flat, especially with the present politics of no tax increases and/or reduction of taxes. The cost of education at state universities has increased faster than inflation and less state support means a much larger cost for the student and/or their parents. Over the long term, the cost of a college education cannot continue to increase faster than inflation.

In my opinion (VN), there are two main problems. Too many students are not willing to sacrifice or to defer other expenses, for example, cars and pickups, for a university education. It must be noted that today, a student cannot earn enough money during the summer to pay for a year at the university. A summer job of \$10/h for 3 months is only \$4,800. So they borrow too much money, and for some careers, they cannot repay the loan. In 2014, the average student-loan debt was \$33,000, and the total student debt was $\$1.2*10^9$, 6% of the U.S. national debt. How did we arrive at this point of the federal government being the main source of student loans? Already some are promoting a bailout for student loans. Remember the bailout for Wall Street and the banks in 2008–2009 and the bailout for credit unions in the 1980s.

The other problem is the increase in administration at universities. For example, in my first year (VN) at WTAMU (1969), the enrollment was around 8,000 students with the following administrators: president, three vice presidents, six deans, and twenty academic department heads (department heads received a 3-h load reduction). In 2014, the fall enrollment was 8,981 with the following administrators: president, provost, five vice presidents, chief diversity officer, chief information officer, five deans, associate VPs and associate deans, fund raiser for each college and school, and nineteen department heads (department heads received a 6-h load reduction). Of course, all those administrators need secretaries and other support. There were 287 full time faculty and 440 full time staff. Then, there is support for the other business of the university, sports, which at regional universities is a black hole for money. At many institutions, sports are partially supported by a mandatory student fee. At major universities, football and basketball are really big business, as salaries for head coaches are in the millions of dollars.

My concerns (KS) about cost for education for students are the way other countries treat education compared to United States. The Alternative Energy Institute has sponsored more than 45 international interns for hands-on experience with the use and operation of renewable energy systems. Generally, they are here for 4–6 months and they were often supported by the university they attended. Some students were fully funded by their country, and other students had to supply all the living expenses in the United States and travel to and from their home, at a cost of $4,000–7,000. If their country supported all the expenses, then they had more disposable income. In United States, we place the burden on student/parents to supply a significant amount for a university education. If we fund a country-wide support for all education, there would be more students staying in class and the result would be a larger pool of educated and skilled workers. Another possibility would be to copy the German program of three paths of training beyond the high school level: technical schools, college (focused toward business and finance), and university (science and philosophical studies). The main thing is the costs are borne by the country and not the student, and so retention is increased.

One of the problems with information on the Internet and web-based textbooks and CDs is the ease of copying material, which make it difficult to cover the cost of developing the material or to make money as with conventional textbooks. Now teachers submit essays and papers to websites or their online platform to check for plagiarized material. The other problem is the lack of review or checks on accuracy or even content on the Internet. Viral nonsense spreads more rapidly than a new peer-reviewed scientific discovery.

The traditional publishing industry for textbooks is changing and other entities are entering the market, for example, iBooks textbooks for iPad (https://www.apple.com/education/ipad/ibooks-textbooks/). Price of textbooks has become expensive, so my opinion (VN) is that university teachers should cooperate (shareware) on developing web-based material by modules and units. With all the photos (and diagrams) available by Google>images there is not much need for photos in textbooks and if copyright problems are resolved they are easily incorporated into web-based modules. As an example, an introductory textbook on environmental science, which is web-based, is available for $50 (http://www.bu.edu/earth/2012/11/08/

world's-first-web-based-env-science-textbook/). Again once an instructor and one student have the CD or web material, then others can copy the material.

16.1 SOLID-STATE ELECTRONICS

For today's students, the Vietnam War (1955–1975; U.S. buildup 1960) is ancient history and Oil War I (Gulf War, 1990) is old history; however, we will start from 1990 with a few earlier excursions. The science of solid-state physics has given rise to solid-state electronics and with integrated circuits (1958) has come the digital age and personal computers (1997), laptops, tablets (Windows XP 2002 and iPad 2010), and smart phones (Nokia 1996; BlackBerry 2003, and iPhone 2007). With the tremendous power of computers, from mainframes to PCs, software packages and tools (sometimes referred to as toolboxes) are now ubiquitous: business (word processing, media presentation, and spreadsheet), graphics and media design, computer-assisted design, math (MATLAB®, Maple, Mathematica), statistics, engineering (computational fluid dynamics and multibody dynamics), geographic information system (GIS), modeling (weather forecasting and climate change), and numerous others. For example, MATLAB® is used for numeric computation, data analysis and visualization, programming and algorithm development, and control systems, test, and measurements. Then, there are the prominent applications, copying songs (iPod 2001), streaming TV and films (Netflix 1997), and games (Game Boy 1989, PlayStation 1995, and X box 2001), which are addictive, especially for boys. However, VN spent one Christmas break playing Frogger until I reached my level of incompetence and KS plays World of Tanks every week. With large bandwidth, multiplayer-interactive games (free and/or purchase) are available online and then there is online poker, where real money is exchanged. Then, one of the latest applications is payment with your iPhone, 2014.

Digital electronics has had a large impact in terms of supervisory control and data analysis (SCADA), which means that renewable energy systems can be controlled from remote sites, monitored for system alerts and faults, curtailment management, and data displayed in real time for the operators. Some companies include the software information for students and even information for the general public, all in real time. National Instruments LabVIEW (Laboratory Virtual Instrument Engineering Workbench) is used for acquisition and analysis of data, instrument control, monitor and embedded control, and automate test and validation systems. There are a number of companies that supply SCADAs for specific applications, for example, renewable energy systems such as wind and solar farms. So check for information on the Internet in your area of interest or field of study. The learning curve for many of the programs is fairly steep, but well worth the effort.

16.2 INTERNET

The beginning was ARPANET (1969) to the World Wide Web (1990) and the first graphical web browser (Mosaic 1993). Then, Google (now also a verb) developed algorithms for search (1997), and the Internet has become the place for most everything that has been digitized, websites even for individuals, blogs, Facebook, Twitter,

YouTube, Instagram, electronic books, music, and who knows all of things that are available, somewhere out there (in the Cloud?). So your mobile device has more power and many more applications and functions than the mainframe computers of the 1970s. The synergistic combination of computing power and software has made a significant difference in almost all aspects of life.

Data loggers for remote sensing are connected to a web portal by cellular or satellite communication. A base program on a PC queries the data logger, records data, performs further analysis, and displays information, sensor values and graphs. Again some programs also provide displays for the public. Most of this has come about due to the increase in reliability and sophistication of the communication network, and the standardization of communication of digital information makes this high speed transfer possible. And the result is the infrastructure (wired and wireless) that has fed the growth of the use of the information at each end of the communication network.

16.3   GEOGRAPHIC INFORMATION SYSTEMS

A GIS is a computer system capable of holding and using data, which is spatially oriented. A GIS typically links different data sets, or a base set is displayed and overlays of other data sets are placed on the base set. Information is linked as it relates to the same geographical area. A GIS is an analysis tool, not simply a computer system for making maps.

There are two general methods of representing the data, raster and vector. Raster based means every pixel has a value and location, and vector based means that the data are represented mathematically—endpoints for lines and groups of lines for polygons. Each pixel can represent an attribute, and the number of attributes depends on the number of bits: 16–256 colors or shades of gray. Therefore, pixels or vectors can have different attributes and are linked to a database, which can be queried. A GIS gives you the ability to associate information with a feature on a map and to create relationships that can determine the feasibility of various locations, for example, a hierarchical system for locating anemometer stations for wind prospecting.

An overlay is a new map with specific features, which is overlaid on the base map. Overlays are one form of database query functions. The overlay can be a raster or vector image, with the base map being a raster or vector image. The number of overlays is generally limited only by the amount of information that can be presented with clarity. Interactive maps are now quite common for displaying data and we googled interactive maps>chapter headings and found a large number of URLs which are at the end of the chapters.

16.4   SATELLITES

Remote sensing in all the different wavelengths from satellites provides huge quantities of data about the Earth, which is useful for assessment of renewable energy resources and for other analysis. Land use, global agriculture information, and seasonal and decade changes (ice cover for Arctic) are examples that were not possible in the past. Your smart phone has a global position system (GPS) and with maps you can find locations, directions, points of interest, and the closest pizza parlor. Google Earth is used for resource analysis, site layout, and interactive maps. In the future, there will be many more applications of interactive maps and GISs in renewable energy information.

16.5 EXAMPLES

The examples provided are a small sample of the large number of applications available for renewable energy. Students are encouraged to visit some of these sites, especially in their area of interest, as in the future they could be better ways of displaying and better ways of analyzing a large amount of data by visual computing and compression and reduction of data. Most interactive maps use the base layer of Google Earth and even the free version allows the addition of icons with attributes. The more detailed maps have multiple levels of query and/or GIS layers.

16.5.1 INTERACTIVE MAPS

International Renewable Energy Agency. Global Atlas. http://globalatlas. irena.org.
> The Global Atlas is an online GIS linked to a number of data centers located around the world. All the information in the catalog can be accessed directly from the Global Atlas GIS interface. It started with solar and wind, and should progressively expand to include other resources: bioenergy, geothermal, and hydropower in 2014, and marine energies in 2015.
> The GIS interface enables users to visualize information on renewable energy resources and to overlay additional information. These include population density, topography, local infrastructure, land use, and protected areas. The GIS interface will progressively integrate software and tools that will allow advanced energy or economic calculations for assessing the technical and economic potential of renewable energy.
> The Map Gallery also features a geo-explorer. The geo-explorer allows users to find the maps available for a specific location. On the GIS interface, users can edit the map, add other datasets from the catalog, and save the map under the user's personal profile.

Renewable Energy Policy Network for the twenty-first century. Renewables interactive map. http://map.ren21.net.
> Technology: Bioenergy (solid, biogas, and biofuels), geothermal, hydropower, ocean, solar (PV, heating/cooling, and CSP), and wind.
> Sector: Policies, targets, shares, installed capacity, energy production, and RE economy.

U.S. National Renewable Energy Laboratory. Interactive mapping tools. http://maps.nrel.gov.
> Renewable Energy Atlas. Resources: hydro, geothermal, biomass, CSP, PV, wind speed, and wave power density.
> Marine and Hydrokinetic Atlas.

Federal Energy Management Program Screening Map. Market potential for various solar technologies. Example queries, payback or savings to investment ratio for solar hot water, PV.

Biofuels Atlas; Biopower Atlas (feedstocks, plant locations, etc.).

Solar Power Prospector (utility scale); PVWatts (estimates energy production and COE, grid connected PV); PVDAQ (solar array efficiency).

Geothermal Prospector (large scale); Wind Prospector; HyDra (hydrogen demand, resources, infrastructure, and cost).

Global RE Opportunity.

U.S. Environmental Protection Agency, RE-Powering American's Land, mapping and screening tools. http://www.epa.gov/oswercpa/rd_mapping_tool.htm.

U.S. Geologic Survey, energy resources program. Energy Vision. http://energy.usgs.gov/Tools/EnVisionSplash.aspx.

Natural Resources Defense Council. Energy map. http://www.nrdc.org/energy/renewables/energymap.asp.

Facilities; wind, solar, advance biofuel, biodigesters, geothermal, low-impact hydroelectric. Energy potential; wind, solar, cellulosic biomass, biogas, enhanced geothermal.

United Kingdom, Map of renewable and alternative energy projects in the UK. http://www.renewables-map.co.uk.

16.5.2   INTERACTIVE DATABASE

U.S. DOE Global Energy Storage Database. http://www.energystorageexchange.org/projects.

Projects, search filters: Technology type, country, state/province, rated power, duration, service/use case, ownership model, status, and grid interconnection.

U.S. Policies, search filters: Policy source, ownership model, service/use case, grid interconnection, and utility type.

U.S. Energy Information Administration. http://www.eia.gov/countries/data.cfm.

International energy statistics: Petroleum and other liquids, natural gas, electricity, coal renewables, total energy, carbon dioxide, and others (population and conversion factors).

U.S. DOE EERE. Transparent cost database. http://en.openei.org/apps/TCDB/.

16.5.3   TOOLS

Sustainable by Design. http://www.susdesign.com.

Design tools: Sun angle tools, window tools, panel shading, and U.S. climate data.

U.S National Renewable Energy Laboratory. http://www.nrel.gov/analysis/models_tools.html.

Homer Energy. http://www.homerenergy.com. Microgrid optimization.

U.S. DOE, EERE. Building energy software tools directory. http://apps1.eere.energy.gov/buildings/tools_directory/subjects.cfm/pagename=subjects/

pagename_menu=whole_building_analysis/pagename_submenu=renewable_
energy.

RETScreen International. http://www.retscreen.net/ang/software_and_data.
php.

Free, available in multiple languages; evaluate energy production, costs, emissions reductions, financial viability; include hydrology and climate databases. RETScreen 4 is an Excel-based energy project analysis tool.

16.5.4 OTHERS

Earth, a graphic look at the state of the world. http://www.theglobaleducation-project.org/earth/index.php.

Global ecology, energy supply, fresh water, development and debt, weapons, human conditions, fishing and aquaculture, food and soil, toxic pollution, and wealth.

BP Statistical review of world energy. http://www.bp.com/en/global/corporate/about-bp/energy-economics/statistical-review-of-world-energy.html. Can download spreadsheet which has historical data.

Hydro Tasmania, King Island Renewable Energy Project. http://www.kingisland-renewableenergy.com.au. Display of real time data.

16.6 PREDICTIONS

Arthur Clarke's, the famous sci-fi writer, three laws of prediction are still applicable [3] which are given as follows:

1. When a distinguished but elderly scientist states that something is possible, he is almost certainly right. When he states that something is impossible, he is very probably wrong.
2. The only way of discovering the limits of the possible is to venture a little way past them into the impossible.
3. Any sufficiently advanced technology is indistinguishable from magic.

VN likes the following example of two unrelated past events and then the actual result in the future. In 1903, on the day of the first flight of the Wright brothers' airplane, the long fourth flight was less than one minute and covered a distance of 260 m (852 ft). In 1905, Albert Einstein developed the special theory of relativity and one of its predictions was the most famous equation for energy and mass, $E = mc^2$. If any scientist at that time had predicted that within 40 years an airplane could deliver a weapon that would destroy a whole city, everyone would say that scientist was crazy, did not know anything, and besides it was theoretically impossible. Compare the round trip of the Enola Gay of 14 h and over 5,000 km with the fourth powered fight of the Wright airplane, a time span of only 40 years from the two events. So you can see the difficulty of prediction, even though now there are lots of articles and books and even a World Future Society site (http://www.wfs. org). Sci-fi has lots of examples of future technologies and science and was much better than almost all scientists in imaging possible future science and technologies. Prediction in the social sciences is much more difficult, although some trends have been identified [4,5] and game theory is used to make predictions in international relations and foreign policy [6]. The impact of technology on social and international policy [7] for 50 years from 1968 was presented for the following areas: weaponry, space, transportation, communication, weather, educational, behavioral, computer technology, energy, food, population, economics, and oceanography. The predicted impacts range from optimism to pessimism and here are some of the predictions still in the future in 2018: control of weather and possibly the climate, control of gravity, waning of nationalism, breeder reactors provide long range hope, and control of population growth. In the chapter on energy, the prediction was that tidal power, geothermal power, and wind can be dismissed as inconsequential.

Here are some interesting predictions by Clarke [3], which were made from 1962–1983.

2030: Lunar settlements, detection of extra-terrestrial intelligence, and weather control
2050: Space colonies, memory recording, artificial life, and space-time control
2100: Planetary colonies, space elevator, machine intelligence exceeds man, climate control, and immortality

16.7 SCIENCE

Our knowledge of the universe is powered by science, so here are a couple of comments about science. There is no absolute truth in science in contrast to religion tenets where there are a large number of different absolute truths, over which there are violent conflicts. However, the laws and models of science are very useful (nobody violates a physical law) and the applications (technology) have been astounding. Hierarchical structures have been very useful in our understanding of what and how in the world; a few examples are nuclear table, periodic table, DNA to genes, and amino acids to proteins. Parts combine to form wholes, which are then used as parts to form new wholes. However, remember that the functions or attributes of the whole are greater than the sum of the parts. For example, chlorine is a deadly gas, and sodium is a metal that bursts into flame when it comes in contact with water; however, we eat salt, $NaCl$, which is necessary to stay alive. The interaction is explained by quantum chemistry, not just knowing the information about the parts.

Another description of phenomena is by particle and waves. The world is made up of particles (description by hierarchical structures), and even for identical particles (electrons and atoms), two particles cannot occupy the same energy state at the same time (in physics they are Fermions). Waves (in physics they are Bosons) obey the principle of superposition as they can be at the same place at the same time. The most common examples are waves that need a medium (sound and water) and electromagnetic waves, which do not need a medium. Place your finger in the air and at that point there are all those different waves, reflected color from different objects in the room, radio, Wi-Fi, and TV. Science and the application of science, technology, have been and will be the major impact on our lives.

16.8 AUTHORS' PREDICTIONS

These are not really predictions but some possible technologies that will have a significant impact in the next 25 years. Predictions of tomorrow's technology from today's science are only limited by our imagination. We are making significant progress in neuroscience and will understand how the brain works, input, output, storage, and retrieval of information (memory), and maybe even consciousness. Knowledge systems (e.g., Watson) are here and will become ubiquitous, even in everyday life. VN always wanted a way to learn calculus without all that hard work, just plug me into a knowledge and teaching unit while I sleep. Or maybe give me some artistic talent or even a little rhythm.

Artificial intelligence has been predicted, it seems like forever, and hard artificial intelligence is still in the future. However, Kurzweil predicts that artificial intelligence will surpass human thinking by 2029 [8], and the singularity, an exponential growth in computing capacity, which will provide a profound and disruptive transformation in human capacity, will occur in 2045 [9]. Asimov, along with other sci-fi writers, explored the psychological and social aspects of intelligent robots and interaction with humans. The problem for humans is that we are presently limited by the size of our brain and the bandwidth for input and output, which is not the case for intelligent computers. Another comment by Arthur C. Clarke, maybe humans are just one step in evolution towards silicon-based intelligence. VN tried to find the quote and was not successful, however I am almost certain he said it, and if not he implied it.

Virtual reality (VR) is the next disruptive technology (Oculus Rift 2014) that will change our world, similar to the impact of the iPhone. Just think how VR might change the tourism industry and airline industry, number of flights, and need for airplanes. VR might even improve education and learning, but our past use of new communication devices for education indicates quite a lot of doubt. When you combine VR with hepatic devices, then the entertainment industry is changed. The adult entertainment business will be an early promoter and developer, teledildonics. Maybe too many people will spend their time playing, much as some boys and young men now spend most of their night and day playing games. Maybe it will be a society where the majority of people are the ultimate couch potatoes. Of course, who does the work and how is energy provided for all that infrastructure?

Nanotechnology is already at the beginning stages of commercialization [10]; however, biophysics will become integrated with nanotechnology to produce novel materials. When coupled with 3D printing, it will be a disruptive technology. Hewlett-Packard came out with a PC that can also scan in 3D, and then you can manipulate the image, and you will be able to send the information to a 3D printer (2014). In 2015, a car company is offering customized vehicles by 3D printing of the body.

The disruptive technology of genetic engineering is already here and many are promoting genetic engineering of crops as one of the solutions to increase production of food for the increase in population. Then, there is the need for modified crops in response to the increased temperatures of climate change. Scientists are working on changing cereal crops from annual to perennial, which would be a major shift. Comments on cloning and other disruptive aspects of genetic engineering due to our increased knowledge will be left to others, especially the cloning of humans.

16.9   ON THE DOWNSIDE

The two major problems in the world are *over consumption* and *over population*, since one projection is an increase of around $2 * 10^9$ people by 2050 and another projection is a little longer, $2 * 10^9$ by 2100. How do you provide food, energy, water, and infrastructure for that many new people, especially as the developing countries change their diets and increase the use of energy per person? In 1985, the number of cars in China was in the thousands, and by 2010, China became the major market in the world for vehicles, with over 20,000,000 sold in 2014. Just think a new phenomenon in the United States is the number of storage units because people have more stuff than they can use or need at their home or apartment. How many container ships are literally bringing stuff (junk) not really needed to the developed world, and then shipped back empty?

Now we are living in the Anthropocene, the era where humans are performing an uncontrolled experiment on the Earth's ecosystem, changing the atmosphere, landscape, and the biome (ecology), for example, deforestation, extinction of species, and overfishing. An estimation of the global or ecological footprint of humans is 1.5 earths [11], where the definition of ecological footprint is the human impact measured in terms of area of land used for food, fiber, energy, and water to produce the material consumed and to assimilate the wastes generated. Within this century, there will be major environmental consequences (VN) from this overuse or overshoot of the global footprint.

What is the carrying capacity of the Earth for humans without major impact on the ecosystems? The mantra of economists and business entities tout growth as the solution for everything, and most politicians in the United States and the rest of the world maintain that public policies should be based on economic growth. Then, the media acquiesce with reports on percent changes and growth is always positive. Nobody really knows for sure at what level humans should be using the Earth's resources; however, if you accept the deniers of global warming and growth is the solution, then it is already too late for prudent action. We will tend to continue along the same path: destruction of the environment, overfishing, and wasting water and energy. Once again, politics, rhetoric, denial, and to forbid the use of the words of climate change and global warming (e.g., governors of Florida and Wisconsin) will not change the physical world. You can label it junk science, spend millions of dollars on a campaign to deny the climate is being affected by humans, but in the end you cannot fool Mother Nature, and the cost will be large as the environment is degraded.

The increase and the continued high rate of emission of CO_2 into the atmosphere due to human activity will be with us for a long time; therefore, global warming will impact everybody. The consequences of higher temperatures on precipitation and sea level will be significant. How will you and governments react to the changes? What adaption, abatement, and mitigation will be put in place and when? In 2014, the European Union set a new goal for CO_2 emissions, 40% below 1990 levels by 2030.

Therefore (VN) sell your ocean front property and move to an elevation at least 10 m above present sea level. Of course, people and business will want mitigation (levees, sea walls, and gates) and subsidies for insurance from the federal level. It will be cheaper to start moving people from those areas today, rather than waiting for the next disaster. Just think of the number of people, around 700 million, living within 10 m of sea level and to where can they move. Also be aware of living in the Southwest United States due to higher temperatures (need for more air conditioning) and scarcity of water. Global climate models are fairly good; however, which regions will experience more drought or precipitation are primarily rough estimates. James Hansen is advocating shutting down all coal plants for generation of electricity [12] to reduce carbon dioxide in the atmosphere to 350 ppm (in 2015, it was 400 ppm).

If higher temperatures decrease crop production, then you will have massive migrations due to lack of food. It will make the peaceful invasion of illegal aliens to the United States and Europe appear as a small problem.

Surveillance of electronic communications is just the beginning as camcorders will be mounted most everywhere and then there are drones. The justification will be for crime prevention, and later, it may be used for control, political and physical. Video of the Boston Marathon bombers led to their arrest. What is the line between public surveillance and individual privacy?

The amount of money for the military expenditures is way too large, as global expenditures in 2013 were $1,747 * 10^9$, around 2.4% of the world GDP [13]. The United States spent the major share, $640 * 10^9$, which is 36.6% of the total, while China and Russia are the next two largest with a combined total of $275 * 10^9$. If you count the United States and allies, their expenditures are 70% of the world total. In my opinion (VN), it is time to reduce the number of bases and troops that the United States has in other countries. A prediction with almost 100% probability is that the United States will be involved in wars and conflicts from now to 2050. Wars and conflicts over resources [14,15] have been fought, Oil War I (Gulf War) and Oil War II (Iraq War), and there will be war and conflict over energy, water, mineral resources, and maybe in the future over food. A side note on war (VN) is that we expect few casualties for our side and combat has become more impersonal, for example, drone pilots are in the United States and observe people (squirters) fleeing the area of a strike. For the United States in the Vietnam War, there were 58,000 killed, while Oil War I had only 148 killed in battle and another 145 in accidents, and Oil War II had 4,491 killed in action (2003–2014). Of course, we are still in the aftermath of Oil War II and nobody knows how much longer it will continue. How many U.S. service people were killed in automobile accidents during the period of those two wars? And those religious conflicts powered by fanatics seem ever so petty and stupid compared to the totality of life on Earth.

How many people can the planet support and how can we use the resources of the planet such that there is a place for all life? Look at the barren photos of Mars from the Curiosity Rover and the barren hot wasteland of Venus and compare that to our amazing planet with life. To continue on the present path will only result in major and costly problems and maybe even catastrophe within this century (VN).

REFERENCES

1. D. J. Suson, L. D. Hewett, J. McCoy, and V. Nelson. 1999. Creating a virtual physics department. *Am J Phys* 67(6), 520.
2. MOOC List. https://www.mooc-list.com.

3. A. C. Clarke. 1984. Hazards of prophecy: The failure of imagination, In: *Profiles of the Future*. Holt, Rinehart, and Winston, New York.

4. J. Naisbitt. 1982. *Megatrends, Ten New Directions Transforming Our Lives*. Warner, New York.

5. A. Toffler. 1970. *Future Shock*. Bantam Books, New York.

6. B. Bueno de Mesquita. 2009. *The Predictioneer's Game: Using the Logic of Brazen Self-Interest to See and Shape the Future*. Random House, New York.

7. *Toward the Year 2018*. 1968. Edited by the Foreign Policy Association. Cowles Education Corporation, New York.

8. R. Kurzweil. 1999. *The Age of Spiritual Machines*. Viking, New York.

9. R. Kurzweil. 2005. *The Singularity is Near: When Humans Transcend Biology*. Viking, New York.

10. K.E. Drexler and C. Peterson with G. Pergamit. 1991. *Unbounding the Future: The Nanotechnology Revolution*. William Morrow, New York.

11. Global Footprint Network. http://www.footprintnetwork.org/en/index.php/GFN/page/world_footprint/.

12. J. Hansen. 2009. *Storms of My Grandchildren: The Truth about the Coming Climate Catastrophe and Out Last Chance to Save Humanity*. Bloomsbury, New York.

13. Stockholm International Peace Research Institute. http://www.sipri.org.

14. M. T. Klare. 2001. *Resources Wars: The New Landscape of Global Conflict*. Metropolitan Books, New York.

15. W. Youngquist. 1997. *GeoDestinies: The Inevitable Control of Earth Resources Over Nations and Individuals*. National Book, Portland, OR.

RECOMMENDED RESOURCES

K. Börner. 2014. *Visual Insights: A Practical Guide to Making Sense of Data*. MIT Press, Cambridge, MA.

Global sea level rise map. http://geology.com/sea-level-rise/.

K. Higgs. 2014. *Collision Course, Endless Growth on a Finite Planet*. MIT Press, Cambridge.

A. Y. Hoestra. 2008. *Globalization of Water: Sharing the Planet's Freshwater Resources*. Wiley-Blackwell, New York.

M. Lima. 2013. *Visual Complexity: Mapping Patterns of Information*. Princeton Architectural Press, Princeton, NJ.

A. Lovins. 2011. *Reinventing Fire: Bold Business Solutions for the New Energy Era*. Chelsea Green Publishing, White River Junction, VT.

P. Sabin. 2013. *The Bet, Paul Ehrlich, Julian Simon, and Our Gamble over Earth's Future*. Yale University Press, New Haven.

P. F. Sale. 2014. *Our Dying Planet: An Ecologist's View of the Crisis We Face*. California University Press, Los Angeles, CA.

S. Solomon. 2011. *Water: The Epic Struggle for Wealth, Power and Civilization*. Harper Perennial, New York.

There are a number of books on the future and forecasts. Economists and others disparaged the *Limits to Growth* because the dire forecasts did not happen within the time frame predicted. However the basic tenet is still applicable as there are limits to growth.

G. Friedman. 2010. *The Next 100 Years: A Forecast for the 21st Century*. Anchor, New York.

D. H. Meadows, J. Randers. and D. L. Meadows. 2012. *Limits to Growth: The 30-Year Update*. Chelsea Green Publishing, White River Junction, VT.

J. Randers. 2012. *2052: A Global Forecast for the Next Forty Years*. Chelsea Green Publishing, White River Junction, VT.

PROBLEMS

16.1. Go to global sea level rise map, use increase of 9 m. Estimate what percent of Florida is now underwater compared to today.

16.2. Ask all students in the class to turn off all their communication devices for two days (Saturday and Sunday); no smart phone, no PC, no TV, no iPod, etc. Have students write down their two main thoughts after that experiment. The first question is how many had the fortitude to do it.

16.3. How many students have taken one or more online courses? What are the two main advantages and disadvantages compared to traditional lecture course?

16.4. Do you prefer to read books online or hard copy? Give short justification for answer.

16.5. Will the bionic human (robo-sapien) happen by 2050? Give short justification for your answer.

16.6. Do you use or will you need to know any renewable energy tools or toolboxes for your major. List two tools and give short explanation.

16.7. Go to any interactive map in Section 1.5.1. List and describe what can be done with that map.

16.8. Go to Earth, a graphic look at the state of the world. Click on global ecology. What does the map show?

16.9. Go to Hydro Tasmania, King Island Renewable Energy Project. What parameters of the project are there in the app?

16.10. Do you think computers will be able to pass the Turing test? When? Give short explanation for your answer.

16.11. How many students have experience or are going to buy a virtual reality unit?

16.12. Are you in favor or against genetic modified food? Give justification for your answer.

16.13. What is your estimate for the population of the Earth in 2050?

16.14. What was the estimated cost for the United States for Oil Wars I and II? Remember casualties include wounded and those costs need to be considered even after they have left the service.

16.15. What is your response to the term global warming? Give short justification for your answer.

Appendix

We are using the SI units: meter, kilogram, and second. For those who are used to English units, it may be somewhat difficult to visualize the size of the quantity; however, SI units makes the problems much easier. Almost all of the problems will have units associated with the answers. Be sure and include the units. Also answer should reflect correct number of significant digits.

A.1 MATHEMATICS

A.1.1 EXPONENTIAL GROWTH

Values of future consumption, r, can be calculated from the present rate, r_0, and the fractional growth per year, k.

$$r = r_0 e^{kt} \tag{A.1}$$

where:
    e is the base of the natural log
    t is the time

Note that the exponent does not have any unit.

Example A.1

Present consumption is 100 units/year and growth rate is 7%. What is consumption after 100 years?

$$r = 100 \ e^{0.07 * 100} = 100 * e^7 = 100 * 1097 = 1 \times 10^5$$

So after 100 years, the consumption is 1000 times greater.

A.1.2 DOUBLING TIME

Doubling time, $T2$ in years, for any growth rate can be calculated from Equation A.1. Final amount is $2 * r_0$, so from Equation A.1.

$$2 r_0 = r_0 e^{kt} \quad \text{or} \quad 2 = e^{kT2}$$

Take the natural ln on both sides of the equation.

$$\ln 2 = kT2, \text{ which is } T2 = 0.69/k$$

or for percent growth rate, R

$$T2 = 69/R$$

In terms of consumption, remember, it is always the last doubling time that is the problem with a finite resource. The amount needed is the sum of all the previous doubling times plus 1. The total sum of the resource used from any initial time to any final time, T, can be estimated by summing up the consumption per year. This can be estimated by using a spreadsheet, or if r is known as a function of time, then the total consumption can be found by integration. The total consumption for exponential growth is given by

$$C = \int rdt = \int_0^T r_0 e^{kt} dt = \frac{r_0}{k}\left(e^{kT} - 1\right)$$ (A.2)

A.1.3 LIFETIME OF A FINITE RESOURCE

If the magnitude of the resource is known, or can be estimated, then the time, T_E, when that resource is used up, can be calculated or estimated by spreadsheet for different growth rates. The size of resource, $S = C$ is put in Equation A.2, and the resulting equation is solved for T_E.

$$S = \frac{r_0}{k}\left(e^{kT_E} - 1\right)$$

$$T_E = \frac{1}{k}\ln\left(k\frac{S}{r_0} + 1\right)$$ (A.3)

If the demand is small enough or is reduced exponentially, a resource can essentially last forever. However, with increased growth, T_E can be calculated for different resources, and the time before the resource is used up is generally much shorter than most people would have estimated. For example in 2008, United States oil consumption was $7.2 * 10^9$ bbl/year; however, U.S. crude oil production was $1.8 * 10^9$ bbl/ year. Simple division of estimated reserves by today's production rate gives only 20 years for domestic oil at today's rate of oil production. However, U.S. domestic oil production has been declining, so the time frame will be somewhat longer.

According to some energy companies, the continued growth in energy use in the United States is to be fueled by our largest fossil fuel resource, coal, and by nuclear. How long can coal last if we continue to increase production to offset decline in domestic production of oil and to reduce the need for importation of oil? The preceding analysis will allow you to make order of magnitude estimates. Also increased or even current production rates of fossil fuels may have major environmental effects. Global warming has become an international political issue.

The lifetime can be estimated for different finite resources (Table 2.1) and in general the time is short, especially if there is increased demand. Remember these are only estimates of resources and other estimates will be higher or lower depending on demand and cost (as cost becomes higher, reserves are usually increased, but there is a limit to finite resources). A good source for energy and exponential growth is the following, A.A. Bartlett, 1987, Forgotten fundamentals of the energy crisis, *Amer. J. Phys.*, 46(9), Sept, p 876.

A.1.4 SIGNIFICANT DIGITS

Answers to problems and estimates cannot be more accurate than the information available for input. With calculators and PCs, it is common to have 8 or even more significant digits displayed; however, the answer can only have the same number of significant digits as the least accurate data input.

Example A.2

Mass = 1.05869 kg. This has 7 significant digits. This body is traveling at 1.53 m/s. This has 3 significant digits. The momentum of the object is mass * velocity.

Momentum = 1.08569 * 1.53 = 1.6611057 kg m/s using a calculator

However, the answer cannot be more accurate than 3 significant digits, so answer is

Momentum = 1.66 kg m/s

For decimals, leading zeros do not count, for example, 0.000152 has 3 significant digits. If you use powers of 10 that would be $1.52 * 10^{-4}$.

A.1.5 ORDER OF MAGNITUDE ESTIMATES

In terms of energy consumption, production, supply and demand, estimates are needed and an order of magnitude calculation will suffice. By order of magnitude, we mean an answer (1 significant or at most 2 significant digits) to a power of 10.

Example A.3

How many seconds in a year? With a calculator it is easy to determine.

365 days * 24 h/day * 60 min/h * 60 s/h = 31,536,000 s

When you round to 1 significant digit, this becomes $3 * 10^7$ s.
  For 2 significant digits, the answer is $3.2 * 10^7$ s.
  For an order of magnitude estimate, round all input to one number with power of 10, then multiply the numbers and add the powers of 10. So without a calculator, the above becomes

$$4 * 10^2 * 2 * 0^1 * 6 * 10^1 * 6 * 10^1 = 4 * 2 * 6 * 6 * 10^5$$
$$= 288 * 10^5$$
$$= 3 * 10^2 * 10^5 = 3 * 10^7$$

A.2   CONVERSION

Conversions are available on web pages or widgets.

length: 1 m = 3.28 ft, 1 km = 0.62 mile, 1 m = 100 cm
mass: 1 kg = 2.2 lbs, 1 metric ton = 1000 kg = 2205 lbs, 1 ton = 2400 lbs,
   1 short ton = 2000 lbs
   Metric tons will be used, unless stated otherwise
area: 1 hectare = 10,000 m^2 = 2.47 acres, 1 km^2 = 100 hectares
volume: 1 L = 1000 cm^3, 1 L = 0.264 gal, 1 m^3 = 1000 L, 1 barrel oil
   (bbl) = 42 gal = 159 L
speed: 1 m/s = 2.24 mph, 1 km/h = 0.62 mph
power: 1 kW = 1.34 horsepower
temperature: Kelvin, K; Celsius, C; Fahrenheit, F
   T(°K) = T(°C) + 273, T (°C) = (5/9) * [T(°F) – 32]
   Freezing point of water = 0 (°C) = 273 (°K) = 32 (°F)
   Boiling point of water = 100 (°C) = 373 (°K) = 212 (°F)
Energy conversion factors
   1 calorie (cal) = 4.12 Joules (J)
   kilocalorie = 1 Calorie (the unit used in nutrition) = 1000 calories
   1 Btu = 1055 J
   1 therm = 10^5 Btu = 100 cu ft of natural gas
   1 quadrillion Btu (quad) = 10^{15} Btu = 1.055 EJ
   1 kWh = 3.6 × 10^6 J = 3.4 * 10^3 Btu
   ton of oil equivalent (toe) = 4 × 10^7 Btu
   For large amounts, quads and Mtoe (million tons oil equivalent) are used
   1 Mtoe will produce around 4.4 TWh of electricity in a steam plant
   R-value conversion, 1 (h ft^2 °F/Btu) = 0.17611 (m^2 °K/W)
   U-values, $U = 1/R$
Average energy content
   oil, 1 metric ton = 7.2 bbls = 42 GJ,
   barrel of oil equivalent (boe), some units for biomass are in boe
   1 barrel of oil (42 gallons) = 6.12 × 10^9 J = 1.7 × 10^3 kWh
   U.S. gal gasoline = 121 MJ
   U.S. gal diesel = 138 MJ
   Note: Energy content/mass is fairly consistent; however, density varies, so
      energy/volume is different
   coal, 1 metric ton = 2.5 × 10^7 BTU = 2.2 × 10^{10} J
   1 cubic foot of natural gas = 1000 Btu
   1 U.S. gallon gasoline = 121 MJ
Bioenergy
   wood, dry, no moisture, 1 metric ton = 18–22 GJ
   wood, air dry, 20% moisture, 1 metric ton = 15 GJ
   charcoal, 1 metric ton = 30 GJ (derived from 6 to 12 tons wood)
   agriculture residue, 1 metric ton = 10–17 GJ

ethanol, 1 metric ton = 7.94 oil bbls = 26.7 GJ (notice energy content/
volume is less than gasoline)
biodiesel, 1 metric ton = 37.8 GJ

A.3 RESISTANCE TO FLOW OF HEAT

Thermos jugs resist the flow of heat as they have a vacuum between two containers
and silver walls so there is no electromagnetic (EM) radiation. From my personal
experience, I think that China makes the best thermos jugs in the world. All water
is boiled in China, and there is always a jug of hot water available, since they drink
a lot of green tea.

Most insulation materials trap pockets of air. Styrofoam is extruded, expanded
beads of polystyrene, which is in the form of boards from 0.5 to 10 cm thickness.
Aerogels have very high R-values, although they have not reached the commercial
building market. Go to the following site for more information, www.mkt-intl.com/
aerogels/. R-values vary widely, so what is used depends on price and R-value per
width of material, especially if there are limitations on space available.

Obtain R-values for blankets and batts, common building materials, surfaces,
and trapped air spaces from the Internet (http://coloradoenergy.org/procorner/
stuff/r-values.htm) or from manufacturer's specifications.

Oakridge National Laboratory has a whole wall R-value calculator, www.ornl.
gov/sci/roofs+walls/AWT/InteractiveCalculators/rvalueinfo.htm.

Index

Note: Locators followed by '*f*' and '*t*' denotes figure and table in the text